39th Heat Transfer Conference (UIT 2022)

Journal of Physics: Conference Series Volume 2509

Gaeta, Italy
20-22 June 2022

ISBN: 978-1-7138-7465-2
ISSN: 1742-6588

Printed from e-media with permission by:

Curran Associates, Inc.
57 Morehouse Lane
Red Hook, NY 12571

Some format issues inherent in the e-media version may also appear in this print version.

Institute of Physics
Dirac House, Temple Back
Bristol BS1 6BE UK

Phone: 44 1 17 929 7481
Fax: 44 1 17 920 0979

techtracking@iop.org

Additional copies of this publication are available from:

Curran Associates, Inc.
57 Morehouse Lane
Red Hook, NY 12571 USA
Phone: 845-758-0400
Fax: 845-758-2633
Email: curran@proceedings.com
Web: www.proceedings.com

TABLE OF CONTENTS

Design of PCM-Based Heat Sinks Through Topology Optimization 1
Nicola Bianco, Andrea Fragnito, Marcello Iasiello, Gerardo Maria Mauro

Accurate RBF-FD Meshless Numerical Simulation of Thermo-fluid Problems for Generic 3D Geometries 9
Riccardo Zamolo, Davide Miotti, Enrico Nobile

Energy Demand in Secondary Steel Making Process: Numerical Analysis to Assess the Influence of the Ladle Working Lining Properties 20
M Neri, A M Lezzi

Numerical Study on Latent Heat Thermal Energy Storage with Phase Change Material and Metal Foam in Shell and Corrugated Tube with External Heat Losses 30
Bernardo Buonomo, Maria Rita Golia, Oronzio Manca, Sergio Nardini, Renato Elpidio Plomitallo

Application of the Porous Media Approach to a Standing Wave Thermoacoustic Engine for CFD Simulations 38
Armando Di Meglio, Nicola Massarotti

Experimental Study of a R290 Variable Geometry Ejector 46
Giorgio Besagni, Fabio Inzoli, Lorenzo Croci, Vittorio Brignoli, Jacek Smolka, Michal Palacz, Michal Haida, Jakub Bodys, Rafal Fingas, Andrzej J. Nowak, Bartlomiej Melka, Michal Stebel, Edyta Piechnik

Lattice Boltzmann Model of a Square Natural Circulation Loop with Small Inner Diameter: Working Fluid Effects 54
Johan Augusto Bocanegra, Mario Misale

On the Solved Turbulent Scales in Turbulent Plume Fires 62
F. S. Ciani, P. Bonfiglio, S. Piva

Accuracy in Evaluating Heat Transfer Coefficient by RANS CFD Simulations in a Rectangular Channel with High Aspect Ratio - Part 1: Benchmark on a Channel with Plane Walls 70
Maria Corti, Pasqualino Gramazio, Damiano Fustinoni, Luigi Vitali, Alfonso Niro

Accuracy in Evaluating Heat Transfer Coefficient by RANS CFD Simulations in a Rectangular Channel with High Aspect Ratio - Part 2: Channel with Ribbed Walls 78
Maria Corti, Pasqualino Gramazio, Damiano Fustinoni, Luigi Vitali, Alfonso Niro

Preliminary Design and Modeling of Mini Channels to Enhance Heat Transfer in a Millimetric Catalytic Combustor 86
Caterina La Terra, Maria Corti, Adelaide Nespoli, Carlo Fanciulli, Alfonso Niro

Local Wall Heating Effects on Aeroacoustic Field Radiated by an Isolated Square Cylinder in a Laminar Flow 94
M. Falone, V. D'Alessandro, L. Giammichele, R. Ricci

Optimised Electro-Osmotic Flow in Rectangular Microchannels with Smoothed Corners 103
Marco Lorenzini, Nicola Suzzi

An Exhaustive Search Optimization of Heat Transfer and Pressure Drop in Kelvin's Open Cell Foams 112
M Iasiello, G M Mauro, N Bianco, A Andreozzi, W K S Chiu, V Naso

Stability Analysis of Dual Viscous Flows Saturating a Vertical Porous Pipe 120
M Celli, A Barletta, P V Brandão, S Da C Hirata, M N Ouarzazi

Boundary Element Method for Contactless Estimation of Spatially Varying Internal Heat Transfer Coefficient in Circular Pipes 128
Alessandra Aimi, Fabio Bozzoli, Luca Cattani, Chiara Guardasoni

Fouling Effect in a Shell-And-tube Heat Exchanger with Twisted Tape Inserts Applied to a Small-scale Biomass Gasification Power Plant 136
Giulio Allesina, Simone Pedrazzi, Filippo Ottani, Paolo Tartarini

Analysis of Variance of the Heat and Mass Transfer Coefficients in an Evaporative Condenser 145
Michael Giovannini, Marco Lorenzini

Transient Analysis of SIRIO Using RELAP5/MOD3.3 System Code 153
Martina Molinari, Vincenzo Narcisi, Marco Caramello, Mariano Tarantino, Fabio Giannetti

The Effect of Different Sediment Conditions on CO_2-CH_4 Replacement in Natural Gas Hydrates 163
Beatrice Castellani, Alberto Maria Gambelli, Giorgio Minelli, Federico Rossi

Microgrid Application of Liquefied Air Energy Storage (LAES) Systems 171
Beatrice Castellani, Parminder Kaur, Adriano Sciacovelli, Federico Rossi, Yulong Ding

Reverse Flow in a Pressure Suppression System Due to Condensation Instabilities 179
G. Giambartolomei, A. Pesetti, L. Berti, M. Raucci, R. Lazzeri, D. Aquaro

Hidden Grid Background Oriented Schlieren in Studying Convective Flows 187
Dario Ambrosini, Tullio De Rubeis, Domenica Paoletti, Jagadesh Ramaiah, Rajshekhar Gannavarpu

Experimental Characterization of the Airflow Within a Car Cabin 194
M. Bertone, A. Sciacchitano, F. Arpino, C. Canale, G. Cortellessa, G. Grossi, L. Moretti

Experiments of Convective Evaporation of Refrigerant R513A in a Horizontal Stainless-Steel Tube 202
A Arcasi, R Mastrullo, A W Mauro, A F Passarelli, I Viscardi

Experimental Analysis of Drop Size Distribution and Nucleation Site Density During Dropwise Condensation from Humid Air Flow 210
M Mirafiori, M Tancon, S Bortolin, A Martucci, D Del Col

Convective Condensation of R134a and R1234ze(E) Inside Microfin Tube 218
A Lucchini, I M Carraretto, T N Phan, P G Pittoni, L Molinaroli, L P M Colombo

Condensation Heat Transfer of Superheated Vapour of R1234ze(E) and R134a Inside a Brazed Plate Heat Exchanger 226
G Cattelan, M Azzolin, A Diani, L Rossetto, D Del Col

Author Index

Preface

The 2022 UIT (Italian Union of Thermo-Fluid Dynamics) International Conference, hereafter referred as 39th UIT 2022, was organized by the Department of Civil and Mechanical Engineering of the University of Cassino (Italy), in collaboration with the UIT, on June 20-22 2022, at the Castello di Gaeta, Italy.

The 39th UIT 2022 was organized by the Organizing Committee of the 38th UIT Heat Transfer Conference, initially scheduled for June 21st-23rd 2021 at the Castello di Gaeta, Italy, and then organized completely online considering the criticalities induced by the spread of COVID-19.

The 39th UIT Heat Transfer Conference 2022 program scheduled two keynote lectures given by Prof. Perumal Nithiarasu from the Swansea University, Swansea, UK, titled "Physics-Informed Neural Networks (PINNs) for solving thermal problems", and by Prof. Wilson K. S. Chiu, from University of Connecticut Storrs, Connecticut, USA, titled "Thermal Transport in Architected Open Cell Foams". A total of 68 abstracts were submitted to the 39th UIT Heat Transfer Conference 2022, 55 of which presented by the Authors in oral sessions, while the remaining 13 abstracts were presented in one poster session.

About 100 researchers participated to the 39th UIT 2022. The Conference was a useful occasion to stimulate discussion, further the understanding of heat transfer and related phenomena, present the state-of-the-art of some topics, discuss emerging trends and promote collaborations. The Organizing Committee hope that the event results constituted significant contribution to the knowledge in the following topics: Computational fluid dynamic and heat transfer; Conduction, radiation, thermophysical properties and porous media; Forced, natural and mixed convection; Heat and mass transfer in nuclear plants and energy systems; Measurement techniques for heat and mass transfer; Multiphase fluid dynamics and heat transfer.

A special tank to UIT, international advisory committee, local organizing committee and all the participants.

With Kind Regards

Paolo Vigo, Università di Napoli "Parthenope", Italy
Marco Dell'Isola, Università degli Studi di Cassino e del Lazio Meridionale, Italy
Fausto Arpino, Università degli Studi di Cassino e del Lazio Meridionale, Italy
Alfonso Niro, Politecnico di Milano, Italy

Committees

Co-chairs
P. Vigo, Università di Napoli "Parthenope", Italy.
M. Dell'Isola, Università degli Studi di Cassino, Italy
F. Arpino, Università degli Studi di Cassino, Italy

International Advisory Committee
F. Asdrubali, Università degli Studi Roma Tre, Italy
P. Asinari, Politecnico di Torino, Italy
N. Bianco, Università degli Studi di Napoli "Federico II" , Italy
G.P. Celata, Centro Ricerche ENEA, Italy
M. Costa, Istituto Motori - CNR, Italy
L. De Santoli, Università degli Studi di Roma "La Sapienza", Italy
D. Del Col, Università degli Studi di Padova, Italy
V. Fernicola, Istituto Nazionale di Ricerca Metrologica - INRiM, Italy
O. Manca, Università degli Studi della Campania Luigi Vanvitelli, Italy
N. Massarotti, Università degli Studi di Napoli "Parthenope", Italy
G. Morini, Università degli Studi di Bologna, Italy
V. Naso, Università degli Studi di Napoli "Federico II" , Italy
A. Niro, Politecnico di Milano, Italy
P. Nithiarasu, Swansea University, Wales
A.J. Nowak, Silesian University of Technology, Poland
B. Panella, Politecnico di Torino, Italy
S. Rainieri, Università degli Studi di Parma, Italy
R. Ricci, Università Politecnica delle Marche, Italy
B. Sarler, University of Ljubljan, Slovenia
M. Sasso, Università degli Studi del Sannio, Italy
R. Vinuesa, KTH Royal Institute of Technology, Sweden
I. Vušanović, University of Montenegro, Montenegro

Local Organising Committee
M. Bertone, Università degli Studi di Cassino, Italy
L. Canale, Università degli Studi di Cassino, Italy
C. Canale, Università degli Studi di Cassino, Italy
G. Cortellessa, Università degli Studi di Cassino, Italy
G. Grossi, Università degli Studi di Cassino, Italy
L. Moretti, Università degli Studi di Cassino, Italy
L. Stabile, Università degli Studi di Cassino, Italy

Journal of Physics: Conference Series **2509** (2023) 011001 doi:10.1088/1742-6596/2509/1/011001

Contact details for the declaration

- ✓ Paolo Vigo, Università di Napoli "Parthenope", Italy, paolo.vigo@uniparthenope.it
- ✓ Marco Dell'Isola, Università degli Studi di Cassino e del Lazio Meridionale, Italy, dellisola@unicas.it
- ✓ Fausto Arpino, Università degli Studi di Cassino e del Lazio Meridionale, Italy, f.arpino@unicas.it
- ✓ Alfonso Niro, Politecnico di Milano, Italy, alfonso.niro@polimi.it

Journal of Physics: Conference Series **2509** (2023) 011002 doi:10.1088/1742-6596/2509/1/011002

Peer Review Statement

All papers published in this volume have been reviewed through processes administered by the Editors. Reviews were conducted by expert referees to the professional and scientific standards expected of a proceedings journal published by IOP Publishing.

- **Type of peer review:** Single Anonymous
- **Conference submission management system:** Morressier
- **Number of submissions received:** 29
- **Number of submissions sent for review:** 29
- **Number of submissions accepted:** 29
- **Acceptance Rate (Submissions Accepted / Submissions Received × 100):** 100
- **Average number of reviews per paper:** 2
- **Total number of reviewers involved:** 32
- **Contact person for queries:**
  Name: Fausto Arpino
  Email: f.arpino@unicas.it
  Affiliation: University of Cassino and Southern Lazio

Design of PCM-based heat sinks through topology optimization

Nicola Bianco[1], Andrea Fragnito[1(c)], Marcello Iasiello[1], Gerardo Maria Mauro[2]

[1]Università degli Studi di Napoli Federico II, Department of Industrial Engineering, Piazzale Tecchio 80, 80125 Napoli (Italy)
[2]Università degli Studi del Sannio, Department of Engineering, Piazza Roma 21, 82100 Benevento (Italy)

[(c)]Corresponding author: andrea.fragnito@unina.it

Abstract. Ever-increasing heat fluxes of electronic components ask for higher performance of devices responsible for their cooling. Against that background, PCM-based heat sinks – because of their compactness, and effectiveness – are well-established solutions for thermal management issues. Even though solutions for the thermal enhancement of PCM devices have been widely presented, new ways to optimize such systems are emerging. Among them, this work investigates the application of density-based topology optimization to define innovative heat sinks design able to minimize thermal resistance under constant wall temperature. A 2D numerical model is developed by means of a finite element tool. The heat equation is solved for the topology optimization problem with the objective of minimizing the average temperature, considering further manufacturing restrictions. Solid-isotropic-material-with-penalization (SIMP) method is applied to link the design variable to material properties. Parameterization of Helmholtz filter's minimum feature size and projection based on the hyperbolic tangent function is performed, showing improved performance as well as feature size decrease. The optimized prototype – with PCM – is then simulated with the enthalpy-porosity model to assess the benefits, *i.e.,* reduction in the melting time, with respect to the baseline. Results show the potential of optimizing heat sinks via a topology-based approach and confirm it as a promising tool for finding new heat sink geometry, whatever the application.

1. Introduction
Thermal management is considered to be one of the main bottlenecks in the growth of innovative microprocessor technologies [1]. In this vein, one of the current research challenges is to find compact and economically feasible devices able to dissipate large amount of heat under appropriate constraints [2]. The need to remove heat fluxes at a rate equal or greater than the generated one is linked to their temperature increase, which can lead to reduction of reliability and performance, and at worst to a failure. In several electronics cooling applications, *e.g.,* temperature stabilization during pulsed operation, short-term thermal storage, and protection from failure during the loss of coolant scenarios, PCM-based heat sinks can be well-suited solutions because of their capability of absorbing the generated waste heat in the absence of high thermal gradients [3]. These passive devices take advantage of the PCM latent heat, at least two orders of magnitude higher than the sensible energy that can be stored by other materials. Moreover, while absorbing thermal energy, PCMs guarantee a minimal temperature rise during the phase transition. Generally, having defined a suitable PCM for an

application, this kind of heat sink can reduce system size, cost, maintenance, and power requirement [4]. Nevertheless, common trade-offs between thermal storage capacity and temperature differences can emerge, *e.g.*, fins number effects thermal gradient and meanwhile the PCM amount, and thus the thermal storage capability. New solutions should be able to survive massive heat production while keeping the space and safety constraints. In this frame, optimizing heat sinks with numerical approaches is a topic of general interest, as proved by several studies that went through potential thermal management solutions in depth. Numerical optimization approaches based on topology optimization would be very useful to find optimum solutions for the additive manufactured heat sinks [5]. For instance, Sigmund *et al.* [6] improved the performance of a liquid-cooled heat sink with topology optimization by a peudo-3D model resulting in a modified airfoil fin shape. Iradukunda *et al.* [7] used topology optimization to generate optimal heat spreading structures that can be coupled with PCM to enhance the heat transfer rate. Specifically, under a pulsed load of 50 W, the hybrid heat sink exhibited the capacity to lower peak temperature by up to 18.9 °C relative to a benchmark plate fin design of similar dimensions. There are still many practical challenges that need to be addressed to achieve volume or mass-optimized, fully functional, and highly reliable PCM heat sink designs. This work points to provide a further example of topology optimization, starting from an existing geometry [8] and proving the benefits of this process. Specifically, the main contribution of this work is to enhance heat transfer mechanisms in the PCM-based heat sinks with an innovative metal structure and contribute to the knowledge gap-filling on thermal analysis of heat sinks for thermal management employing innovative – but well established – optimization tools.

2. Problem formulation

A half plate-fin (*PF*) PCM-based heat sink for electronic cooling is illustrated in Fig. 1, together with a topology-optimized one (*TO*). Both geometries are made of a conductive metal structure, *i.e.*, aluminium. The traditional one is taken from a reference work [8], as well as the materials thermophysical properties, while the topology-based is achieved through steps described later. The investigated domain is 11 mm wide and 12 mm high, with a base of 2 mm and fins of 1 mm thickness. All the walls of the PCM-based heat sink are adiabatic except for the heat source area where a constant temperature (T_w) of 44 °C is applied [8]. Therefore, heat is acquired from the hot wall (T_w) to which the plate fins are attached, and then transferred into the PCM for temporal heat dissipation. The initial temperature of the computation domain is set at 20 °C.

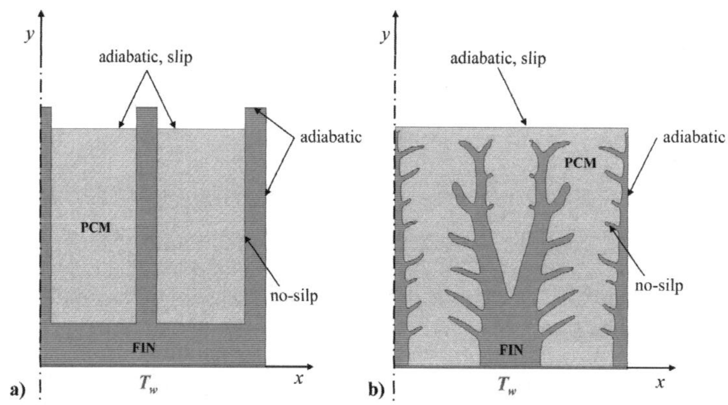

Figure 1. Heat sink geometry: a) plate fin (*PF*), topology optimized (*TO*)

The PCM is demanded to absorb heat from a metal structure – aimed at enhancing heat diffusion – and then to overcome the low thermal conductivity of PCM. Under a fixed metal volume fraction, the optimization effort is worthwhile in designing a highly efficient conductive thermal structure in a PCM-based heat sink.

2.1. Topology optimization formulation

The aim of this optimization problem is to design a conductive heat sink (Fig. 1) that enhances the heat transfer via diffusion between the domain and the constant temperature boundary. To balance the heat rate in the design domain, an artificial volumetric heat source term ($\dot{Q}$) is introduced on the right-hand side of Eq. (1) to mimic the heat absorbed by the PCM during the phase change process:

$$-k\nabla^2 T = \dot{Q} \quad on\ \Omega \tag{1}$$

To simulate the electronic heat source, a constant wall temperature (T_w) – above the PCM melting temperature – is prescribed, while remaining boundaries are assumed to be adiabatic:

$$T = T_w \quad on\ \Gamma_D \tag{2}$$

$$\nabla T \cdot \boldsymbol{n} = 0 \quad on\ \Gamma_N \tag{3}$$

To find the optimal topology for given objective functions and constraints, the density method is used [9]. For this purpose, a design variable, *i.e.,* design density (γ) is defined, which can assume for each finite element values between 0, *i.e.,* no material or fluid, and 1, *i.e.,* solid material. The solid isotropic material with penalization (SIMP) method is used to influence the material distribution such that intermediate design values are repressed, and a well-defined material distribution is achieved, *i.e.,* replaces the discrete variables with continuous ones. In this frame, using the SIMP method, the thermal conductivity of (1) is replaced by:

$$k_{SIMP}(\gamma) = k_{min} + (k_{max} - k_{min})\gamma^n \tag{4}$$

where k_{min} = 0.2 W/mK and k_{max} = 202.4 W/mK are the thermal conductivities of the poor and the high conductors, respectively, while n is the penalization factor, set equal to 5. The limit on the solid fraction is defined based on the reference case, *i.e.,* 35% of the total area. From a topological point of view, the design objective is to find the optimized material distribution that minimizes the global temperature distribution of the domain. At the same time, to avoid geometries with low-scale parts – to be hardly achievable with 3D printing techniques – manufacturing constraints are also considered in the optimization approach [10]. The mathematical problem can be drawn as following:

$$\begin{cases} minimize\colon\ f(\gamma) = \int_\Omega \nabla(k(\gamma)\nabla T)d\Omega + \dfrac{h_0 h_{max}}{A}\int_\Omega |\nabla\gamma(x)|^2 d\Omega \\ subject\ to\colon \int_\Omega \gamma d\Omega < \delta A \\ \qquad 0 \le \gamma \le 1 \end{cases} \tag{5}$$

Where h_0=0.3 mm and h_{max}=0.05 mm are the parameters that govern the size of details in the solution and the mesh size, respectively. A is the domain surface, and δ is the solid material fraction, *i.e.,* 35 %. Several papers performed an analysis with a topology-based structure considering the PCM during the optimization process. However, the resulting geometry bears a close resemblance to the TO structure obtained considering pure conduction without PCM motion. Therefore, in the present study, the optimized structure is derived neglecting the PCM presence. Two conductive metal structures, *i.e.,* the conventional plate-fin and optimized tree shape structures deriving from the topology optimization process are investigated.

2.2. Phase change material model

The following assumptions are set for the PCM thermal model definition: PCM is isotropic but not homogeneous; phase change during solidification/melting occurs in a temperature range established in [8]; volume reduction/expansion during the phase change is ignored, together with air gaps caused by this; laminar flow model for PCM liquid phase, which is treated as a Newtonian fluid; Boussinesq approximation to include the buoyancy effects. The phase change modelling is based on three distinct regions, *i.e.,* a totally liquid region, a solid one and a mushy zone, where both phases coexist. To model the above-mentioned phase change, the three basic conservation equations have to be solved throughout the computational domain, *i.e.*, mass (6), momentum (7) and energy conservation (8). Therefore, the model equations are written as follows:

$$\nabla \cdot \boldsymbol{v} = 0 \tag{6}$$

$$\rho_{PCM}\left(\frac{\partial \boldsymbol{v}}{\partial t} + \boldsymbol{v}\nabla \cdot \boldsymbol{v}\right) = -\nabla p + \mu_{PCM}\nabla^2\boldsymbol{v} + \bar{F}_g + \bar{F}_v \tag{7}$$

$$(\rho c)_{PCM}\left(\frac{\partial T}{\partial t} + \boldsymbol{v} \cdot \nabla T\right) = \nabla \cdot (k_{PCM}\nabla T) + S_h \tag{8}$$

where $(\rho c)_{PCM,}$ is the product between the density and the specific heat at constant pressure of the PCM, that represents the volumetric heat capacity, k_{PCM} is the thermal conductivity, t the time, T the temperature, p the pressure and μ_{PCM} the dynamic viscosity. The Boussinesq hypothesis is adopted to consider these buoyancy effects, and to reduce the complexity of the Navier-Stokes equations. The related Boussinesq term (9) has been added in the momentum conservation equation (7), as follows:

$$\bar{F}_g = \rho_{PCM}\boldsymbol{g}\beta(T - T_{ref})\varphi(T) \tag{9}$$

where $\boldsymbol{g}$ is the gravitational acceleration, β is the volumetric thermal expansion coefficient, T_{ref} is the reference temperature, fixed equal to the melting point temperature T_M, and φ is the melt fraction, *i.e.,* the non-dimensional parameter that quantifies the percentage of liquid phase contained in the mushy zone, evaluated as follows:

$$\varphi(T) = \frac{T - T_S}{T_L - T_S} = \frac{T - T_M + \Delta T_M}{2\Delta T_M} = \begin{cases} 0 & for \quad T < T_S \\ 0 < \varphi < 1 & for \quad T_S \le T \le T_L \\ 1 & for \quad T > T_L \end{cases} \tag{10}$$

In (10) the definition of the melt fraction is based on: the phase change temperature interval, ΔT_M– assumed symmetrical with respect to the T_M –; the maximum temperature at which the PCM is completely solid, T_S; and the minimum temperature at which the PCM is completely liquid, T_L. These temperatures have been evaluated as a function of the PCM melting point T_M, and the phase change interval ΔT_M as follows:

$$T_S = T_M - \Delta T_M \tag{11}$$

$$T_L = T_M + \Delta T_M \tag{12}$$

To force all velocities in solid regions to be zero, a volume force source term S (Eq. (13)) is defined in the enthalpy-porosity approach, based on the Carman-Kozeny equation [11], and to be multiplied by the fluid velocity to define the source term defined in the momentum equation:

$$\bar{F}_v = S\boldsymbol{v} = S(T)\boldsymbol{v} = A_{mush}\frac{(1 - \varphi(T))^2}{\varphi(T) + \delta}\boldsymbol{v} \tag{13}$$

where A_{mush} is an arbitrary constant that influences the effective viscosity into the mushy zone. Its value – set within the range of 10^3-10^7 – in this study is assumed to be 10^4. Moreover, δ is an arbitrary constant, usually set equal to 10^{-3}, introduced to avoid a null denominator. Furthermore, to overcome the issues of the evaluation of heat and mass transfer near the phase change layer, the basic approach consists of defining an appropriate generation term. In detail, for the enthalpy-porosity approach the latent heat influence is accounted by the definition of the energy equation generation term (S_h), as follows:

$$S_h = -\rho_{PCM} L_f \frac{\partial \varphi}{\partial t} \tag{14}$$

For all PCM thermophysical properties refer to [8].

2.3. Numerical resolution and model validation

Each numerical simulation solves the enthalpy-porosity approach for the PCM numerical analysis. In particular, the energy and momentum equations are discretized using the fully coupled solver. The convergence criteria for the continuity and momentum equations and energy equation are applied, performing mesh independence study and time step evaluation to ensure the numerical accuracy, as in Fig. 2.

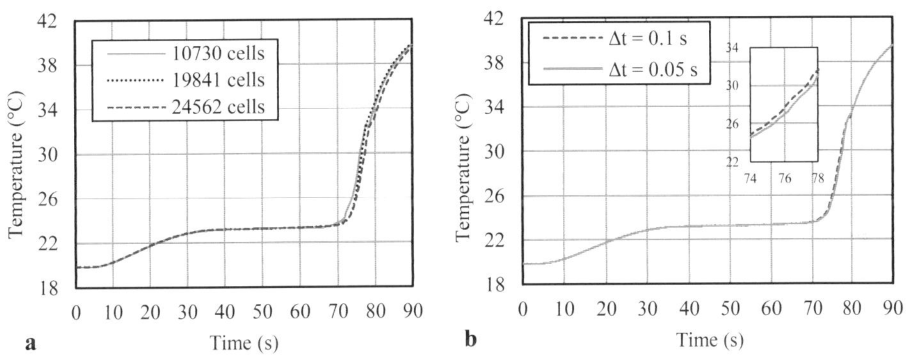

Figure 2. a) grid convergence test at time step of 0.1 s; b) time step convergence test at cells number of 24562

Looking at these trends, further increasing the numbers of elements – with respect to 10730 cells – has negligible effects on the simulation results. At the same time, the time steps independence is assessed, selecting at first a Δt = 0.05 s and then equal to 0.1 s. The second one represents a suitable trade-off between acceptable computational costs and model reliability. Once optimal grid and time step have been defined, a validation step is performed with respect to the reference case (Fig. 3). The comparisons are made in terms of temperature evolution – at the point P = ($7.8e^{-3}$, $5e^{-3}$) with reference to the axis position in Fig. 1 – and PCM melt volume fraction, varying the phase change interval ΔT_{M}. In both cases, except for a slight difference in the PCM response to the liquid phase transition, the model confirms the reference trends. However, the case with ΔT_M seems to be more accurate.

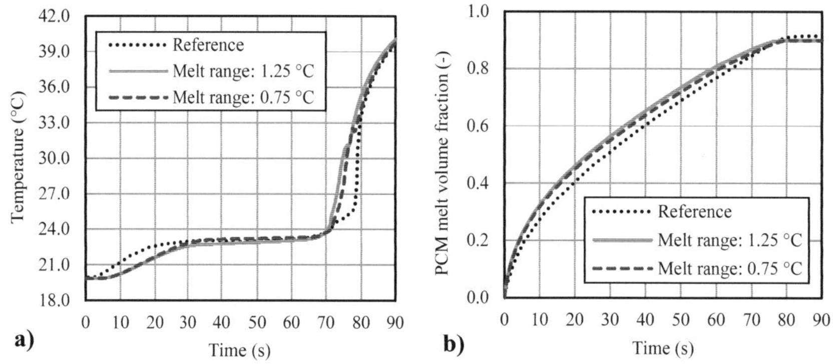

Figure 3. Model validation varying PCM melt range: a) temperature profile at P = (7.8e^{-3}, 5e^{-3}) m; b) PCM melt volume fraction

3. Optimized and reference heat sink comparison

After the model validation, the optimized tree-shaped structure has been obtained according to the density-based topological optimization approach. The performance of this shape, supposed to have the so-called optimal heat conduction path from the heat source into the PCM domain, has been compared with the conventional one, looking at temperature and melt fraction evolution. Finally, remarks on the heat flux spread by the two heat sinks are proposed.

3.1. Temperature evolution

Fig. 4 presents the temporal evolution of the heat sink temperature, for 3 time instants, *i.e.,* 10, 20 and 40 s. These values are selected to catch the main differences between the heat diffusion inside the heat sink until the PCM of the *TO* case is fully melted. As can be seen, the temperature evolution inside the *TO* PCM-structure is strongly affected by the shape of fin structures. They act as heat spreaders due to their high thermal conductivity and high surface area and mitigate temperature gradients faster than the conventional geometry. Conversely, the thickness of the PCM in the plate-fin heat sink makes the heat diffusion slower.

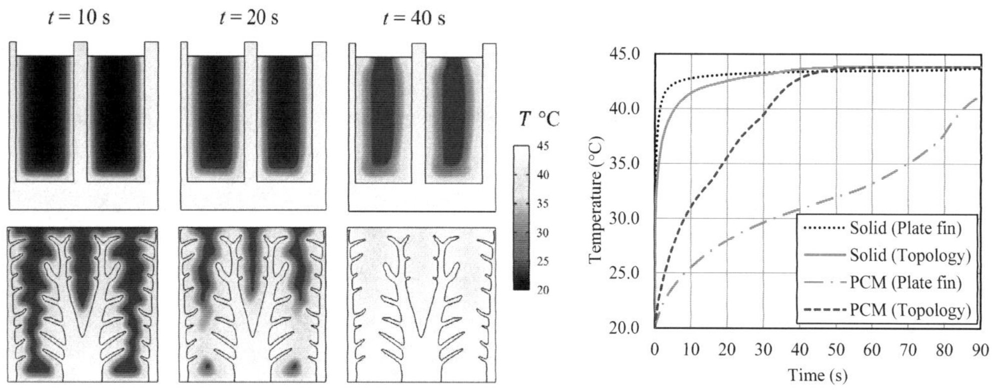

Figure 4. Comparison of heat sink temperature field at 3 time instants: t = 10 s, 20 s and 40 s; average temperature evolution for solid (aluminium) and PCM

After 40 s, the PCM inside the *TO* heat sink has reached an almost uniform temperature distribution, while the one in the *PF* heat sink is still changing phase. At the same time, looking at the average fin temperature (Solid), the *TO* configuration exhibits a slower temperature rising due to the absence of a base plate, which conversely guarantees in the *PF* geometry a lower thermal resistance.

3.2. Melt fraction evolution

By comparing the two cases in terms of melt fraction evolutions (Fig. 5), convection flow patterns appear during the PCM melting process in the conventional geometry rather than in the optimized one, thus depending on the shape of the metal structure. Fig. 5 shows that the PCM phase change process around the tree-shaped heat sink is deeply influenced by the multiple ramifications, resulting in a complete melting even before 40 s.

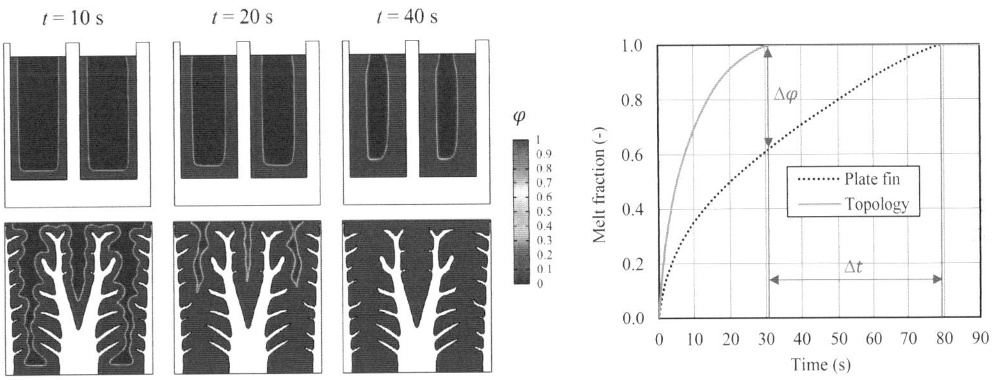

Figure 5. Comparison of heat sink melt fraction field at 3 time instants: t = 10 s, 20 s and 40 s; melt fraction evolution

Looking at Fig. 5, at $t = 31$ s – for which the PCM in the TO configuration is fully melted – the PCM inside the plate-fin heat sink has an average melt fraction value of 0.62. Therefore, adopting *TO* fins allows reducing the time to melt by about 50 s, as highlighted in Fig. 5.

3.3. Heat removal performance

Finally, Fig. 6 displays the comparison of heat fluxes resulting from the different geometries. Until the PCM fully melts in the *TO* heat sink *i.e.,* after 31 s, the heat flux is substantially higher than the plate-fin heat sink. At $t = 10$ s the heat flux related to the plate-fin heat sink is about 2.42 W/cm^2, while for the topology one 4.7 W/cm^2. Therefore, at a fixed temperature, the *TO* configuration allows it to dissipate up to 2 times the heat flux of a traditional one. This benefit lasts until the full exploitation of the phase change, after which the difference between the heat fluxes tends to zero.

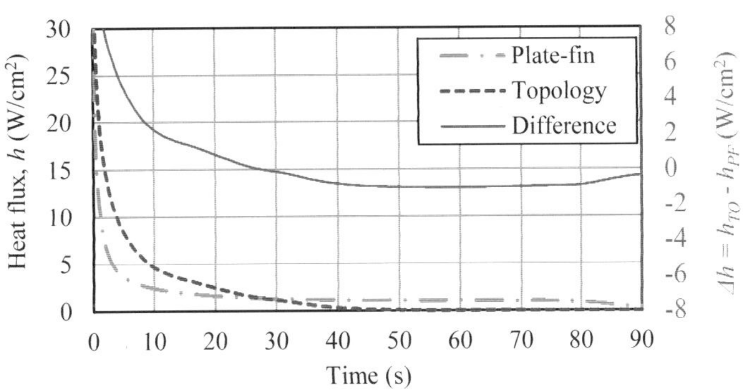

Figure 6. Comparison of plate-fin and topology heat sink heat fluxes

4. Conclusions

This work presents a brief analysis of the benefits of the adoption of topology optimization to a passive PCM-based heat sink. In detail, starting from a reference case, *i.e.,* plate-fin PCM-based heat sink, a topology-optimized structure has been derived and thermally analysed. The comparisons are made by fixing the wall temperature on the heat sink base and considering the same PCM and solid fraction value, *i.e.,* aluminium. The density method is here performed to assess the topological structure, while the enthalpy-porosity method is performed for the PCM simulation. According to the mathematical formulation of the optimization problem, results exhibit a faster heat conduction mechanism inside the topology-based heat sink, leading to a faster thermal transient. The adoption of the tree-shaped heat sink allows it to spread up to 2 times the heat flux of the conventional one, *i.e.,* from 5 to 10 W/cm^2, exploiting all the PCM storage potential in less than half time.

References

[1] Z. He, Y. Yan, and Z. Zhang, “Thermal management and temperature uniformity enhancement of electronic devices by micro heat sinks: A review,” *Energy*, vol. 216, p. 119223, Feb. 2021, doi: 10.1016/j.energy.2020.119223.

[2] N. Bianco, S. Busiello, M. Iasiello, and G. M. Mauro, “Finned heat sinks with phase change materials and metal foams: Pareto optimization to address cost and operation time,” *Appl. Therm. Eng.*, vol. 197, p. 117436, Oct. 2021, doi: 10.1016/j.applthermaleng.2021.117436.

[3] W. Hua, L. Zhang, and X. Zhang, “Research on passive cooling of electronic chips based on PCM: A review,” *J. Mol. Liq.*, vol. 340, p. 117183, Oct. 2021, doi: 10.1016/j.molliq.2021.117183.

[4] S. K. Sahoo, M. K. Das, and P. Rath, “Application of TCE-PCM based heat sinks for cooling of electronic components: A review,” *Renew. Sustain. Energy Rev.*, vol. 59, pp. 550–582, Jun. 2016, doi: 10.1016/j.rser.2015.12.238.

[5] T. Dbouk, “A review about the engineering design of optimal heat transfer systems using topology optimization,” *Appl. Therm. Eng.*, vol. 112, pp. 841–854, Feb. 2017, doi: 10.1016/j.applthermaleng.2016.10.134.

[6] S. Yan, F. Wang, J. Hong, and O. Sigmund, “Topology optimization of microchannel heat sinks using a two-layer model,” *Int. J. Heat Mass Transf.*, vol. 143, p. 118462, Nov. 2019, doi: 10.1016/j.ijheatmasstransfer.2019.118462.

[7] A.-C. Iradukunda, A. Vargas, D. Huitink, and D. Lohan, “Transient thermal performance using phase change material integrated topology optimized heat sinks,” *Appl. Therm. Eng.*, vol. 179, p. 115723, Oct. 2020, doi: 10.1016/j.applthermaleng.2020.115723.

[8] X.-Q. Wang, C. Yap, and A. S. Mujumdar, “A parametric study of phase change material (PCM)-based heat sinks,” *Int. J. Therm. Sci.*, vol. 47, no. 8, pp. 1055–1068, Aug. 2008, doi: 10.1016/j.ijthermalsci.2007.07.016.

[9] E. M. Dede, S. N. Joshi, and F. Zhou, “Topology Optimization, Additive Layer Manufacturing, and Experimental Testing of an Air-Cooled Heat Sink,” *J. Mech. Des.*, vol. 137, no. 11, Nov. 2015, doi: 10.1115/1.4030989.

[10] F. Lange *et al.*, “Numerical optimization of active heat sinks considering restrictions of selective laser melting Topology Optimization considering the restrictions of Additive Manufacturing View project ShipLight View project Numerical optimization of active heat sinks co,” no. November, 2018, [Online]. Available: https://www.researchgate.net/publication/328738213.

[11] M. Iasiello, M. Mameli, S. Filippeschi, and N. Bianco, “Metal foam/PCM melting evolution analysis: Orientation and morphology effects,” *Appl. Therm. Eng.*, vol. 187, no. September 2020, p. 116572, 2021, doi: 10.1016/j.applthermaleng.2021.116572.

Accurate RBF-FD meshless numerical simulation of thermo-fluid problems for generic 3D geometries

Riccardo Zamolo, Davide Miotti and Enrico Nobile

Dipartimento di Ingegneria e Architettura, Università degli Studi di Trieste, via A. Valerio 10, 34127 Trieste, Italy

E-mail: rzamolo@units.it

Abstract. In contrast to traditional mesh-based methods for the numerical solution of boundary value problems, e.g., Finite Element (FEM) and Finite Volume (FVM), in the recent period many meshfree approaches have been proposed in order to avoid those typical issues due to the mesh. For example, the quality of the mesh greatly affects the reliability of the final solution in the case of CFD problems and the human intervention of a professional is often still needed when dealing with complex-shaped domains. This in turn increases both cost and time required for the reliable simulation of problems of engineering relevance. Meshless methods, on the other side, usually rely on a simpler distribution of nodes and do not require the storage of connectivity information. Among others, one of the most promising meshless methods in terms of accuracy and flexibility is the one based on the Radial Basis Function – Finite Difference (RBF-FD) scheme. RBF-FD methods, however, are usually affected by severe ill conditioning issues when Neumann boundary conditions are employed. This fact is the main responsible for the appearance of large discretization errors near the boundary and for the lack of stability of traditional time integration schemes. In order to address this issue, some new algorithms for the robust treatment of boundary conditions have been developed and successfully employed to solve fluid flow problems with heat transfer. Furthermore, it is well acknowledged that the efficient resolution of boundary layers arising in this class of problems requires an adequate spatial discretization in the neighbourhood of the boundary, i.e., increased node/mesh density along the direction of large gradients only. This result is achieved by employing anisotropic node distributions, which is a novelty in the context of the RBF-FD method to the best of the authors' knowledge. The method described above is successfully employed for the accurate solution of a representative 3D heat transfer problem with incompressible fluid flow.

1. Introduction

Meshless methods applied to Computational Fluid Dynamics (CFD) problems are at the center of an increasingly animated activity of research. The reason for such an interest may be looked for in the practical shortcomings of traditional mesh-based methods adopted in this field, namely Finite Element (FEM) [1] and Finite Volume (FVM) [2] methods.

Indeed, the generation of a mesh over complex 3D geometries is a very delicate process which often requires human intervention by an adequately experienced professional. If done improperly, either the final results end up being unreliable, for example when the mesh is not appropriately refined where required, or the computational cost becomes unreasonable, such as when the refinement is excessive. Automatic mesh generation is not always possible, especially in CFD where proper refinement of the mesh is needed at specific locations of the domain in order to capture local phenomena with adequate accuracy, e.g., boundary layers.

On top of that, even when the mesh is generated properly, the resulting data structure can not be flexible enough to allow large geometric deformations of the domain without any remeshing. This can happen, for instance, when the solver is paired with some optimization algorithm and the geometry needs to be deformed several times.

When meshless methods are used, the mesh generation is avoided and usually replaced by a simple set of nodes which are properly distributed over the computational domain. Afterwards, an interpolation scheme is employed to derive the discrete form of the governing equations of the problem of interest. In our case the Radial Basis Function-Finite Difference (RBF-FD) scheme is adopted [3–8]. Its main advantage lies in the fact that there is no need to store explicitly the usual geometric connectivity information required by a mesh, e.g., those quantities related to faces, elements, etc. This theoretically permits greater freedom in the discretization procedure, and also in the case of node movements during the early stage of node generation or during the simulation, in order to account for domain deformations. The solution is computed only at the node locations and therefore, in order to achieve an adequate accuracy, a corresponding adequate node density must be employed throughout the whole domain and especially near the boundaries, where additional requirements may be necessary to guarantee stable schemes.

In this paper, an implementation of the RBF-FD meshless method is employed to accurately solve a steady-state, incompressible fluid flow problem with heat transfer on a simplified 3D domain. The feasibility of using anisotropic node distributions, i.e., in which the spacing between the nodes depends upon the direction, is investigated in order to assess the capability of the RBF-FD method to face practical CFD problems characterized by large gradients at the boundary layers. Finally, the results are compared with the ones obtained using a commercial FVM-based solver, highlighting an excellent degree of accuracy.

The required computations for the numerical simulation of a CFD problem, i.e., discretization of the domain, discretization of the governing equations and solution of the resulting algebraic equations, are all performed using the Julia programming language [9], which allows extensive code reuse and excellent computational performance already at the development stage.

2. Governing equations

The thermo-fluid problem of interest is described by coupled conservation equations of mass, momentum and energy. An incompressible fluid of density ρ, kinematic viscosity ν, thermal diffusivity α, thermal conductivity k and volumetric temperature expansion coefficient β is considered. The Boussinesq approximation is employed to model the buoyancy effects. The aforementioned equations thus take the following nondimensional form:

$$\nabla \cdot \boldsymbol{u} = 0, \tag{1}$$

$$\frac{\partial \boldsymbol{u}}{\partial t} + (\boldsymbol{u} \cdot \nabla)\boldsymbol{u} = -\nabla p + \frac{1}{\sqrt{\mathrm{Gr}_d}} \nabla^2 \boldsymbol{u} + \hat{\boldsymbol{z}} T, \tag{2}$$

$$\frac{\partial T}{\partial t} + \boldsymbol{u} \cdot \nabla T = \frac{1}{\mathrm{Pr}\sqrt{\mathrm{Gr}_d}} \nabla^2 T, \tag{3}$$

where $\hat{\boldsymbol{z}}$ is the unit vector along the vertical direction z. In the above equations, length, velocity $\boldsymbol{u} = (u, v, w)$, time t, pressure p and temperature T are made dimensionless by taking respectively d, $U_0 = \sqrt{g\beta d \Delta T}$, d/U_0, ρU_0^2 and $\Delta T = T_h - T_c$ as reference quantities (see §4), where g is the gravitational acceleration. $\mathrm{Gr}_d = g\beta d^3 \Delta T / \nu^2$ is the Grashof number, $\mathrm{Pr} = \nu/\alpha$ is the Prandtl number and $\mathrm{Ra}_d = \mathrm{Gr}_d \cdot \mathrm{Pr}$ is the Rayleigh number. For the presented computations we consider $\mathrm{Pr} = 0.71$ and three different values $\mathrm{Ra}_d = 1,000$, $5,000$ and $10,000$, leading to steady-state solutions for the presented geometry. Slightly larger values of Ra_d, e.g., above $\mathrm{Ra}_d = 15,000 \div 20,000$, lead to periodic solutions.

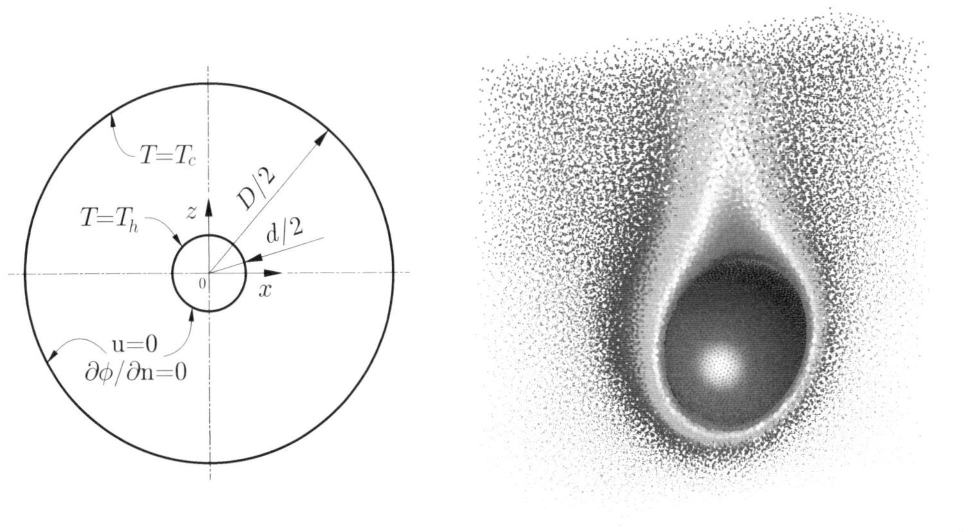

Figure 1. $x - z$ section of the computational domain and adopted boundary conditions (left) and an example of a meshless temperature distribution over the internal nodes, i.e., nodes inside the computational domain (right, 320k nodes, $\text{Ra}_d = 10^4$).

3. Numerical method

Here follows a brief description of the adopted RBF-FD method. An interested reader might refer to more extensive descriptions reported in [3–8].

The whole simulation process is composed by three phases, namely node generation, RBF-FD discretization and final solution, which are briefly described as follows.

3.1. Node generation

Node distributions are represented by a certain number N of points fulfilling a prescribed spacing function $s(\boldsymbol{x})$ scattered within the domain Ω and on its boundary $\partial\Omega$.

Node generation can be further divided in two stages:

- generation of a volumetric node distribution within Ω that satisfies a certain spacing s on average; $\partial\Omega$ is not taken into account yet,
- iterative refinement of the initial node placement through a node-repel approach until a sought node distribution is reached.

During the second stage nodes are moved according to mutual repulsion forces that arise between neighbors. Whenever one of them is pushed outside Ω, it is projected onto the nearest point on $\partial\Omega$. An example of 3D node distribution for the domain described in §4 is depicted in Figure 1 where the quality of the boundary distribution on the sphere due to the iterative refinement process can be appreciated. If the domain is defined implicitly or explicitly by means of analytic functions, the implementation of the projection is straightforward. In the generic case where the domain is defined by a CAD geometry, e.g., by means of a `.stl` (stereolithography) file, the projection operation can be efficiently performed by exploiting an octree data structure for the boundary surface [10–12].

Isotropic node distributions are obtained when the mutual repulsion forces that arise between neighbor nodes during the iterative refinement are equally weighted regardless of their direction.

Smoothness	Type	Definition $\Phi(r)$	Positive definite, order
Infinitely	Gaussian (GA)	$e^{-(\varepsilon r)^2}$	strictly
	Multiquadric (MQ)	$\sqrt{1+(\varepsilon r)^2}$	conditionally, 1
	Inverse Multiquadric (IMQ)	$1/\sqrt{1+(\varepsilon r)^2}$	strictly
	Inverse Quadratic (IQ)	$1/(1+(\varepsilon r)^2)$	strictly
Piecewise	Monomial PHS	$r^{2k+1},\ k \in \mathbb{N}$	conditionally, $k+1$
	Thin Plate Spline PHS	$r^{2k} \log r,\ k \in \mathbb{N}$	conditionally, $k+1$

Table 1. Most common Radial Basis Functions.

If the component of the repulsion force along a particular direction $\boldsymbol{e}_{an}$ has a different weighting than the orthogonal directions, the resulting node distribution will be anisotropic. Such anisotropy can then be described by two quantities: the anisotropic direction $\boldsymbol{e}_{an}(\boldsymbol{x})$ and the anisotropic factor $\kappa(\boldsymbol{x}) \geq 1$, both functions of the position $\boldsymbol{x}$. The corresponding linear transformations, which are mutually inverse, are simply given by:

$$\boldsymbol{v} \rightarrow \boldsymbol{v} + (\kappa - 1)(\boldsymbol{v} \cdot \boldsymbol{e}_{an})\boldsymbol{e}_{an} \tag{4}$$

$$\boldsymbol{w} \rightarrow \boldsymbol{w} - (1 - \kappa^{-1})(\boldsymbol{w} \cdot \boldsymbol{e}_{an})\boldsymbol{e}_{an} \tag{5}$$

representing, respectively, an expansion and a contraction of the space along $\boldsymbol{e}_{an}$ by a factor $\kappa \geq 1$. The effect of this anisotropy on the node distribution is that the node spacing along $\boldsymbol{e}_{an}$ will be $s(\boldsymbol{x})/\kappa(\boldsymbol{x})$, i.e., more nodes along $\boldsymbol{e}_{an}$, while the spacing along the orthogonal directions will be unchanged, i.e., $s(\boldsymbol{x})$. The resulting node density, i.e., nodes per unit volume, is therefore:

$$\delta(\boldsymbol{x}) = \frac{2\kappa(\boldsymbol{x})}{\zeta s^D(\boldsymbol{x})} \tag{6}$$

where $\zeta = \sqrt{5 - D}$ in D dimensions.

In order to enhance the stability of the RBF-FD scheme employed to solve the time dependent equations (1)-(3), after the iterative refinement phase a single layer of internal nodes is added from the boundary nodes in the opposite direction to the outer normals. Such strategy has a beneficial effect when dealing with Neumann boundary conditions for the auxiliary variable ϕ, equation (19).

3.2. RBF-FD collocation method

The starting point in the RBF-FD method is the construction of a suitable function that interpolates nodal values. Such interpolant is a linear combination of m Radial Basis Functions (RBFs) $\Phi(||\boldsymbol{x} - \boldsymbol{x}_i||_2)$, and the set $\{\boldsymbol{x}_i\}_1^m$ is the interpolation stencil. The most commonly used RBFs are reported in Table 1, where they are defined in terms of $r = ||\boldsymbol{x} - \boldsymbol{x}_i||_2$, for $\boldsymbol{x}_i$ fixed. In this paper the Multiquadric (MQ) radial basis function is adopted. In order to improve accuracy [13], a polynomial basis $\{p_j\}_1^q$ of the multivariate polynomial space Π_P^d of total degree P in d dimensions is also included in the RBF expansion as follows:

$$u^h(\boldsymbol{x}) = \sum_{i=1}^{m} \alpha_i \varphi_i(\boldsymbol{x}) + \sum_{j=1}^{q} \beta_j p_j(\boldsymbol{x}) \tag{7}$$

where the notation $\varphi_i(\boldsymbol{x}) = \Phi(||\boldsymbol{x} - \boldsymbol{x}_i||_2)$ is used.

The dependence of the weights $\{\alpha_1, \ldots, \alpha_m\}$ and $\{\beta_1, \ldots, \beta_q\}$ on the nodal values is obtained by considering the following interpolation conditions:

$$u^h(\boldsymbol{x}_i) = u(\boldsymbol{x}_i), \quad i = 1, \ldots, m \tag{8}$$

When boundary nodes are included in the stencil, the corresponding interpolation conditions of equation (8) are replaced by the respective boundary conditions. In the case of Robin b.c., i.e., $au + b\,\partial u/\partial \mathbf{n} = g$, applied to the ith node of the stencil, the corresponding equation becomes:

$$a(\boldsymbol{x}_i)\, u^h(\boldsymbol{x}_i) + b(\boldsymbol{x}_i)\,\frac{\partial u^h}{\partial \mathbf{n}}(\boldsymbol{x}_i) = g(\boldsymbol{x}_i), \quad \boldsymbol{x}_i \in \partial\Omega \tag{9}$$

where a, b, g are all known functions defined on the boundary.

In order to make the system solvable, other q equations are attained by enforcing the following additional orthogonality conditions:

$$\sum_{i=1}^{m} \alpha_i p_j(\boldsymbol{x}_i) = 0, \quad j = 1, \ldots, q \tag{10}$$

By substituting equation (7) into (8) and (9), the unknown values of α_i and β_j can be calculated by solving the following linear system:

$$\underbrace{\begin{bmatrix} \varphi_{BC} & \mathbf{P}_{BC} \\ \mathbf{P}^T & \mathbf{0} \end{bmatrix}}_{\mathbf{M}} \begin{Bmatrix} \boldsymbol{\alpha} \\ \boldsymbol{\beta} \end{Bmatrix} = \begin{Bmatrix} \boldsymbol{u} \\ \boldsymbol{g} \\ \mathbf{0} \end{Bmatrix} \tag{11}$$

where the block matrix $\boldsymbol{P}^T$ comes from equation (10), $\{\alpha_1, \ldots, \alpha_m, \beta_1, \ldots, \beta_q\}$ is the vector of unknowns and both $\boldsymbol{\varphi_{BC}}$ and $\boldsymbol{P_{BC}}$ come from equations (8) and (9).

Given a linear partial differential equation $\mathcal{L}(u) = f$ in the unknown field u, the RBF expansion (7) is made valid at each node $\boldsymbol{x}_i$ which does not lie on the boundary. The following sparse linear system is thus obtained:

$$\mathbf{C}\left\{\boldsymbol{u}^h\right\} = \boldsymbol{q} - \boldsymbol{f} \tag{12}$$

where $\boldsymbol{u}^h = \{u^h(\boldsymbol{x}_1), \ldots, u^h(\boldsymbol{x}_{N_I})\}$ is the vector of unknown scalar values $u^h(\boldsymbol{x}_i)$ evaluated at all N_I inner nodes, i.e. those contained in Ω, while the vector $\boldsymbol{q}$ comes from the enforcement of boundary conditions at the interpolant level. Each row $\mathbf{c}_i$ of the matrix $\mathbf{C}$ therefore correspond to the inner node at $\boldsymbol{x}_i$ and is attained through the solution of the local linear system (13), associated to the stencil built around that node.

$$\mathbf{M}^T\,\mathbf{c}_i = \begin{Bmatrix} \boldsymbol{\Psi}(\boldsymbol{x}_i) \\ \boldsymbol{\Pi}(\boldsymbol{x}_i) \end{Bmatrix} \tag{13}$$

where $\boldsymbol{\Psi}(\boldsymbol{x}_i) = \{\mathcal{L}\varphi_i(\boldsymbol{x}_i) \ldots \mathcal{L}\varphi_m(\boldsymbol{x}_i)\}$ and $\boldsymbol{\Pi}(\boldsymbol{x}) = \{\mathcal{L}p_1(\boldsymbol{x}_i) \ldots \mathcal{L}p_q(\boldsymbol{x}_i)\}$ are column vectors and $\mathbf{M}$ is the same as in equation (11).

3.3. Solution procedure

The computation of velocity, pressure and temperature through equations (1)-(3) is decoupled at each time step by using a projection scheme with a three-level Gear scheme for the time discretization. A tentative velocity $\boldsymbol{u}^*$ is computed from the linearized nondimensional momentum equation:

$$\frac{3\boldsymbol{u}^* - 4\boldsymbol{u}^l + \boldsymbol{u}^{l-1}}{2\Delta t} + \boldsymbol{u}^l \nabla \boldsymbol{u}^* = -\nabla p^l + \frac{1}{\sqrt{\mathrm{Gr}_D}} \nabla^2 \boldsymbol{u}^* + \hat{\boldsymbol{z}} T^l, \tag{14}$$

where l is the time level and Δt is the nondimensional time step size, which is chosen to be $\Delta t = 0.25$ for the presented results.

The tentative velocity $\boldsymbol{u}^*$ is then forced to satisfy the continuity equation (1) by means of an irrotational correction $\boldsymbol{u}^{l+1} = \boldsymbol{u}^* - \nabla\phi$, leading to the following Poisson equation in the auxiliary variable ϕ:

$$\nabla^2 \phi = \nabla \cdot \boldsymbol{u}^* \tag{15}$$

The pressure is then updated as $p^{l+1} = p^l + \phi/\Delta t$ and the temperature is computed from the discretized nondimensional energy equation:

$$\frac{3T^{l+1} - 4T^l + T^{l-1}}{2\Delta t} + \boldsymbol{u}^{l+1} \nabla T^{l+1} = \frac{1}{\mathrm{Pr}\sqrt{\mathrm{Gr}_D}} \nabla^2 T^{l+1}. \tag{16}$$

The previous equations are discretized in space with the RBF-FD scheme presented in §3.2. Explicit hyperviscosity ∇^4 [14] is employed to stabilize the time evolution of the variables $\boldsymbol{u}$, p and T. Sparse linear systems like the one in Equation (12) are attained for each component of velocity, for the temperature and for the auxiliary variable ϕ. Such sparse linear systems are preconditioned with an Incomplete LU factorization (ILU) [15] (package `IncompleteLU` in `Julia`). In the case of the Poisson equation for the velocity correction, equation (15), the ILU factorization can be performed only once at the beginning of the simulation. The Biconjugate Gradient Stabilized Method [16] (package `IterativeSolvers` in `Julia`) is then employed as iterative solver using a relative tolerance of 10^{-10}. The computational time required for each time step ranges from 2 seconds for $N \approx 100,000$ nodes to approximately 9 seconds for $N \approx 500,000$ nodes on a laptop equipped with a quad-core Intel i7 2.6GHz processor.

3.4. Convergence criterion

The convergence to steady-state through the time stepping procedure explained in §3.3 is declared when the average of the root-mean-square (RMS) time derivatives for the velocity components and temperature become less than $tol = 5 \cdot 10^{-6}$. The RMS time derivative $\dot{f}_\Omega$ for a generic field f over the domain Ω is defined as:

$$\dot{f}_\Omega = \sqrt{\frac{1}{\mu(\Omega)} \int_\Omega \left(\frac{\partial f}{\partial t}\right)^2 d\Omega} \approx \sqrt{\frac{1}{\sum_i s_i^3} \sum_i \left(\frac{f_i^{l+1} - f_i^l}{\Delta t}\right)^2 s_i^3} \tag{17}$$

where $\mu(\Omega)$ is the measure of the domain Ω and the notation $f_i = f(\mathbf{x}_i)$ for $\mathbf{x}_i \in \Omega$ is used. Therefore the convergence criterion can be explicitly written as:

$$\sqrt{\frac{\dot{u}_\Omega^2 + \dot{v}_\Omega^2 + \dot{w}_\Omega^2 + \dot{T}_\Omega^2}{4}} < tol \tag{18}$$

4. Geometry, boundary conditions and generic definitions

A 3D sphere of diameter d, assumed as the reference length, is enclosed in a spherical cavity with diameter $D = 5d$, i.e, a spherical shell. Both spheres share the same center point, which is also the origin of the considered cartesian reference frame, as shown in Figure 1. The following boundary conditions are enforced:

$$\begin{cases} \text{no-slip condition } \mathbf{u} = \mathbf{0} \text{ on all boundaries,} \\ T = T_h \text{ on the surface of the internal sphere of diameter } d, \\ T = T_c \text{ on the surface of the external sphere of diameter } D, \\ \dfrac{\partial \phi}{\partial \mathbf{n}} = 0 \text{ on all boundaries.} \end{cases} \tag{19}$$

The last condition in (19) is due to the fact the domain is a cavity, i.e., it has no inlets or outlets. Therefore equation (15) must satisfy the compatibility condition $\int_\Omega \nabla \cdot \mathbf{u}^* = 0$ in order to be solvable. At the PDE level such condition is always met in a cavity because of the divergence theorem. However, when considering the discretized version of the Poisson equation (15), i.e., the linear system $\mathbf{A}\boldsymbol{\phi} = \boldsymbol{b}$, the compatibility condition is:

$$\boldsymbol{w}^T \boldsymbol{b} = 0 \tag{20}$$

where $\boldsymbol{w}$ is any vector of the 1D nullspace of $\mathbf{A}^T$, i.e., $\mathbf{A}^T \boldsymbol{w} = \mathbf{0}$. The compatibility condition (20) is not generally met when using non-conservative schemes like the bare RBF-FD method. In order to meet such condition, each entry of the right hand side vector $\boldsymbol{b}$ of the discretized Poisson equation can be corrected by subtracting the value $\boldsymbol{w}^T \boldsymbol{b} / \sum w_i$.

The mean Nusselt number over the surface Γ of the internal sphere is given by:

$$\overline{\text{Nu}}_d = \frac{d}{\Delta T \cdot \mu(\Gamma)} \int_\Gamma \frac{\partial T}{\partial \mathbf{n}} d\Gamma \approx \frac{d}{\Delta T \sum_b s_b^2} \sum_b \left. \frac{\partial T}{\partial \mathbf{n}} \right|_b s_b^2 \tag{21}$$

where the index b runs over each boundary node on Γ.

The spacing function $s(\boldsymbol{x})$ and the anisotropic factor $k(\boldsymbol{x})$ are defined as follows:

$$s(\boldsymbol{x}) = s_{min} + (s_{max} - s_{min}) \Delta r \frac{1 + \lambda_s}{1 + \lambda_s \Delta r} \tag{22}$$

$$\kappa(\boldsymbol{x}) = \kappa_{max} - (\kappa_{max} - 1) \Delta r^2 \frac{1 + \lambda_\kappa^2}{1 + \lambda_\kappa^2 \Delta r^2} \tag{23}$$

where s_{min} and s_{max} are the minimum and the maximum values of the spacing function, respectively, and κ_{max} is the maximum value of the anisotropic factor. $\Delta r = (2r - d)/(D - d)$ where r is the radius from the center of the spheres and the following parameters are employed: $s_{max}/s_{min} = 3$, $\lambda_s = 3$ and $\lambda_k = 10$. These definitions are such as to obtain node distributions with a small spacing s_{min} at the surface of the internal sphere and a large spacing s_{max} at the surface of the external one. The anisotropic factor grows from $\kappa = 1$ at the external sphere to $\kappa = \kappa_{max}$ at the internal one while the anisotropic direction $\boldsymbol{e}_{an}$ is chosen to be radial, i.e., it points towards the center of the spheres.

5. Results

5.1. Preliminary considerations on a 2D Poisson equation

In order to highlight the advantages and the basic feasibility of the use of an anisotropic node distribution in the context of the RBF-FD method, let us consider the following 2D Poisson equation:

$$\nabla^2 u = b \tag{24}$$

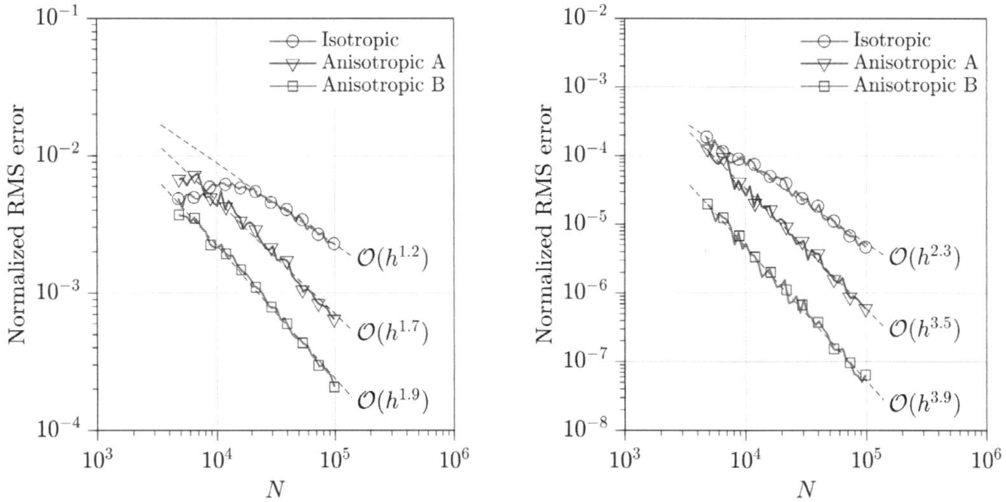

Figure 2. Convergence curves for a 2D Poisson equation in the case of a solution with large gradients, solved by RBF-FD with polynomial degree $P = 2$ (left) and $P = 4$ (right) and by using isotropic and non-isotropic node distributions.

with Dirichlet b.c. on the unit square $[0, 1]^2$. The chosen analytical solution is $u = e^{-50x}$, which has a very large gradient at $x = 0$ which requires a localized increase in the node density in order to be accurately resolved, for example by using the following spacing function:

$$s(\boldsymbol{x}) = s_{min} + (s_{max} - s_{min})\frac{1 - e^{-3x}}{1 - e^{-3}} \tag{25}$$

where $s_{max} = 3s_{min}$ is chosen. The resulting node distribution is isotropic and more refined near $x = 0$, although the increase in the node density along y is unnecessary since the large gradient of u is along x whereas u is constant along y. Therefore it would be much more effective to employ an anisotropic node distribution with increased node density only along x, i.e., with a spacing along x defined by equation (25), $s_x = s(\boldsymbol{x})$, and a constant spacing along y, $s_y = s_{max}$. The convergence curves for these two approaches are shown in Figure 2 for polynomial degrees $P = 2$ and $P = 4$ and $n = 2q$ local nodes. In strategy A the space for the interpolant is scaled with shape factors $\varepsilon_x = 0.5/s_x$ along x and $\varepsilon_y = 0.5/s_y$ along y, i.e., the interpolant "sees" isotropic nodes, while in strategy B the interpolant is not scaled, i.e., $\varepsilon = 0.5/s_y$ as usual. From Figure 2 we see that the anisotropic node distributions provide a significant increase in accuracy over the isotropic distributions. We can also see that strategy B seems to be the best in terms of both rate of convergence (1.9 for $P = 2$ and 3.9 for $P = 4$, as expected) and absolute error. This is due to the fact that the scaling employed in strategy A is equivalent to an increase in the shape factor ε, which is known to affect the accuracy of the RBF interpolation.

This example shows that, as expected, the most effective way to accurately capture a rapidly varying solution is to increase the node density only in the direction of the largest gradient.

5.2. Natural convection in a spherical shell

In order to highlight the properties of the employed RBF-FD approach for the solution of equations (1)-(3), several simulations are carried out for different node distributions using a total number of nodes ranging from $N \approx 100,000$ nodes to $N \approx 500,000$ nodes. Polynomial degree $P = 3$ is employed with $n = 2q = 40$ local nodes and $\varepsilon = 0.5/s(\boldsymbol{x})$.

Table 2. Mean Nusselt number.

	N	$\overline{\mathrm{Nu}}_d$ $\mathrm{Ra}_d = 10^3$	$\mathrm{Ra}_d = 5 \cdot 10^3$	$\mathrm{Ra}_d = 10^4$
Isotropic	$100k$	4.0336	5.3949	6.1342
	$130k$	4.0340	5.4000	6.1352
	$165k$	4.0321	5.3879	6.1198
	$320k$	4.0393	5.3958	6.1262
	$455k$	4.0399	5.3964	6.1262
Anisotropic	$120k$	4.0300	5.3858	6.1144
	$150k$	4.0324	5.3896	6.1176
	$190k$	4.0366	5.3938	6.1245
	$360k$	4.0361	5.3921	6.1223
	$515k$	4.0351	5.3905	6.1204
	Fluent	4.0415	5.3994	6.1300

The results are summarized in Table 2, where the computed mean Nusselt number $\overline{\mathrm{Nu}}_d$ is reported for the three chosen values of the Rayleigh number $\mathrm{Ra}_d = 1,000$, $5,000$ and $10,000$, and for the two strategies employing isotropic and anisotropic node distributions. In the last case, the anisotropic factor is chosen to be $\kappa_{max} = 2$ for which the interpolation scaling is chosen to be halfway between strategies A and B, briefly presented in §5.1. This choice is due to the fact that strategy B turns out to be the most unstable, despite it seems to be the most accurate.

Table 2 shows also the values of $\overline{\mathrm{Nu}}_d$ computed with Ansys Fluent 2022R1 as reference, where the axial symmetry of the problem is exploited in order to provide a reliable and accurate solution. From the comparison of the values of this table we can observe a very strong agreement between the presented results and the reference ones for each of the three values of Ra_d, even for the smallest number of nodes, i.e., $N \approx 100k$ nodes, with errors always much lower than 1%.

From these values it can also be observed that the anisotropic results are slightly worse than the isotropic ones, when compared to the reference values computed by Fluent. This can be due to the interpolant scaling strategy which is not the optimal one, i.e., strategy B presented in §5.1. In other words, the nodal spacing is reduced anisotropically along the radial direction $\boldsymbol{e}_{an}$ when approaching the surface of the internal sphere, while at the same time the shape factor for the RBF interpolant increases along the same direction. This last effect seems to be predominant in the presented cases, where the gradients of the field variables in the boundary layers are such as to be sufficiently resolved without the need of anisotropic distributions.

Nonetheless, these results confirm that the RBF-FD approach can operate with consistency also with anisotropic node distributions, and therefore it can be employed to face practical engineering problems at large values of Rayleigh or Reynolds numbers. The consistency of the presented anisotropic RBF-FD approach is also confirmed by the fact that, in the presented cases, the computed velocity and temperature profiles, like the ones presented in Figure 3, are indistinguishable from the isotropic ones.

Figure 3 shows different comparisons between the profiles of some variables of interest, obtained with an isotropic distribution and $N \approx 320k$ node, for each of the considered values of Ra_d. Reference values obtained with Fluent are also shown for comparison. Figures 3a and b show, respectively, the temperature T and the vertical velocity w along the horizontal line $y = z = 0$. Figure 3c shows the local Nusselt number Nu_d at the surface of the internal sphere as a function of the angle α taken from the horizontal line $y = z = 0$.

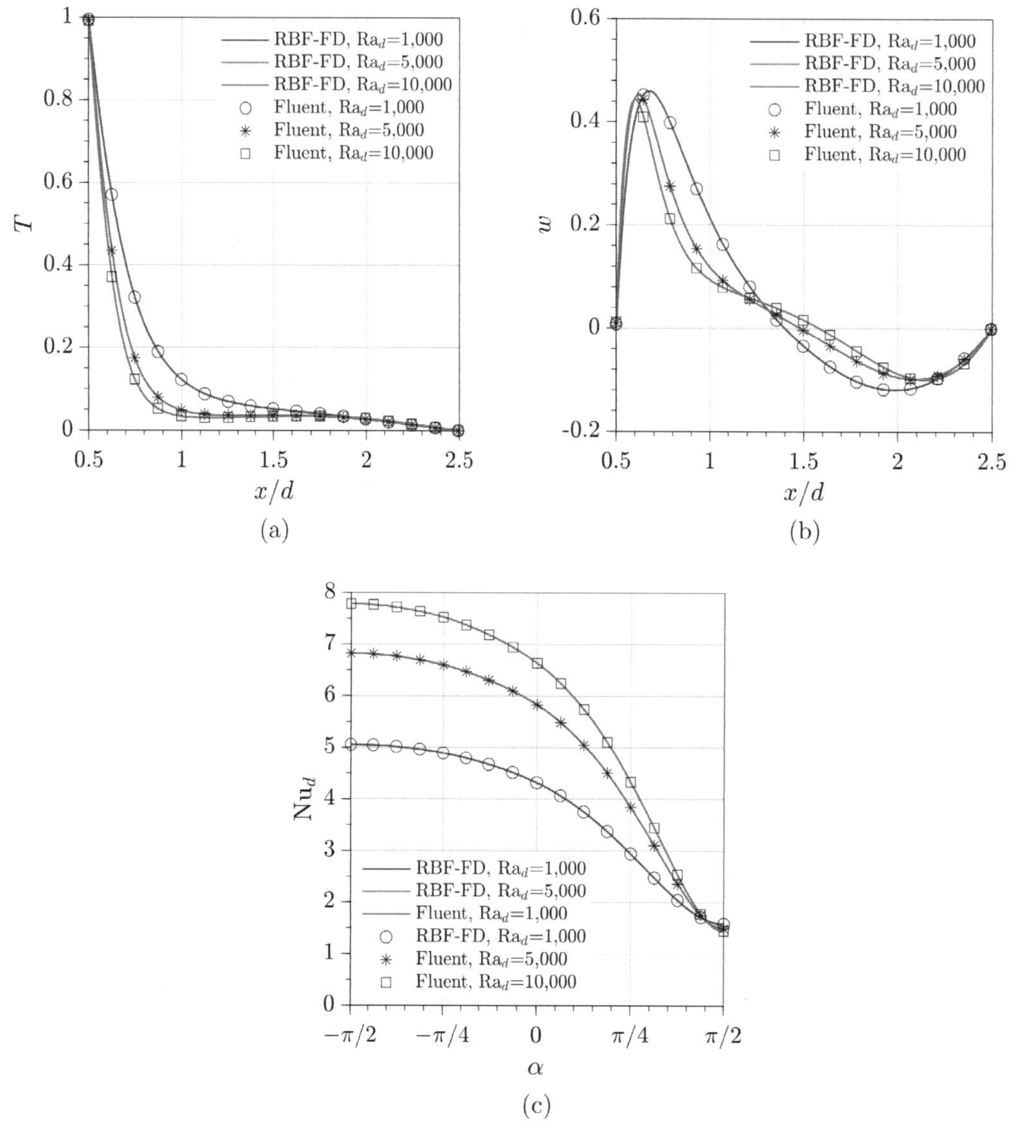

Figure 3. Comparison of temperature T (a) and vertical z-velocity w (b) profiles along the horizontal line $y = z = 0$. Comparison of local Nusselt number Nu_d (c) on the internal sphere.

Again, all the previous graphical comparisons show an excellent agreement between the computed profiles and the reference ones.

6. Conclusions

In this work the RBF-FD meshless method is applied to the simulation of a natural convection problem with laminar flow in a spherical shell at Rayleigh numbers $\mathrm{Ra}_d = 1,000$, $5,000$ and $10,000$. Excellent agreements with reference results are consistently obtained. Particular attention was reserved to the development and testing of a new method for attaining anisotropic node distributions. As a matter of fact, while node generation algorithms and meshless methods in general are theoretically capable of coping with very complex shapes, isotropic node placements might lead to unnecessary node concentrations at certain spots of the domain, leading to a waste of computational resources. The adoption of anisotropic node distributions addresses this issue by increasing the node density only along a single meaningful direction in order to provide an efficient resolution of large gradients, e.g., boundary layers.

Future developments will insist at perfecting the algorithms for the optimal node placement and for stable anisotropic RBF approaches. For instance, the natural evolution of the anisotropic node generation lies in its integration within an adaptive node refinement algorithm. Indeed, it appears clear that, in order to increase both convenience and accuracy of collocation meshless methods, a lot of effort needs to be invested in the development of efficient and totally autonomous node generation algorithms.

The attained results suggest that the RBF-FD method with isotropic and even anisotropic node distributions has the potential to provide both high accuracy and ease of use for the numerical solution of generic heat transfer problems of engineering interest, and therefore represents an ideal candidate for the emerging needs of the industry.

References

[1] Zienkiewicz O C 1977 *The finite element method* vol 3 (McGraw-hill London)
[2] Versteeg H K and Malalasekera W 2007 *An introduction to computational fluid dynamics: the finite volume method* (Pearson education)
[3] Fornberg B and Flyer N 2015 *A primer on radial basis functions with applications to the geosciences* (SIAM)
[4] Divo E and Kassab A J 2007 *J HEAT TRANSF* **129** 124–136
[5] Zamolo R and Nobile E 2019 *NUMER HEAT TR B-FUND* **75** 19–42
[6] Šarler B and Vertnik R 2006 *COMPUT MATH APPL* **51** 1269–1282
[7] Waters J and Pepper D W 2015 *NUMER HEAT TR B-FUND* **68** 185–203
[8] Kosec G and Slak J 2020 *AIP Conference Proceedings* **2293** 420094
[9] Bezanson J, Edelman A, Karpinski S and Shah V B 2017 *SIAM REV* **59** 65–98
[10] Frey P and George P 2013 *Mesh Generation: Application to Finite Elements* ISTE (Wiley) ISBN 9781118623824
[11] de Berg M, Cheong O, van Kreveld M and Overmars M 2008 *Computational Geometry: Algorithms and Applications* (Springer Berlin Heidelberg) ISBN 9783540779742
[12] Zamolo R and Nobile E 2021 *J PHYS CONF SER* **2116**
[13] Bayona V, Flyer N, Fornberg B and Barnett G A 2017 *J COMPUT PHYS* **332** 257–273
[14] Fornberg B and Lehto E 2011 *J COMPUT PHYS* **230** 2270–2285
[15] Saad Y 2003 *Iterative Methods for Sparse Linear Systems* (Philadelphia, Pennsylvania: SIAM) chap 10: Preconditioning Techniques, pp 297–368 2nd ed
[16] van der Vorst H 1992 *SIAM J SCI COMPUT* **13**(2) 631–644

Energy demand in secondary steel making process: numerical analysis to assess the influence of the ladle working lining properties

M Neri and A M Lezzi

University of Brescia, via Branze 38, 25123 Brescia, Italy

E-mail: manuela.neri@unibs.it and adriano.lezzi@unibs.it

Abstract. In secondary steel making process, steel scrap is melted in an electric arc furnace, then poured into a ladle where ferroalloy are added to obtain the desired steel composition. Energy is also supplied in LF (Ladle Furnace) to guarantee steel temperature required in the teeming phase. In this paper a simplified, but sound numerical model of the process the ladle undergoes is presented: it is meant to provide a way to compare energy demand for different choices of materials used in the ladle or for changes in the ladle work cycle. As an example of application of the model, energy demand in LF are evaluated and compared for two different choices of the refractory material used as working lining: magnesia and alumina. To take into account the reduction of the working lining due to wear processes, energy demand has been calculated for three different thicknesses. In addition, the beneficial effect of a lid during the waiting phase after cleaning is investigated.

1. Introduction

In the secondary steel making process, steel scrap is rapidly melted in an electric arc furnace (EAF) and, then, transformed into final products through a refining process that takes place in a ladle. The process can be schematized as a sequence of phases as shown in figure 1. The first phase is the treatment in the Electric Arc Furnace (EAF), where scrap is melted by supplying energy through an electric arc and gas burners. Once the steel is completely melted and it has reached a threshold temperature, it is poured into a ladle jointly with a certain amount of slag, slag formers and deoxidizers (tapping phase). The presence of the slag is beneficial since it reduces the heat losses from the steel surface, and guarantees a better control of the steel temperature. To limit thermal shocks due to the sudden contact with the hot molten steel, the ladle is previously heated for several hours by means of gas burners. After the tapping phase, the ladle is transferred to the ladle furnace (LF) where ferroalloys are added to adjust the steel composition, and energy is supplied by means of electrodes immersed in the molten steel. In a standard cycle, the LF is the last phase when temperature is controlled through the supply of energy to overcome temperature drop due to cold ferroalloys addition, endothermic chemical reactions, and heat losses towards the environment. In the following phases, the ladle is closed with a lid to limit the liquid temperature decrease. At this point, the ladle may follow two different paths: it can be sent directly to the Continuos Casting (CC) where the molten steel is drained from the ladle to be transformed into final products or, before going to CC, it can undergo to the Vacuum Oxygen Decarburizing (VOD) - a process to dissolve undesired gases

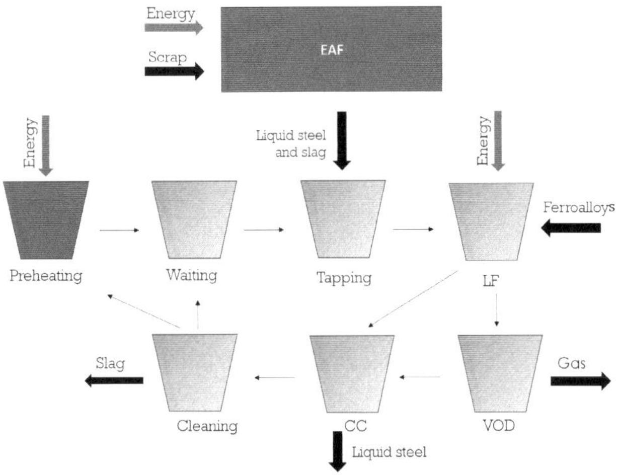

Figure 1. Scheme of the secondary steel making process.

in high quality steels. Finally, the ladle is sent to a cleaning station where residuals of steel and slag are removed. The set of these phases is called a cycle. Between each couple of phases described above, the ladle is made waiting a variable period of time, called in the following *holding* or *waiting* phase. The temperature of the steel and of the ladle decreases progressively and, for this reason, the greater the cycle time duration, the greater the energy required to heat the steel. Heat losses depend on the ladle thermal properties and the cycle time duration that is variable depending on delays that may occur in the following phases of the process such as potential failures (failure of the ladle opening in the CC), replacement of electrodes, etc. The waiting time of the ladle to reenter the cycle may cause a significant cooling of the ladle and, in this case, it is heated again.

A ladle is a slightly cone-shaped vessel internally lined with refractory bricks resistant to wear process. As shown in figure 2, walls and base are made of an external carpentary, an intermediate permanent lining, and an internal working lining. At the top of the ladle there is the *slag lining* that is highly resistant to corrosion. The permanent lining protects the metallic carpentry in case of a working lining break and its thickness does not vary over time, while the working lining undergoes a progressive reduction in thickness as a consequence of the periodic contact with the molten steel and, for this reason, it is periodically renovated. The number of cycles of use between two renovations determines the ladle service life: it may span between 70 and 90 cycles. The working lining material affects the ladle life-duration; refractory materials such as magnesia and alumina guarantees a longer ladle life but they are sensitive to thermal shocks and, for this reason, they need to be preheated. To remix the steel and homogenize its temperature, argon is insuflated through the bottom of the ladle and this process is called *stirring*. The induced agitation results in thermal homogenization of the molten steel, removal of alumina inclusion, but, at the same time, greater heat losses towards the ambient.

In the literature, the secondary steel making process has been investigated from several points of view: some studies focused on the entire process like [1–3], while others analyzed a single phase, such as preheating [3], holding [4–6], continuos casting [6, 7], tapping [7], and the desulfurization phase [8]. The analysis of the entire process show that the larger energy losses occur in the waiting phase between two cycles, and in the VOD facility [1]. According to the study in [4], the preheating phase, that is the initial ladle temperature, strongly affects the temperature distribution in the molten steel. On the contrary, the distance between lid and ladle

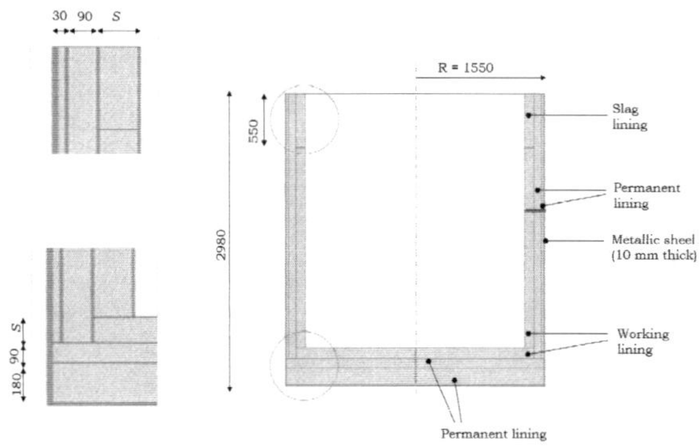

Figure 2. Geometry of the ladle defined in the numerical model. The working lining thickness S has been varied to consider different ladle conditions, and the horizontal red line represents the position where the ladle temperature reported in the paper has been evaluated. Dimensions are in millimeters.

does not affect the temperature distribution [3]. The preheating duration could be reduced for a thinner working lining. Also the heat losses through the slag strongly affect the temperature distribution in the molten steel [4]. By varing the heat flux through the slag, it was shown that stratification occurs with a temperature variation of 6 °C [6]. Thermal stratification and inclusions trajectories during the holding period are evaluated in [7]. Other studies focused on the influence of the ladle material properties. It was shown that alumina has better performances than magnesite [8], but very insulating refractory materials may cause a stronger corrosion and thermal shocks [2,9]. A thermal and thermomechanical lining optimization was proposed in [10]. Other studies investigated the casting process from the exergy point of view [11–13]. For example, the second law analysis has been applied to the ladle furnace (LF) and it was found that it is possible to reduce the irreversibilities by improving the insulation of the LF, by using high quality refractors [14].

The study presented in this paper was prompted by a steel plant company based in Brescia (Italy) interested in understanding how the ladle characteristics affect the energy demand in the secondary steel production process. The analysis has been done by means of a simplified axisimmetric unsteady numerical model of the entire process that the ladle undergoes - from the preheating (before the first tapping) to the cleaning (after CC) and the waiting before next tapping. This has been repeated for several cycles until a stabilized periodic condition is reached. At each cycle the required output is an estimate of the energy consumption which is taken equal to the amount of energy supplied in the LF. Although the physical and numerical model proposed is rather crude, it is sound enough to predict energy demand trends and to allow quantitative comparisons between different choices of ladle characteristics or process parameters. As an example of application of the model, different scenarios are considered such as different working lining materials – magnesia and alumina – or different thicknesses to investigate the effect of wear on energy demandand thickness. In addition, the beneficial effect of a lid during the waiting phase after cleaning is investigated.

2. Model description

In this study a 2D axi-symmetrical numerical model of the ladle has been developed to estimate the energy demand in the LF as a function of the thickness and thermal properties of the ladle working lining (see figure 1). In addition, energy savings due to the presence of a lid in the final phase of the process are evaluated. As reported in table 1, 12 cases have been considered: the thickness S in figure 1 has been set equal to 150 mm (S_{max}) representative of the situation in the cycles after renovation of the working lining, 115 mm (S_{int}) representative of lining thickness at half service-life, and 80 mm (S_{min}) representative of the condition just before the lining renovation. Therefore, the ladle external radius and height do not vary, but the inner radius and depth do. Since for all the three thickenesses the initial steel mass is the same (83846 kg, including the slag), the initial steel level changes as thickness varies. To assess the influence of the working lining properties, two refractory materials - magnesia and alumina - have been considered, whereas to assess the effect of a lid in the final phase, two scenarios - with (WL) and without (NL) lid - have been considered.

The steel-making process has been divided into cycles (identified by letter C) and phases (identified by letter F) shown in figure 3. Letters i and f identify the phase initial and final time, respectively; for example, $C1fF2$ is the final time of the second phase $F2$ in the first cycle. A standard cycle is composed by the following phases: filling ($F2$), LF ($F3$), holding ($F4$), emptying in CC ($F5$), cleaning ($F6$), and waiting ($F7$). In this study we do not consider steels that require VOD. In the preheating phase $F1$, the ladle is prepared to receive the molten steel and it is preheated with burners placed inside. Preheating is carried out only when the ladle enters the production process; after the first cycle the ladle arriving from the CC is still hot enough to avoid thermal shock for the working lining. In this study, only one preheating is considered. Since phases $F1$ and $F1_{bis}$ are present in the first cycle $C1$, the first cycle $C1$ counts 8 phases while the other ones count 6 phases. Then, the ladle waits about 30 minutes (phase

Table 1. Cases analyzed. WL and NL are representative of the scenario with and without lid respectively.

With lid (WL)						Without lid (NL)					
Magnesia			Alumina			Magnesia			Alumina		
S_{max}	S_{int}	S_{min}	S_{max}	S_{int}	S_{min}	S_{max}	S_{int}	S_{min}	S_{max}	S_{int}	S_{min}

Table 2. Material properties set in the model. Data without reference have been provided by producers who wish to remain anonymous.

Material	ρ (kg/m^3)	c_p (kJ/(kg·K))	λ (W/(m·K))	ε (-)
Magnesia	3030	1.30	8 (0 °C), 6 (400 °C), 4.8 (800 °C), 4 (1200 °C)	0.45
Alumina	3150	1.49	3.2 (0 °C), 2.4 (400 °C) 2.75 (800 °C), 2.54 (1000 °C)	0.45
Slag	2800 [8]	0.52 [8]	0.48 [8]	0.6 [9]
Perm. lin. - wall	2700	1.05	2.3	-
Carpentery	7200 [8]	0.79 [8]	52 [8]	0.8 5
Perm. lin. - bottom	2100	1.05	2	-
Molten steel	6981	0.78	41	-
Lid internal	2390	1.00	1.35	0.45
Lid external	2390	1.00	1.35	0.85
Slag lining	3030	0.89 (200 °C), 1.06 (700 °C), 1.14 (1200 °C)	8 (0 °C), 6 (400 °C), 4.8 (800 °C), 4 (1200 °C)	0.45

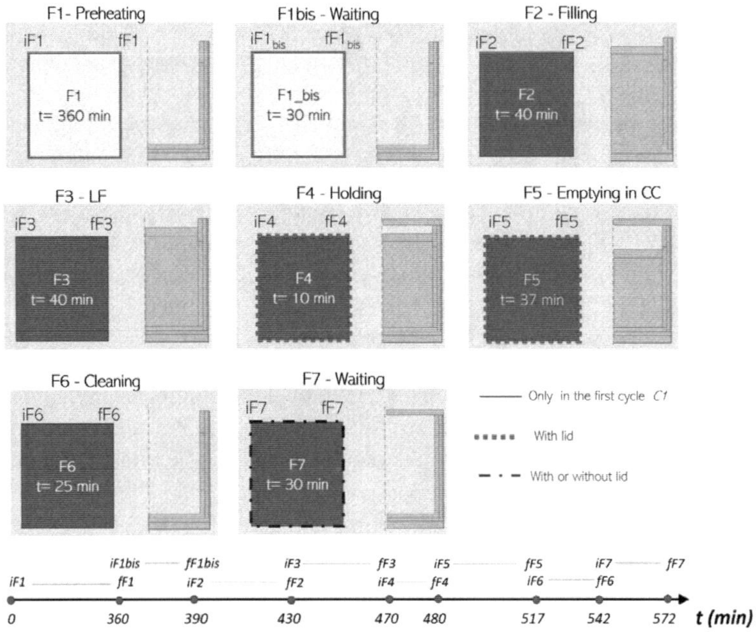

Figure 3. Scheme of the steel production process and geometry considered in the different phases of the study. Letters i and f indicate the phase initial and final time, respectively.

$F1_{bis}$) before being filled with steel. In phase $F2$, the ladle is quickly filled with steel and slag at 1600 °C from the EAF and waits 40 minutes. In the LF - phase $F3$ - ferroalloy and energy are supplied to the ladle and the molten steel is heated up to 1595 °C and the process lasts 40 minutes. Then, in phase $F4$, the ladle is closed with a lid and waits 10 minutes before entering the CC - phase $F5$ - where it is progressively emptied. To remove the slag and residuals of steel, in phase $F6$, the ladle is open. Finally, in phase $F7$, the ladle waits 30 minutes before entering the next cycle, and it may be open (scenario indicated as NL) or closed (scenario WL). The phase time durations reported in figure 3 have been agreed with the steel making company that prompted the study and are representative of average times during normal operations.

2.1. Boundary conditions

Each of the phases shown in figure 3 has been analyzed by means of one or more numerical simulations: the computed temperature distribution at the end of each simulation has been used as initial condition for the following one. The mesh cell size is about 10 mm. On the ladle surfaces, both convection and radiation have been considered:

$$q = h(T_{surf} - T_{amb}) + \sigma\varepsilon(T_{surf}^4 - T_{amb}^4) \tag{1}$$

where h is the convective heat transfer coefficient (see table 3), the ambient temperature T_{amb} is 30 °C and, unless otherwise stated, the air temperature inside the ladle is 400 °C. Convective motions in the molten steel and in the slag have been neglected: due to the intense stirring action, steel temperature is forced to be uniform (but not constant) and thermal contact between steel and ladle is assumed perfect.

During the preheating phase $F1$, the contribution of gas burners has been modeled by

Table 3. Convective heat transfer coefficients in $W/(m^2\,K)$ imposed on the ladle surfaces during each phase.

Surface	F1	$F1_{bis}$	F2	F3	F4	F5	F6	F7 (NL)	F7 (WL)
A	-	7	-	-	-	-	7	7	7
B	-	5	-	-	-	-	5	5	5
C	2	2	9	0	-	-	9	9	-
D	5	7	7	7	7	7	7	7	7
E	2	3	3	3	3	3	3	3	3
F	-	-	9	0	3	3	-	-	-
G	-	-	7	0	5	5	-	-	-
H	-	-	-	-	9	9	-	-	8
I	-	-	-	-	6	6	-	-	3

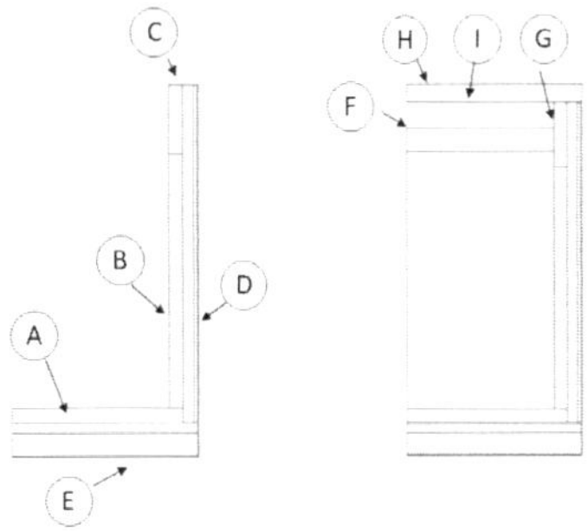

imposing a time dependent Dirichlet condition on the internal surface:

$$T_{\mathrm{br}} = 25°\mathrm{C} + \frac{1100°\mathrm{C} - 25°\mathrm{C}}{11\,\mathrm{h}} \cdot t \qquad 0\,\mathrm{h} < t < 11\mathrm{h} \tag{2}$$

$$T_{\mathrm{br}} = 1100\,°\mathrm{C} \qquad 11\,\mathrm{h} < t < 13\,\mathrm{h} \tag{3}$$

where t is the time in hours. Since *tapping* is a quite fast process, it is considered instantaneous in the model, so simulation of phase $F2$ starts with the ladle full of steel at 1600 °C. In the LF, real processes are characterized by multiple and complex positive and negative energy flows due to the electrodes, to the addition of ferroalloy and to chemical reactions. These aspects are taken into consideration in phase $F3$ by imposing in the steel a source term q''' constant for the entire phase duration. This term is determined in a direct way. A few simulations of $F3$ are performed for slightly different values of q''' and final steel temperature is recorded. Then by interpolation the value of q''' that takes steel at 1595 °C at the end of $F3$ is determined. The E_{LF} energy to be supplied is:

$$E_{LF} = q''' \cdot V \cdot \Delta t \tag{4}$$

where V is the volume of steel and slag, and Δt is the duration of $F3$. In phase $F4$, the lid is initially at 250 °C, and the air temperature inside the ladle is:

$$T_{\mathrm{air}} = 1.2064 \cdot t^3 - 29.584 \cdot t^2 + 249.68 \cdot t + 45.349 \tag{5}$$

where T is in °C and t is measured in hours. Since the emptying of the ladle in CC is a relative long process, phase $F5$ has been divided into 11 numerical simulations. In each simulation, the steel level is lowered of 200 mm, and the portion of freed ladle is replaced by air at 800 °C. The average temperature of the slag at the end of each simulation has been imposed as initial condition in the next one. In phases $F6$ and $F7$, the air in ladle is at 400 °C [28], and the initial temperature of the lid (if present) is 250 °C. The steady condition, considered achieved when the difference in the energy supplied in the LF does not change over cycles, has been reached after 8 cycles. Therefore, a total of 116 numerical simulations has been performed.

3. Results and discussion

The influence of the working lining properties and the presence of a lid in the final phase ($F7$) is analyzed by comparing the energy E_{LF} estimated for phase $F3$.

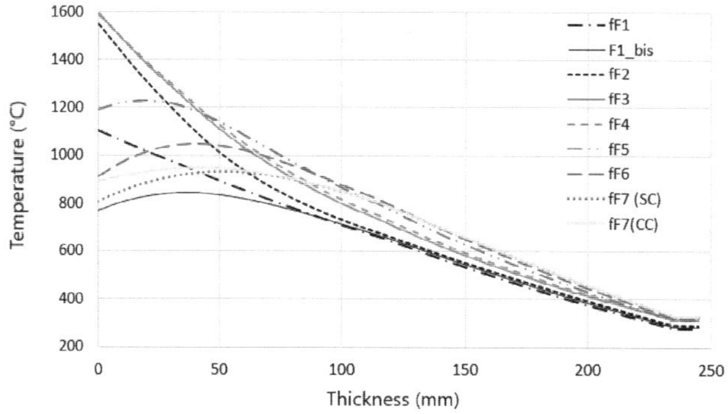

Figure 4. Computed temperature profile across the wall of the ladle for maximum thickness ($S_{\max}$) and alumina working lining. Temperature distributions refer to the final time of phases. For phase $F7$ the temperature profile both in cases with (WL) and without (NL) lid are reported.

Numerical model verification To verify the numerical model, the ladle wall temperature measured with a thermocamera has been compared with that estimated numerically, and they have shown good agreement. Moreover, the cooling of the steel in the ladle detected in the production plant has been compared with data provided by the company.

Temperature profiles in the ladle wall To understand how the ladle temperature varies throughout the process, figure 4 shows the temperature detected across the ladle wall thickness (red line in figure 1) at the end of each phase. Indeed, during the steelmaking process the ladle undergoes both heating - due to preheating at the beginning of $C1$, to contact with hot molten steel during and after tapping and to energy supply in $F3$ - and cooling - due to convection and radiation towards the environment. It can be seen that internally the temperature varies considerably, while externally the variation is limited to a hundred degrees. At the end of the preheating phase ($fF1$), the temperature on the internal surface is 1100 °C that is the temperature imposed as boundary condition in the last two hours of $F1$. During the waiting phase $F1_{bis}$, the internal surface cools due to convection and radiation, but the temperature of the external layers increases. The time duration of phase $F1$ (360 minutes) does not allow to reach a steady condition. During phases $F2$, $F3$ and $F4$, the temperature increases as a result of the contact with the molten steel and the energy supplied in LF. During phase $F5$, the ladle is progressively emptied and internally cooled of about 400 °C. During phases $F6$ and $F7$ the ladle is cooled further depending on the presence of the lid that guarantees an internal temperature 90 °C higher.

Influence of the working lining thickness When the ladle that has to enter in the production cycle is pre-heated ($F1$ and $F1_{bis}$), lining thickness plays two counteracting effects. On one hand for smaller S the ladle inner surface increases and the ladle capacity decreases: that results in higher temperatures in the ladle at time $tF1_{bis}$ as clearly shown in figure 5. As a consequence during the waiting time after tapping (phase $F2$), molten steel cools down less for S_{min} than for S_{max}: this in turn results in smaller energy demand in LF (phase $F3$). On the other hand, for smaller S the ladle wall thermal resistance decreases and the external wall temperature is higher: that causes larger heat loss to the environment and increases energy demand in LF. There is a critical thickness that depends on the lining material for which the energy demand during the first cycles is maximum. That is clear in figures 6 and 7, especially for magnesia. As

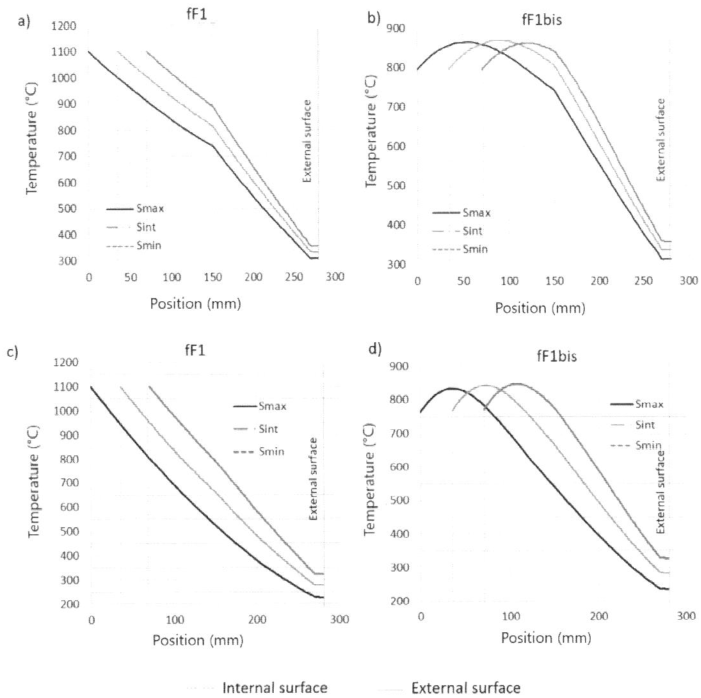

Figure 5. Computed temperature profiles across the ladle wall at the end of the preheating $F1$ and the waiting $F1_{\text{bis}}$ phases for alumina (a,b) and magnesite (c,d).

the number of cycles increases the temperature field tends to stabilize and the energy demand E_{LF} converges to a constant value: the asymptotic values is attained after 7 or 8 cycles. At this point the effect of the ladle wall thermal resistance prevails and E_{LF} increases for decreasing thickness S. However the increment is not large, from cycle $C8$, for example, for alumina case with no lid each reduction in thickness determines an energy demand increase of 3%. In the case with lid, the increases is of 4% from S_{max} to S_{int}, and of 3% from S_{int} to S_{min}. For magnesia and for the lid scenario, the variation is 5.5% from S_{max} to S_{int}, and 3% from S_{int} to S_{min}.

Influence of the working lining material In figure 6, it can be seen that the ladle made of alumina requires a minor quantity of energy. From the eighth cycle (considered to be representative of the steady condition), the energy saving is of 12% for S_{max}, and about 13.5% for the other two thicknesses for the case without lid, and about 11.4% in the scenario with lid.

Influence of the lid in phase F7 The energy demand E_{LF} estimated with and without lid is compared in figure 7. In both scenarios, the presence of the lid reduces the energy demand because the ladle internal surface exchanges energy by radiation with the lid initially at 250 °C. For the case of working lining made of alumina, a energy reduction of about 7% is estimated, while in the case of working lining made of magnesite, the energy reduction is a function of the working lining thickness and it is about 7% for the S_{max}, and 4.5% for the other two thicknesses.

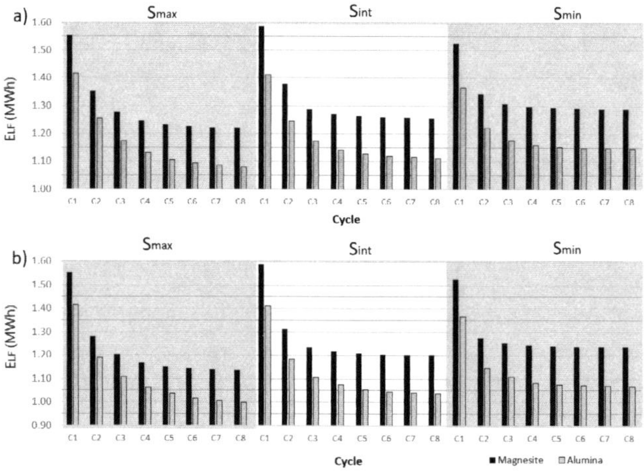

Figure 6. Comparison of the energy E_{LF} supplied in phase $F3$ for different cycles, different materials and different thicknesses, for ladle working lining made of alumina and magnesite. Case a) without lid, and case b) with lid.

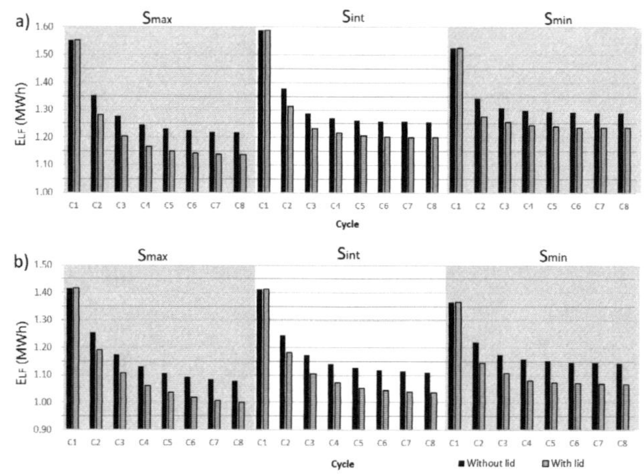

Figure 7. Comparison of the energy E_{LF} supplied in LF with and without lid for the ladle working lining made of magnesite a) and alumina b).

4. Conclusions

In this study by using a simplified, but sound numerical model, it was studied relative energy consumptions in LF between different working lining material for different thicknesses. The influence of closing the ladle with a lid was also analysed. It was observed that the preheating phase has a strong influence on the amount of energy required in LF. Furthermore, when comparing the estimated consumption for the magnesite and alumina ladle, it was observed that the latter average saves is 10%. Althought thickness influced consumption, the presence of a lid at the end of each cycle proved quite effective effective in saving energy.

References

[1] Zimmer A *et al* 2008 Heat transfer in steelmaking ladle *Journal of Iron and Steel Research International,* **15** 3 pp11-60
[2] Santos M F *et al* 2018 Enhanced numerical tool to evaluate steel ladle thermal losses *Ceramics International* **44** pp 12831-12840
[3] Glaser B, Gornerup M and Sichen D 2011 Thermal modelling of the ladle preheating process *Steel research international* **82** pp 1425-1434
[4] Xia J L and Ahokainen T 2001 Transient flow and heat transfer in a steelmaking ladle during the holding period *Metallurgical and Materials Transactions B* **32** pp 733-741
[5] Haojian D, Ying R and Lifeng Z 2019 Thermal stratification and inclusion motion during holding period in steel ladles *Metallurgical and Materials Transactions B* **50** pp 1476-1489
[6] Chakraborty S and Sahai Y 1992 Effect of slag cover on heat loss and liquid steel flow in ladles before and during teeming to a continuous casting tundish *Metallurgical Transactions B* **23** pp 135-151
[7] Glaser B, Gornerup M and Sichen D 2011 Fluid flow and heat transfer in the ladle during teeming *Steel research international* **82** pp 827-835
[8] Farrera-Buenrostro J E *et al* 2019 Analysis of temperature losses of the liquid steel in a ladle furnace during desulfurization stage *Transactions of the Indian Institute of Metals* **72** pp 899-909
[9] Santos M F *et al* 2017 Steel ladle energy saving by refractory lining design *Proc. Conf. UNITECR2017* (Santiago, Chile)
[10] Aidong H *et al* 2018 A method for steel ladle lining optimization applying thermomechanical modeling and Taguchi approaches *Refractory Materials for Metallurgical Uses* **70** pp 2449-2456
[11] Hernandez A G, Paoli L and Cullen J 2017 Resource efficiency in steelmaking: energy and materials combined *Energy Procedia* **142** pp 2429-2434
[12] Neri M *et al* 2019 Energy- and exergy- based analysis for reducing energy demand in heat processes for aluminum casting *Journal of Energy Resources Technology* **141** pp 1045011- 10450110
[13] Camdali U and Tunc M 2003 Energy and exergy analysis of a ladle furnace *The Canadian Journal of metallurgy and materials science* **42** pp 439-446
[14] Camdali U, Tunc M and Dikec F 2001 A thermodynamic analysis of a steel production step carried out in the ladle furnace *Applied Thermal Engineering* **21** pp 643-655
[15] Shigaki N, Akiyama T and Tsukihashi F 2002 Exergy analysis of steel production processes *Materials Transactions* **43** pp 379-384
[16] Kor G J W and Glaws P C 1998 *The Making, Shaping and Treating of Steel* United States Steel Corp 11th edition - Steelmaking and Refining Volume, (Pittsburgh, Pa: AISI Steel Foundation) chapter 11 pp 661-664

Journal of Physics: Conference Series **2509** (2023) 012004 doi:10.1088/1742-6596/2509/1/012004

Numerical study on latent heat thermal energy storage with phase change material and metal foam in shell and corrugated tube with external heat losses

Bernardo Buonomo, Maria Rita Golia, Oronzio Manca, Sergio Nardini, Renato Elpidio Plomitallo

Dipartimento di Ingegneria, Università degli Studi della Campania "Luigi Vanvitelli" via Roma 29, 81031 Aversa (Italy),

bernardo.buonomo@unicampania.it mariarita.golia@unicampania.it oronzio.manca@unicampania.it sergio.nardini@unicampania.it renatoelpidio.plomitallo@unicampania.it

Abstract. In the present work, a numerical investigation of the Latent Heat Thermal Energy Storage system (LHTESs) is performed on a vertical shell and tube geometry, made of a porous medium filled with a phase change material (PCM). The LHTESs is a cylinder with a corrugated inner surface at a constant temperature above the PCM melting temperature; on the external surface there are heat losses to the outside and the top and bottom surfaces are adiabatic. The PCM used is a pure paraffin wax, the metal foam, instead, is made of aluminum. The phase change process is modelled with the enthalpy-porosity theory, while the Local Thermal Non-Equilibrium (LTNE) and the Darcy-Forchheimer models are adopted to analyse the heat transfer between the paraffin and the metal foam. The solutions of the governing equations are computed with Ansys-Fluent commercial code. The survey considers different conductivity variations on the external wall and distinct corrugated wall geometric parameters, with different wavelengths and wave amplitudes. The results of the numerical simulations, concerning the LHTESs charging phase, are reported as a function of time and compared in terms of melting time, enthalpy stored, and energy loss.

Key Words: thermal energy storage, phase change material, metal foam, local thermal non-equilibrium

1. Aim of work

Energy storage systems have gained significant importance to reduce the environmental impact and solve the mismatch in time and method between energy demanded and energy produced. Thermal Energy Storage systems have high TES capacities per unit mass, and they operate with a small temperature range, as the heat interaction occurs at quasi-constant temperature. Among these systems, one of the most promising is based on PCM, which is characterized by a high energy density. The use of PCM in LHTESs, in combination with metal foams, enhances the thermal energy storage system; using metal foam as matrices undoubtedly improves the low thermal conductivity of PCM and fully exploits their high energy density [1,2]. In this way, a thermal storage system, with a fast-absorbing energy zone characterized by a low thermal resistance, is realized, although the heat storage ability of

metal foam-PCM composite is slightly reduced.
Diani et al. [3] investigated the interaction between three different paraffins, as phase change materials, and a porous matrix. The porous matrix is a modified aluminum alloy BCC cell with ad edge of 5 mm, and it was made by additive manufacturing. Mecieb et al. [4] studied numerically, with a 2D simulation, a LTES combined with a solar collector and with phase change material. Sakhaei and Valipour [5] analysed the effects of PCM combined with a helical corrugated heat tube. They noted that the heat losses decrease by 39.8% in comparison to plain heat collecting ones.
Since it appears that there are some gaps from the examination of LHTES with a shell and tube configuration with corrugated internal tube and heat transfer coefficients, the aim of this work is filling those gaps and analyse thermal storage in different configurations in order to assess how different geometries of corrugation and different external heat transfer coefficients can influence the parameters such as liquid fraction, melting rate and stored energy. In this paper the metal foam is considered to be a porous medium and the numerical analysis is carried out in the transient regime under the Local Thermal Non-Equilibrium (LTNE).

2. Physical model

The present work is a numerical investigation of Latent Heat Thermal Energy Storage on a vertical shell and tube geometry, made of a porous medium filled with a phase change material (PCM) with a corrugated internal cross-section. The space between the external shell and the corrugated internal tube is composed of an aluminum foam with a porosity of 0.9546 and a pore density of 20 PPI, filled with the paraffin wax RT50. The 3D axisymmetric geometrical configuration is reported in Figure 1.

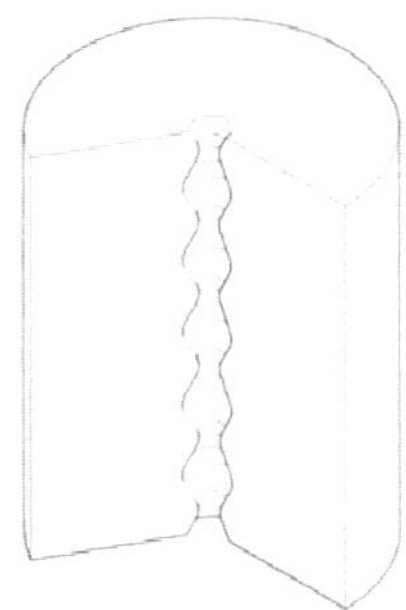

Fig.1. Latent thermal energy storage system (LHTESs): 3D configuration.

The investigated geometry is also used by Buonomo et al.[6] and the 2D configuration, shown in Figure 2, represents the same geometry simulated and solved using Ansys-Fluent commercial code. The determined geometric parameters are the cylinder height L=100 mm, the average internal radius $r_{i,avg}$ = 6 mm and the external one r_e = 50 mm. The inner face is assumed to be heated and is modelled as a cosine wave, with the following analytical expression:

$$r = r_{i,avg} + A\cos\left(2\pi n\frac{z}{L}\right) \tag{1}$$

where r is the radial coordinate, A is the wave half-amplitude and n stands as the wave number. In this investigation, the wavelength, the wave amplitude, and the heat loss at the external wall were varied to analyze the charging phase in a different configuration. In the analyzed geometric configuration, the parameter that identifies the cosine wave has been given the values A=2 mm and A=4 mm, while the parameter that identifies the wavelength *n* has been given the values n=6, n=12 and n=16.

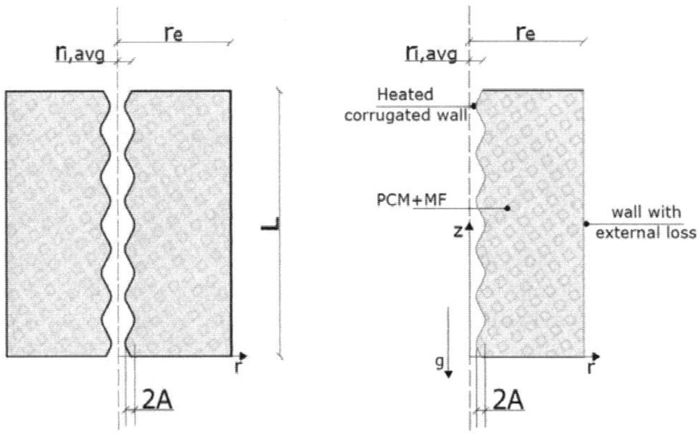

a) b)

Fig. 2. LHTESs geometry: a) middle cross section and b) computational domain.

The thermophysical properties are considered temperature-independent and summarized in Table 1 [7,8]. The properties of the aluminium foam are evaluated following the work of Battacharya et al. [9], especially using sample 6. Table 2 summarizes the parameters of the metal foam.

Table 1. Thermal properties of the paraffin RT 50 and aluminum.

Properties	RT 50	Aluminum
Density [kg/m^3]	820	2719
Specific Heat [J/kg K]	2000	871
Thermal Conductivity [W/m K]	0.2	202.4
Dynamic Viscosity [kg/ m s]	0.0038	-
Thermal Expansion Coeff. [W/mK]	0.0006	-
Melting Heat [J/kg]	168000	-
Solidus Temperature [K]	318	-
Liquidus Temperature [K]	324	-

Table 2. Metal foam parameters

Properties	
PPI	20
ε	0.9546
d_f [m]	0.0003
d_p	0.0027
C_F	0.093
K x10^7 [m^2]	1.3
k_{eff} [W/mK]	3.42

This study is performed starting from the following assumptions: an unsteady and laminar two-dimensional axisymmetric flow. The internal cylindrical surface represents the heat source and is supposed to be at the uniform temperature of 350 K.
The temperature is considered uniform because the convective heat-transfer coefficient related to the water flows in the internal tube is of order of magnitude 10^3 and the temperature variation of water is negligible. Moreover, the internal wall is assumed metallic with a high thermal conductivity.
At the beginning of the thermal transient, the temperature is equal to 293 K. To simulate the natural convection in the PCM the Boussinesq approximation is used, due to the buoyancy phenomenon, which is assumed to be triggered when the temperature becomes higher than the RT50 solidus temperature. The PCM thermal properties are assumed to be constant and evaluated in the liquid phase, at 300 K, except for the density, which is assumed to be variable with temperature, according to the Boussinesq approximation. The metal foam is isotropic and homogenous and is simulated as a porous medium. The external wall has a heat loss to the outside and its overall conductance varies from 0.1 to 1 W/m^{2}K; and the Heat Transfer Coefficients (HTCs)= 0.1, 0.5, and 1 W/m^{2}K.

Furthermore, the local thermal non-equilibrium (LTNE) hypothesis and the Forchheimer-extended Darcy model have been adopted to model the heat transfer between the paraffin and the aluminum foam. Consequently, the continuity and momentum equations of the physical model are given in the work of Buonomo et al. [7]:

$$\frac{1}{r}\frac{\partial(ru)}{\partial r}+\frac{\partial w}{\partial z}=0 \tag{2}$$

with u and w being, respectively, the radial and axial components of the fluid velocity in the cylindrical coordinates r and z.
Momentum equations[7]:

$$\begin{aligned}&\frac{\rho_p}{\varepsilon}\left(\frac{\partial u}{\partial t}+\frac{u}{\varepsilon}\frac{\partial u}{\partial r}+\frac{w}{\varepsilon}\frac{\partial u}{\partial z}u\right)=-\frac{\partial p}{\partial r}+\frac{\mu_p}{\varepsilon}\left(\frac{\partial^2 u}{\partial r^2}+\frac{1}{r}\frac{\partial u}{\partial r}-\frac{u}{r^2}+\frac{\partial^2 u}{\partial z^2}\right)+\\&-S_i u-\left(\frac{\mu_p}{K}+\frac{C_F}{\sqrt{K}}\rho_p\sqrt{u^2+w^2}\right)u\end{aligned} \tag{3}$$

$$\begin{aligned}&\frac{\rho_p}{\varepsilon}\left(\frac{\partial w}{\partial t}+\frac{u}{\varepsilon}\frac{\partial w}{\partial r}+\frac{w}{\varepsilon}\frac{\partial u}{\partial z}u\right)=-\frac{\partial p}{\partial r}+\frac{\mu_p}{\varepsilon}\left(\frac{\partial^2 w}{\partial r^2}+\frac{1}{r}\frac{\partial w}{\partial r}-\frac{u}{r^2}+\frac{\partial^2 w}{\partial z^2}\right)+\\&-S_i w-\left(\frac{\mu_p}{K}+\frac{C_F}{\sqrt{K}}\rho_p\sqrt{u^2+w^2}\right)w+\rho_p g\gamma(T-T_0)\end{aligned} \tag{4}$$

Where ρ_p and μ_p stand for the paraffin density and dynamic viscosity, p represents the static pressure and ε is the foam porosity. The buoyancy effect is considered through the last term of the equation (4), where γ is the volumetric expansion coefficient and T_0 is a reference temperature equal to 318K and corresponding to the RT 50 solidus temperature. In the momentum equations, there are the Darcy and Forchheimer terms, that simulate the pressure losses caused by the viscous and inertial effects related to the coexistence of the PCM and the metal foam. The foam permeability (K) and the drag factor (C_F) are given by the work of Calmidi et al. [10]. In this work has been assigned the experimental value of d_p which is provided by the reference [9,10]. Concerning equations (3) and (4), the presence of the solid matrix in the mixed zone and the enthalpy-porosity model is linked to the S_i term, which can be expressed as follows [11]:

$$S_i=\frac{(1-\beta)^2}{(\beta^3+\omega)}A_{mush} \tag{5}$$

The term A_{mush} has been fixed at $1\mathrm{x}10^6$ [kg m^{-3} s^{-1}] and it represents a constant that governs the amount of velocity release due to phase change processes. The term β represents the volume liquid fraction and it is an index that measures the ratio between the liquid volume and the PCM total volume; it can have a value between 0 and 1. The term ω is added to prevent the division of zero when β is next to zero. According to local temperature, β is defined as shown in the work of Voller et al.[12]. The LTNE assumption is implemented for these simulations, which means that PCM and metal foam are featured by distinct temperature fields; therefore, two energy equations must be solved.
Energy Equations are given by [13]:

$$\begin{aligned}&\frac{\rho_p c_p \partial(\varepsilon T_p)}{\partial t}+\rho_p c_p\left(u\frac{\partial T_p}{\partial r}+w\frac{\partial T_p}{\partial z}\right)=k_{eff,p}\left[\frac{\partial}{\partial r}\left(\frac{1}{r}\frac{\partial}{\partial r}(rT_p)\right)+\frac{\partial^2 T_p}{\partial z^2}\right]+\\&-\varepsilon\rho_p H_L\frac{\partial\beta}{\partial t}+h_{sf}a_{sf}(T_s-T_f)\end{aligned} \tag{6}$$

$$\frac{\rho_s c_s \partial[(1-\varepsilon)T_s]}{\partial t} = k_{eff,s}\left[\frac{\partial}{\partial r}\left(\frac{1}{r}\frac{\partial}{\partial r}(rT_s)\right)+\frac{\partial^2 T_s}{\partial z^2}\right] - h_{sf}a_{sf}(T_s - T_f) \tag{7}$$

The fourth term of equation (6) is associated with the phase change, and it represents the unsteady term; moreover, it is proportional to the PCM melting latent heat H_L.

The terms $k_{eff,p}$ and $k_{eff,s}$ in the energy equations are the conductivity of the PCM and metal foam. These properties are evaluated using the formulations below [14] :

$$k_{eff,p} = \varepsilon k_{RT50} + k_d \tag{8}$$

$$k_{eff,s} = 0.181\left[(1-\varepsilon)^{0,763}\right]\cdot k_{Al} \tag{9}$$

where the conductivity of the paraffin wax RT50 is k_{RT50} and the aluminum one is k_{Al}, both at 300 K (provided in Table 1). Additionally, in equation (8) the contribution of the dispersion thermal conductivity k_d has been considered, which is a coefficient that increases linearly with velocity, as it is illustrated by the correlation provided by Calmidi and Mahajan [10]. The parameter k_d can be neglected because it is two orders of magnitude smaller than k_{RT50}. As it can see in the energy equations the term h_{sf} appears, which is the convective heat transfer coefficient and indicates the convective heat transfer between metal foam and PCM. To determine this coefficient several correlations can be used, but in our case, the Churchill correlation has been chosen [15], employed with a User-Defined Function in C++ language. At last, a_{sf}, the interface surface per unit volume of the metal foam, was calculated as follows [10]:

$$a_{sf} = \frac{3\pi d_f (1 - e^{-((1-\varepsilon)/0.04)})}{(0.59 d_p)^2} \tag{10}$$

3. Numerical model

For these simulations, the finite volume method has been used, which employs SIMPLE algorithm for the pressure-velocity coupling and PRESTO algorithm for the pressure calculation, solving all the governing equations above with Ansys-Fluent commercial code. The convergence errors are set to values of 10^{-5}, for the continuity and momentum equation, and of 10^{-8}, for the energy equation. To verify the accuracy of the results, reference was made to the work of Buonomo et al. [6], who used a geometry almost equal to the reported one and verified that among the three analyzed meshes, respectively made of 738, 2916 and 11431 cells, the first two, which have a lower cost of calculation, present a modest difference both in terms of average temperature and of liquid fraction (less than 1%), and that especially the mesh with 2916 cells is acceptable in terms of accuracy and costs of calculation.

4. Results and discussion

In the present study, the basic configuration with no heat loss on the external wall was compared with systems characterized both by different geometric characteristics (amplitude and wavelength) and different heat transfer coefficient of the external surface, to evidence how these variables influence the results of the simulations. The following indicators were evaluated to characterize the LHTESS system analyzed: average temperature and stored enthalpy (for both the PCM and the metal foam zones), liquid fraction, inlet, and outlet storage rates (T_{avg}, enthalpy, β, q_{int}, q_{ext}). Several simulations have been performed to investigate the influence of the cosine wave amplitude and of the wavelength on the thermal transient heat flux during the charging phase. The cases analyzed are six and they are a combination of wave amplitudes A=2 mm and A=4mm, with wave numbers n=6, n=12 and n=16. These cases have been compared to highlight how the different combinations performed in terms of melting time, stored enthalpy, and heat flux. The results provided in Figure 3 show the evolution of liquid fraction during the time. The melting rate, that is the liquid fraction derivative in time, is poorly influenced by the different HTC on the external wall, instead, it varies more when the geometry of the section varies. The increase of wavelength leads to a heat transfer surface area; therefore, greater waves numbers generate a higher PCM melting rate, which results in a more efficient charging time for the TESS. Anyway, after a specific number of waves, the improvements in the heat transfer and in the reduction of the melting time are negligible in proportion to the difficulty of realizing a geometry with

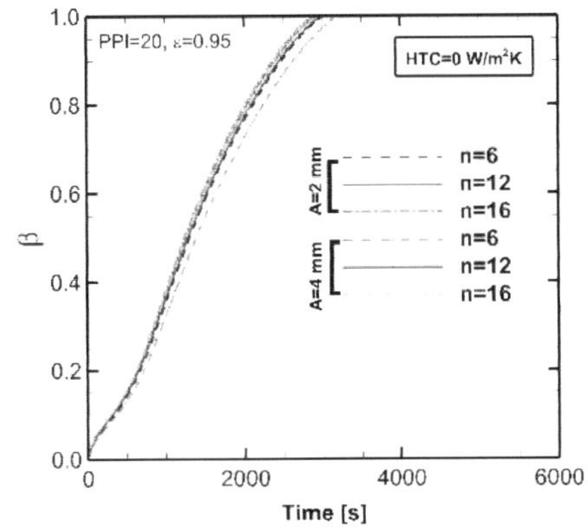

a)

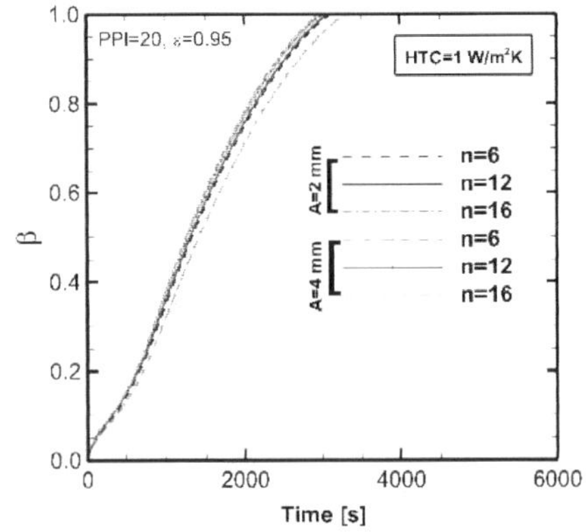

b)

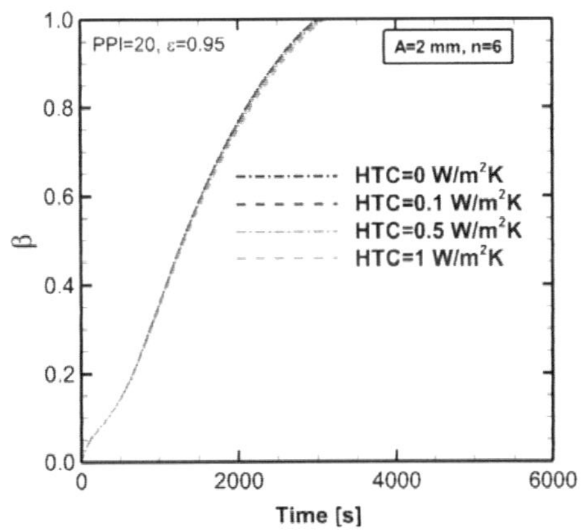

c)

Fig. 3. Liquid fraction for: a) HTC=0 W/m²K, b) HTC=1 W/m²K, c) case A=2 mm, n=6.

this number of waves. In fact, as it can be noticed, the curves of n=12 and n=16 are nearly overlapping in all cases The results provided in Figure 4 show the time evolution of the enthalpy of the solid part and fluid part, as it can be seen, the case (A=2 mm, n=16) that achieves the value of the maximum system temperature also achieves the maximum value of enthalpy storage. The last parameter under investigation is the effect of different heat transfer coefficients of the external wall on the heat flux. The results shown in Figure 5, as was expected, proved that by increasing the HTC the losses on the external wall increase and greater values of amplitude and wavelength correspond to higher losses because these values are linked to a higher heat storage rate on the internal surface. The different HTC influences the heat transfer rate of the external of the considered configurations. Since the selected HTCs are low, the values of average temperature, liquid fraction, and stored enthalpy depend only on the geometry of the section. It should be noted that by choosing higher values of HTCs, their influence could be higher, which is why the authors believe that it is also necessary to investigate this phenomenon with higher values of the heat transfer coefficient. Referring to Figure 5 it can be noted that the first section of the plotted profiles has a significant slope because the difference in temperature between the outside and the inside is big at the beginning. The next section has a less pronounced slope and represents the accumulation of latent heat when the PCM melt, at the end, the plotted curves have an asymptotic value that is reached when all the PCM is melted.

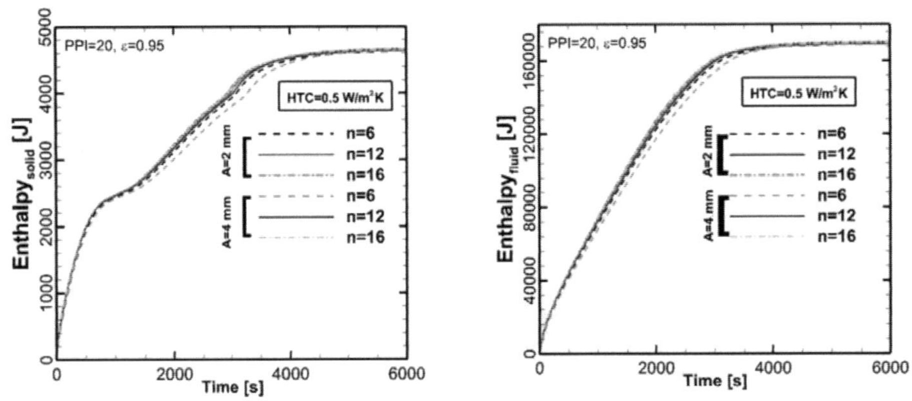

a) b)

Fig. 4. Enthalpy for HTC=0.5 W/m^{2}K: a) Enthalpy of solid part, b) Enthalpy of fluid part

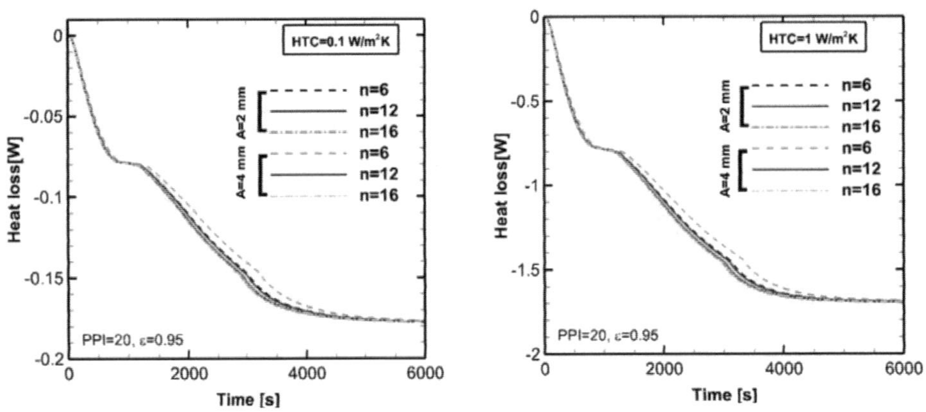

a) b)

Fig. 5. Heat transfer rate on the external surface as function of time for: a) HTC=0.1 W/m^{2}K, b) HTC=1 W/m^{2}K.

5. **Conclusions**

This study intended to evaluate the effect of structural and geometrical parameters of an LHTESS, composed of a high conductivity aluminum foam filled with RT 50, during the charging phase. The survey was performed with a constant pore density of 20 PPI, two different waves amplitudes (A=2 mm and A=4 mm), three different waves number (n=6, n=12 and n=16) and different external convective heat transfer coefficient variations (setting HTCs=0, 0.1, 0.5 and 1 W/m^2K.). The heat source was located on the corrugated inner surface of the cylinder and the initial temperature was assumed to be higher than the melting temperature range of the PCM. Since there are few studies on corrugated configurations and with external loss, it was chosen to analyze this configuration. The paper shows how the different considered HTCs affect the heat transfer rate and described the trend of heat loss on the external surfaces for different geometries. Meanwhile, the HTCs have little influence on the melting rate, average temperature and enthalpy stored. The simulation has been conducted in the melting phase and could be expanded by conducting it with HTC greater than 1 or extending it to the solidification phase or to multiple cycles.

References

[1] Cabeza L F 2021 *Advances in Thermal Energy Storage Systems: Methods and Applications* (Elsevier)

[2] Dincer I and Rosen M A 2011 *Thermal Energy Storage: Systems and Applications* ed John Wiley & Sons (New York)

[3] Diani A, Moro L and Rossetto L 2021 Melting of Paraffin Waxes Embedded in a Porous Matrix Made by Additive Manufacturing *Applied Sciences* **11** 5396

[4] Mecieb F Z, García-Bermejo F, Solano J P, Herrero-Martín R and Laouedj S 2022 Solar-driven melting dynamics in a shell and tube thermal energy store: A numerical analysis *J Energy Storage* **52** 104924

[5] Sakhaei S A and Valipour M S 2021 Thermal behavior of a flat plate solar collector with simultaneous use of helically heat collecting tubes and phase change materials *Sustainable Energy Technologies and Assessments* **46** 101279

[6] Buonomo B, di Somma F, Manca O, Nardini S and Plomitallo R E 2021 Numerical investigation on latent thermal energy storage in shell and corrugated internal tube with PCM and metal foam *E3S Web of Conferences* **312** 03003

[7] Buonomo B, Celik H, Ercole D, Manca O and Mobedi M 2019 Numerical study on latent thermal energy storage systems with aluminum foam in local thermal equilibrium *Appl Therm Eng* **159**

[8] Chen X, Li X, Xia X, Sun C and Liu R 2021 Thermal storage analysis of a foam-filled PCM heat exchanger subjected to fluctuating flow conditions *Energy* **216** 119259

[9] Bhattacharya A, Calmidi V V and Mahajan R L 2002 Thermophysical properties of high porosity metal foams *Int J Heat Mass Transf* **45** 1017–31

[10] Calmidi V v and Campmode R L M 2000 *Forced Convection in High Porosity Metal Foams*

[11] Yang J, Yang L, Xu C and Du X 2015 Numerical analysis on thermal behavior of solid–liquid phase change within copper foam with varying porosity *Int J Heat Mass Transf* **84** 1008–18

[12] Voller V R and Prakash C 1987 A fixed grid numerical modelling methodology for convection-diffusion mushy region phase-change problems *Int J Heat Mass Transf* **30** 1709–19

[13] Chen X, Li X, Xia X, Sun C and Liu R 2019 Thermal Performance of a PCM-Based Thermal Energy Storage with Metal Foam Enhancement *Energies (Basel)* **12** 3275

[14] Calmidi V v. and Mahajan R L 1999 The Effective Thermal Conductivity of High Porosity Fibrous Metal Foams *J Heat Transfer* **121** 466–71

[15] Churchill S W and Chu H H S 1975 Correlating equations for laminar and turbulent free convection from a vertical plate *Int J Heat Mass Transf* **18** 1323–9

Application of the porous media approach to a Standing Wave Thermoacoustic Engine for CFD simulations

Armando Di Meglio* and Nicola Massarotti

Department of Engineering, University of Naples "Parthenope" Centro Direzionale, Isola C4, 80133.

*armando.dimeglio001@studenti.uniparthenope.it

Abstract. In this article, an application of the macroscopic porous media approach, suitable for CFD simulations in oscillating flows, is proposed for a simplified Standing Wave Thermo-Acoustic Engine (SWTAE), composed of a hot buffer tube, a stack - where the energy conversion takes place - and the rest of the resonator. While for a Travelling Wave Thermo-Acoustic Engine (TWTAE) a macroscopic model for porous media has been successfully applied to both regenerators (similar to the stack in a SWTA) and heat exchangers, for SWTAE this is not true. The results illustrate that a Local-Thermal Non-Equilibrium model is required to start up the SWTAE, otherwise (with the Local Thermal Equilibrium model) the thermoacoustic instability cannot arise. Furthermore, the comparison between the simulation conducted at the microscopic scale and that one at the macroscopic level, depicts that a purely Darcy-linear model employed for the macroscopic model, characterized for oscillating flows, overpredicts the pressure amplitude at periodic steady-state. For this reason, a Forchheimer-like coefficient needs to be implemented to fit the macroscopic solution with the microscopic one.

1. Introduction

Thermoacoustics is a multidisciplinary science that combines acoustics, heat transfer and fluid dynamics [1]. According to the thermoacoustic effect, if a temperature difference is applied along a porous material, pressure oscillation, exploitable to obtain acoustic work, can arise. Vice-versa, if acoustic work is provided along the stack a temperature difference can be generated [2]. According to the type of phase shift between pressure and velocity, thermoacoustic devices can be classified into Standing Wave (SW) and Travelling Wave (TW) devices. In a SW device, the phase shift between pressure and velocity is approximately 90°, while in a TW device these quantities are almost in phase [2].

Computational Fluid Dynamic is a powerful tool able to overcome the intrinsic shortcomings of the thermoacoustic linear theory with which these devices are generally designed [3]. However, being a multiscale physics, high-fidelity simulation of a complete device cannot be always affordable from a computational cost perspective. For this reason, the macroscopic porous media approach can be used to simulate the porous core, called stack in a SW thermoacoustic device. While for a TW Thermo-Acoustic Engine (TAE) this approach has been almost always considered, for a SWTAE instead it cannot be claimed the same.

In this work, a macroscopic porous media model, derived from a previous article in case of isothermal case [4] and extended in subsequent work [5], has been applied for the first time to the stack of a SWTAE.

NOMENCLATURE

c_p	Specific heat ($1000\ J/(kgK)$)	**Greek letters**	
e	Internal energy (J/kg)	δ	Viscous/Thermal penetration depths (m)
f	Frequency, (Hz)	ϵ_s	Complex heat capacity ratio
h	Enthalpy (J/kg)	$\tilde{\kappa}_c$	Complex Permeability (m^2)
k	Thermal conductivity ($0.0256\ W/(mK)$)	λ	Wavelength (m)
k_w	Wavenumber ($1/m$)	μ	Dynamic viscosity ($1.8 \cdot 10^{-5}\ Pa \cdot s$)
i	Imaginary unit	ρ	Density (kg/m^3)
Ls	Stack length (30 mm)	$\boldsymbol{\tau}$	Viscous stress tensor (Pa)
$\widetilde{Nu}_c$	Complex Nusselt number	ϕ	Porosity (0.5)
p	Pressure (Pa)	ω	Angular frequency (Hz)
Pr	Prandtl number (0.7)	**Subscript**	
r_h	Hydraulic radius (0.5 mm)	1	First order variables
t	Time (s)	D	Darcian/macroscopic velocity
T	Temperature (K)	eff	effective
		f	Fluid
		k	Thermal
		p	Porous media
		s	Solid, stack
		v	Viscous
		w	Wall

The second section briefly reviews the governing equations of the "microscale" and "macroscale", and the third one describes the numerical setup employed in this work. In the fourth one, the results are discussed, and in the last one some conclusions are drawn.

2. Theoretical background

2.1. Microscopic governing equations

Thermoacoustic phenomena can be fully described by the set of Navier Stokes Equations (NSE). The numerical solution of such equations can be defined as "microscopic", because, in this case, any surrogated model is not employed to capture the phenomena. For this reason, it will be considered a benchmark solution when compared with the solution based on a macroscopic approach to the stack.

$$\frac{\partial \rho}{\partial t} + \nabla \cdot (\rho \boldsymbol{v}) = 0 \tag{1}$$

$$\frac{\partial (\rho \boldsymbol{v})}{\partial t} + \nabla \cdot (\rho \boldsymbol{v}\boldsymbol{v}) = -\nabla p + \mu \nabla^2 \boldsymbol{v} + \frac{\mu \nabla (\nabla \cdot \boldsymbol{v})}{3} \tag{2}$$

$$\frac{\partial(\rho e+\frac{1}{2}\rho|\boldsymbol{v}|^2)}{\partial t}=-\nabla\cdot\left[\rho\boldsymbol{v}\left(\frac{1}{2}|\boldsymbol{v}|^2+h\right)-k_f\nabla T-\boldsymbol{v}\cdot\boldsymbol{\tau}\right] \tag{3}$$

2.2. Macroscopic governing equations

By (volumetrically) averaging the previous governing equations over a Representative Elementary Volume, the macroscopic equations that can be obtained need to be closed by specifying both the interaction terms in the momentum and the energy equations [6], [7]. In the framework of a linear problem, for oscillating flows application, this can be done from the analytical solution of the linearized problem in the frequency domain [1].

2.2.1. Macroscopic momentum equation. The macroscopic momentum equation differs from its microscopic version for two terms. The first one, on the right-hand side of the following eq. (4) resembles the steady-state Darcy friction term. On the left-hand side, there is a virtual mass correction to take into account the eventual phase shift between pressure gradient and velocity in oscillating flow. They are respectively related to the real and imaginary parts of the complex permeability.

$$\frac{1}{\phi}\rho\frac{\partial\boldsymbol{v_D}}{\partial t}\left\{\phi\Im\left[\frac{\delta_v^2}{\tilde{k}_c}\right]+1\right\}+\frac{1}{\phi^2}(\boldsymbol{v_D}\cdot\nabla)\boldsymbol{v_D}=\nabla\cdot\left[-\langle p\rangle\boldsymbol{I}+\frac{\langle K\rangle}{\phi}\right]+\Re\left[-\frac{\mu}{\tilde{k}_c}\right]\boldsymbol{v_D},$$
$$K=\mu\left(\nabla\boldsymbol{v_D}+\nabla\boldsymbol{v_D}^{\mathrm{T}}\right)-\frac{2}{3}\mu\,(\nabla\cdot\boldsymbol{v_D}) \tag{4}$$

2.2.2. Macroscopic energy equation. For the macroscopic energy equation, two approaches can be used. The Local Thermal Equilibrium (LTE), as the nomenclature suggests, assumes the thermal equilibrium between the fluid and solid media and is based on a single following equation:

$$\left(\rho c_p\right)_{eff}\frac{\partial\langle T_p\rangle}{\partial t}+\rho_f c_{p,f}\boldsymbol{v_D}\cdot\nabla T+\nabla\cdot\left(-k_{eff}\nabla\langle T_p\rangle\right)=\frac{\partial\langle p\rangle}{\partial t}+\boldsymbol{v_D}\cdot\nabla p \tag{5}$$

where "⟨ ⟩" is the volume average operator.

On the other hand, the Local Thermal Non-Equilibrium considers the irreversible heat transfer exchanged between the two media introducing in both the fluid and energy equations a source term (with a clear opposite sign) to describe it, as a function of a volumetric heat transfer coefficient or an interfacial Nusselt number $\widetilde{Nu}_c$ as in this case. Similarly, for the macroscopic momentum equation, there are two additional terms on the right and left-hand sides. The fluid and solid energy equations can be written as follows:

$$\phi\rho_f c_p\left(1+\frac{\delta_k^2\Im[\widetilde{Nu}_c]}{2\phi r_h^2}\right)\frac{\partial\langle T\rangle}{\partial t}+\rho_f c_p\boldsymbol{v_D}\cdot\nabla\langle T\rangle-\nabla\cdot(\phi k\nabla\langle T\rangle)=\frac{\partial\langle p\rangle}{\partial t}+\boldsymbol{v_D}\cdot\nabla p+\frac{\Re[\widetilde{Nu}_c]}{r_h^2}(\langle T\rangle-\langle T_s\rangle) \tag{6}$$

$$\frac{1-\phi}{\phi}\rho_s c_s\left(1+\frac{\phi}{2(1-\phi)}\left(\frac{\delta_s}{r_h}\right)^2\Im[\widetilde{Nu}_c]\right)\frac{\partial\langle T_s\rangle}{\partial t}-\frac{1-\phi}{\phi}\nabla\cdot(k_s\nabla\langle T_s\rangle)=-\frac{\Re[\widetilde{Nu}_c]}{r_h^2}(\langle T\rangle-\langle T_s\rangle) \tag{7}$$

where r_h is half the gap between two solid plates.

3. Numerical setup

Both macroscopic and microscopic models have been solved with the Finite Element Method. The geometrical characteristics of the computational domain are introduced in Table 1, inspired by the work of Chen et al. [8]. A finer mesh size in the stack is required to solve the oscillating thermal and viscous boundary layer for the microscopic model compared to the case in which the porous media approach has been used. Second-order spatial schemes have been used for the velocity and temperature, as well as a second-order backward scheme has been considered for the time discretization. The computational domain and meshes are pictured in the following Figure 1.

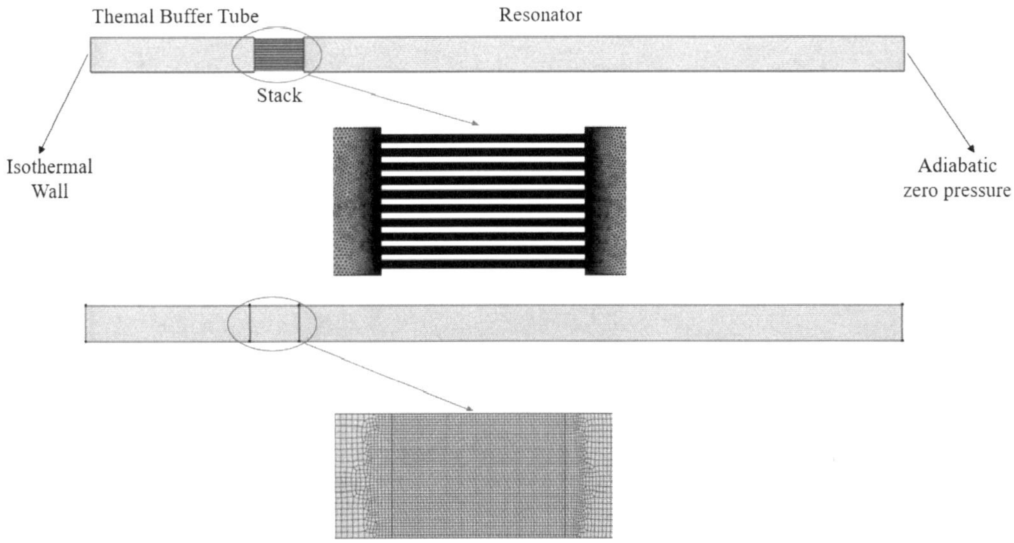

Figure 1 Computational domain and zoom of meshes for both microscopic (top) and macroscopic (bottom) setups

The boundary conditions are isothermal wall and adiabatic zero pressure for the left and the right boundaries respectively. A linear temperature profile, from 550 K to 300 K is prescribed along the stack walls to trigger the thermoacoustic instability. The solid medium is considered to have a much higher thermal capacity compared to the fluid one so solving an additional energy equation can be avoided [9]. In order to accelerate the reach of the periodic steady-state, a co-sinusoidal pressure distribution and the above linear temperature profile have been imposed along the longitudinal coordinate.

4. Results and discussion

At the microscopic scale, the time history of pressure pictured in Figure 2 depicts an exponential growth, well predicted by a complex frequency (linear regime), whose real part is the frequency of the pressure oscillation and its imaginary part is related to the pressure peak-peak ratio [10]. The pressure growth ratio decreases in the so-called "saturation phase" and then pressure reaches an almost constant amplitude oscillation (limit cycle), as illustrated qualitatively in Figure 2.

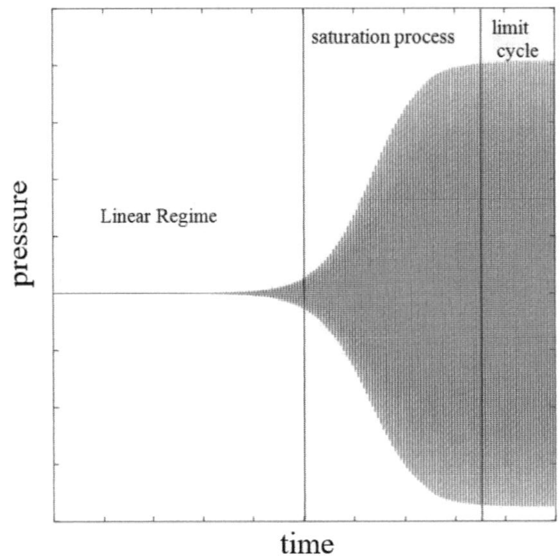

Figure 2 Pressure evolution from the quiescent state to the limit cycle

On the other hand, for a macroscopic approach based on the LTE model, even though the same cosinusoidal pressure distribution and the same initial temperature profile have been used, pressure decreases with time until the quiescent state is obtained (Figure 3). This can be physically explained by the heat transfer mechanism within the stack of a SWTAE. In this case, the height channel, required to be of the same order of magnitude as the thermal penetration depth for such types of devices, should allow a finite and irreversible heat transfer between the two media. This clearly violates the assumption under which a LTE model is built. This finding is in agreement with what was also reached by Yu et al. [11] and analyzed in a previous review work focusing on the CFD modelling of thermoacoustic devices [12].

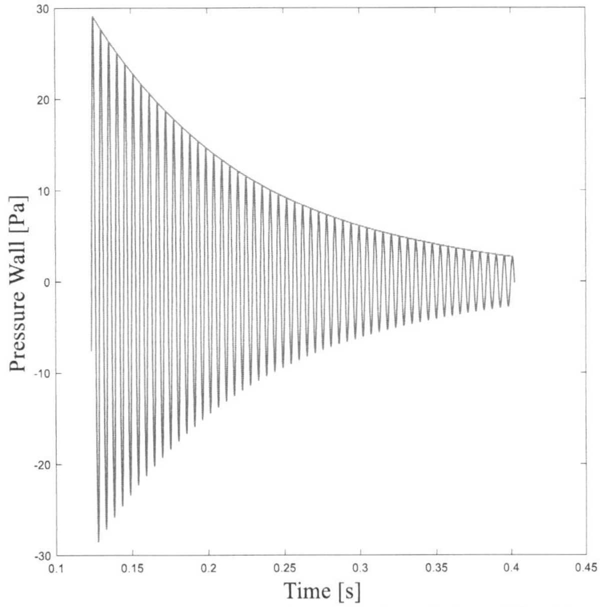

Figure 3 Pressure at Thermal Buffer Tube wall in case of Local Thermal Equilibrium (LTE)

A LTNE approach enables to describe the oscillating heat transfer between the fluid and solid media thanks to the definition of an interfacial Nusselt number as shown in eqs. (6) and (7). The Nusselt number, as well as the permeability, slightly depends on the x-coordinate because the viscous and thermal penetration depths change with mean density/temperature. The complex permeability and Nusselt number, calculated as shown in previous work [5] and used in the macroscopic equations, are pictured in the following Figure 4.

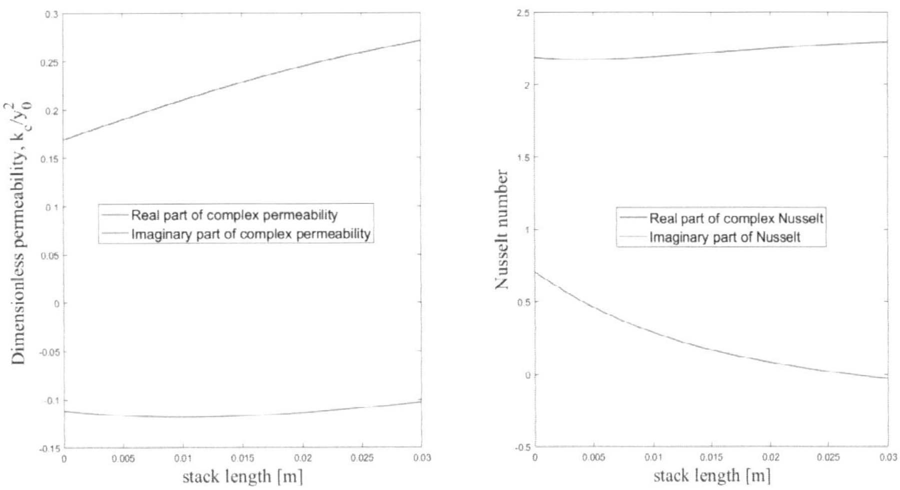

Figure 4 Real and imaginary part of complex permeability (left) and Nusselt number along the stack length

The results confirm that such a macroscopic model is suitable for modelling the thermoacoustic instability in a SWTAE, as proved by the comparison with microscopic results. Pressure oscillations match both in terms of frequency and growth ratio factor. However, when the saturation process starts (until the periodic steady-state condition), the pressure amplitude of the microscopic model becomes higher than the macroscopic one, as illustrated in Figure 5. Such a departure is intrinsically related to the fact that the effects on the fluid flow of geometry discontinuities at stack ends are not taken into account in the macroscopic model.

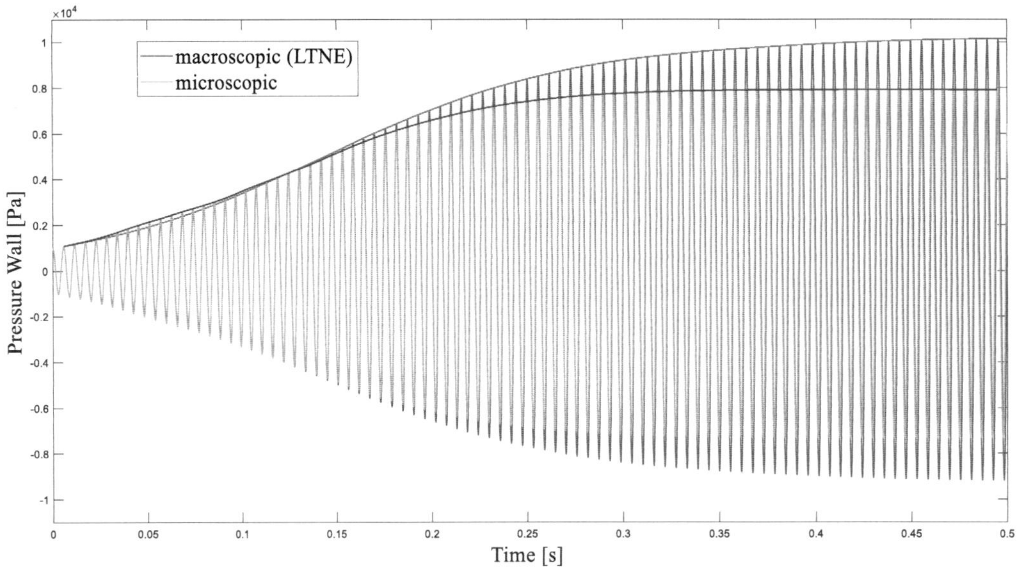

Figure 5 Comparison, in terms of pressure at the thermal buffer tube wall, between microscopic and macroscopic (LTNE) results

5. Conclusions

In this paper, a CFD simulation of a simple Standing Wave Thermoacoustic Engine has been carried out by using a developed porous media approach in oscillating flows. After reporting the governing equation of both microscopic and macroscopic scales, the numerical setup of the thermoacoustic engine has been shown, by specifying the geometry and mesh details, the boundary and initial conditions. With regards to the macroscopic scale, two possible approaches are usually considered for the energy equation in porous media: the Linear Thermal Equilibrium (LTE) and the Local-Thermal Non-Equilibrium (LTNE). The results have shown that the with a LTE approach, the engine was not able to start up because the irreversible heat transfer between the solid and fluid media is not modelled. On the other hand, adopting a LTNE approach allows one to reproduce such a phenomenon. In particular, a good agreement has been found between the micro and macro-scale in terms of frequency oscillation and growth ratio of the pressure at the solid wall of the engine. However, at high amplitude, where the saturation process takes place, the nonlinearities, not taken into account in such a Darcy-like porous media approach, determine a discrepancy between the pressure level reached in the engine. Therefore, further studies are required to consider such nonlinear losses macroscopically, by introducing a Forchheimer-like term in the source term of the momentum equation.

References

[1] G. W. Swift, "Thermoacoustic engines," *J. Acoust. Soc. Am.*, vol. 84, no. 4, pp. 1145–1180, 1988, doi: 10.1121/1.396617.
[2] G. W. Swift, *Thermoacoustics: A Unifying Perspective for Some Engines and Refrigerators*. .
[3] J. P. Clark, W. C. Ward, and G. W. Swift, "Design environment for low-amplitude thermoacoustic energy conversion (DeltaEC)," *J. Acoust. Soc. Am.*, vol. 122, no. 5, p. 3014,

2007, doi: 10.1121/1.2942768.

[4] A. Di Meglio, N. Massarotti, and V. Trombetta, "Macroscopic model for porous media in oscillating flow for CFD applications," *J. Phys. Conf. Ser.*, vol. 2177, no. 1, p. 012012, 2022, doi: 10.1088/1742-6596/2177/1/012012.

[5] A. Di Meglio, E. Di Giulio, R. Dragonetti, and N. Massarotti, "A novel model for macroscopic simulation of oscillating heat and fluid flow in porous media," *Int. J. Therm. Sci.*, vol. 181, no. August 2021, p. 107758, 2022, doi: 10.1016/j.ijthermalsci.2022.107758.

[6] K. Vafai, *Handbook of porous media, second edition*. CRC Press, 2005.

[7] S. Whitaker, "Flow in porous media I: A theoretical derivation of Darcy's law," *Transp. Porous Media*, vol. 1, no. 1, pp. 3–25, 1986, doi: 10.1007/BF01036523.

[8] G. Chen, L. Tang, and B. R. Mace, "Bistability and triggering in a thermoacoustic engine: A numerical study," *Int. J. Heat Mass Transf.*, vol. 157, p. 119951, 2020, doi: 10.1016/j.ijheatmasstransfer.2020.119951.

[9] A. Di Meglio, E. Di Giulio, R. Dragonetti, and N. Massarotti, "Analysis of heat capacity ratio on porous media in oscillating flow," *Int. J. Heat Mass Transf.*, vol. 179, 2021, doi: 10.1016/j.ijheatmasstransfer.2021.121724.

[10] M. Guedra and G. Penelet, "On the use of a complex frequency for the description of thermoacoustic engines," *Acta Acust. united with Acust.*, vol. 98, no. 2, pp. 232–241, 2012, doi: 10.3813/AAA.918508.

[11] G. Yu, W. Dai, and E. Luo, "CFD simulation of a 300 Hz thermoacoustic standing wave engine," *Cryogenics (Guildf).*, vol. 50, no. 9, pp. 615–622, 2010, doi: 10.1016/j.cryogenics.2010.02.011.

[12] A. Di Meglio and N. Massarotti, "CFD Modelling of Thermoacoustic Energy Conversion: A Review," *Energies*, 2022.

Experimental study of a R290 variable geometry ejector

Giorgio Besagni[1], Fabio Inzoli[1], Lorenzo Croci[2], Vittorio Brignoli[2], Jacek Smolka[3], Michal Palacz[3], Michal Haida[3], Jakub Bodys[3], Rafal Fingas[3], Andrzej J. Nowak[3], Bartlomiej Melka[3], Michal Stebel[3], Edyta Piechnik[3]

[1]Politecnico di Milano, Department of Energy, via Lambruschini 4a, 20156, Milano (Italy)
[2]Ricerca sul Sistema Energetico – RSE S.p.A., Power System Development Department, via Rubattino 54, 20134, Milano (Italy)
[3]Silesian University of Technology, Department of Thermal Technology, Konarskiego 22, 44-100, 20156, Gliwice (Poland)

giorgio.besagni@polimi.it

Abstract. Ejectors are classified as fluid-dynamics controlled devices where the "*component-scale*" performances are imposed by the local-scale fluid dynamic phenomena. For this reason, ejector performances (measured by the pressure-entrainment ratio coordinate of the critical point) are determined by the connection of operation conditions, working fluid and geometrical parameters. Given such a connection, variable geometry ejector represents a promising solution to increase the flexibility of ejector-based systems. The present study aims to extend knowledge on variable geometry systems, evaluating the local and global performances of the R290 ejector equipped with a spindle. The prototype ejector was installed at the R290 vapour compression test rig adapted and modified for the required experimental campaign. The test campaign considered global parameter measurements, such as the pressure and the temperature at inlets and outlet ports together with the mass flow rates at both inlet nozzles, and the local pressure drop measurements inside the ejector. In addition, the experimental data were gathered for different spindle positions starting from fully open position the spindle position limited by the mass flow rate inside the test rig with the step of 1.0 mm.

1. Introduction

Ejector is a device constituted by a primary nozzle, a suction chamber, a mixing chamber, and a diffuser, as displayed in Figure 1. An "*high energy*" primary flow accelerates and expands through the primary nozzle creating a low-pressure region nearby the nozzle exit; subsequently, the secondary flow is entrained into the mixing chamber because of the vacuum-effect and the shear action between the primary and secondary flows. The primary and the secondary flows mix within the mixing chamber and the resulting stream moves into a diffuser where the high velocity fluid is gradually decelerated and increased in pressure due to subsonic conditions.

The entrainment, the pressure recovery, and the mixing effects provided by the ejector, makes it suitable to be employed in ejector refrigeration systems (*ERSs*; Figure 2a presents the layout of a standard ejector refrigeration system, *SERS*) [1]. In general, *ERSs* are promising alternative compared with compressor-based technologies owing to their reliability, limited maintenance, low initial and operational costs, and no working fluid limitations [2]; also, the generator of an *ERS* might exploit low-grade heat energy, making *ERSs* valuable in contributing towards reducing electricity consumption in the residential sector [3].

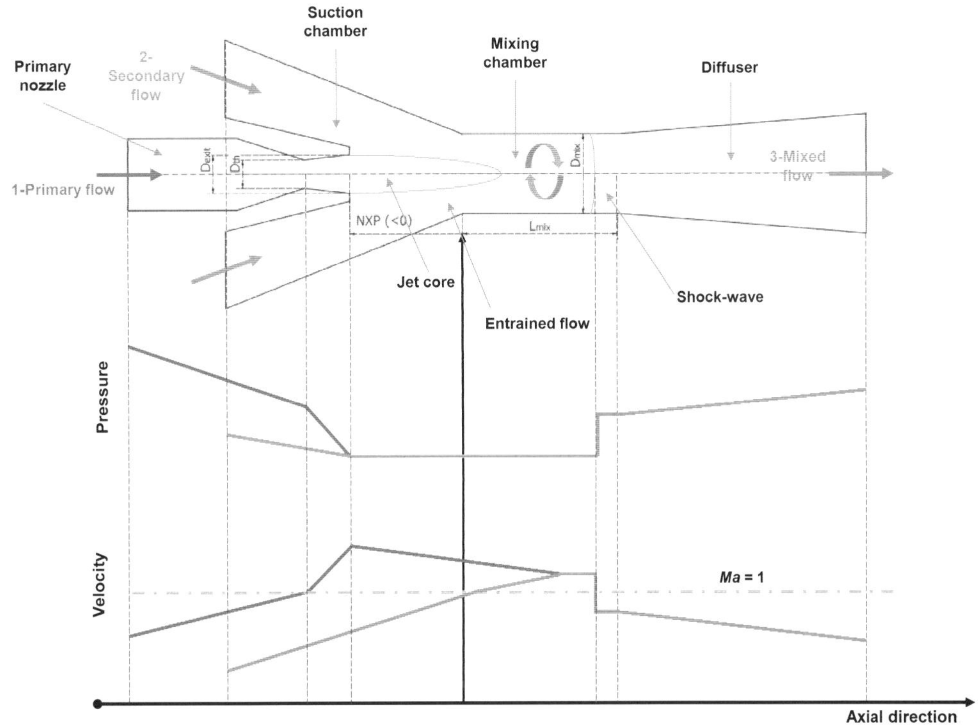

Figure 1. Ejector design and qualitative axial pressure and velocity trends

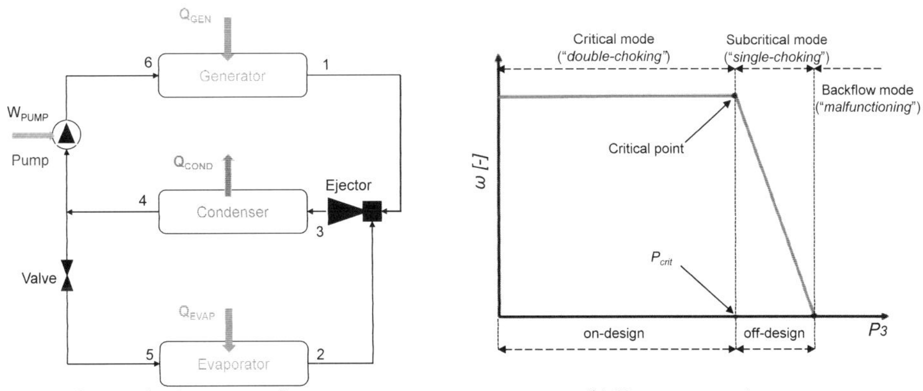

(a) Layout of a standard ejector refrigeration system *(b) Ejector operating curve*

Figure 2. Ejector component (a) and ejector refrigeration system (b)

Unfortunately, the large-scale deployment of *ERSs* is hindered owing to two main drawbacks: the low coefficient of performance (in the range of 0.1–0.7) and the relevant influence of ejector operation on the performance of the whole system. The latter can be easily explained by the fact that the ejector is a fluid-dynamics controlled device, where the fluid-dynamic interactions at the "*local-scale*" impact on the performances at the "*component-scale*", namely the entrainment ratio ω (viz- the ratio between secondary and primary mass flow rates, Figure 2b). These multi-scale relationships can be easily explained considering that the ejector operation relies on two concurrent physical phenomena: (i) the low-pressure fluid entrainment process caused by the primary flow expansion and (ii) the compression effect, provided by the diffuser, which raises the secondary flow pressure from the evaporator to the condenser pressure. These two effects are contrasting, and an improvement of the former would deteriorate the latter. Hence, for a given nozzle area ratio, the ejector operating curve is imposed by the primary and secondary flow boundary conditions. To this end, several attempts have been made to provide a viable ejector control system which could improve the system COP and at the same time overcome the above-described operating limitation. In this direction, variable geometry ejectors (VGE) represents a promising solution to provide *ERS* with the flexibility to dynamically adapt to the boundary conditions, increasing the performances compared to fixed-geometry ejectors. In this respect, a short summary of the presented literature review has been also provided in Table 1 of Ref. [4]. Indeed, Variable Geometry Ejectors (*VGE*) represent a promising solution to increase the flexibility and operation range of this component. In a *VGE*, the spindle acts on the nozzle area, and primary flow rate, and changes the ejector's response by adjusting its entrainment ratio accordingly with the requirements of the system (i.e., temperature set point, thermal load, …). The present-day discussion regarding ejector technology is even more challenging owing to the current transition in refrigerants; given the European regulation aimed to limit the emission of fluorinated greenhouse gases, the existing market is expected to change sharply in the next years, and 3rd generation refrigerants (i.e., R134a) will be most likely replaced by natural (i.e., hydrocarbons) and 4th generation fluids (i.e., hydrofluoroolefins) [5]. Among new refrigerants, *R290* (propane) is promising owing to its favourable thermodynamic properties, which make it suitable for refrigeration applications in the medium-long views. Besides, few recent literature studies investigated the performance of R290 as working fluid in heat-pump systems with encouraging outcomes, as for example Refs [6, 7, 8].

This paper aims at extending knowledge on *VGE* systems, which is mostly limited to theoretical studies. To this end, this paper experimentally evaluates the impact of a spindle-provided ejector operated with R290 on the performance of an *ERS*. In a previous study proposed by the authors, a *VGE* ejector operated with R290 has been designed and numerically studied [4]. Thus, the proposed study is an advancement compared with the present body of knowledge as it aims at experimentally evaluating the local and global performances of the R290 *VGE* ejector equipped with a spindle. The test campaign considered global parameter measurements, such as the pressure and the temperature at inlets and outlet ports together with the mass flow rates at both inlet nozzles, and the local pressure drop measurements inside the ejector.

2. Methods

2.1. Tested ejector

The tested ejector has been derived by Besagni and Cristiani [4] starting from Del Valle et al. [6]. It should be noted how, compared with Del Valle et al. [9], the mixing chamber diameter has been reduced to 4 mm to extend the on-design operating mode. The present ejector has been provided of a spindle (Figure 3) which can move axially within the primary nozzle.

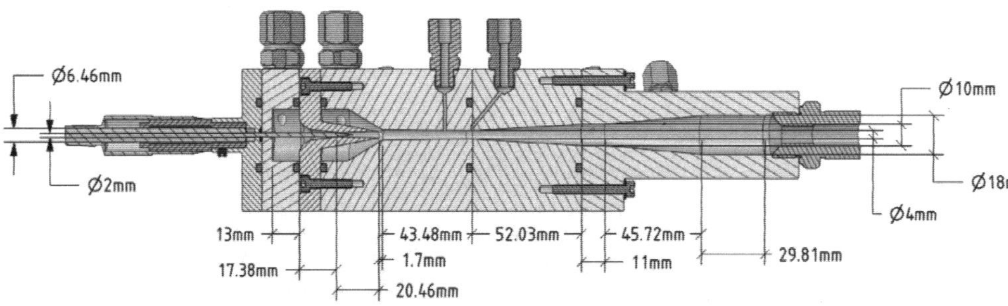

Figure 3. Ejector equipped with the spindle.

The neutral spindle position (*SP*) in which the tip of the spindle is placed right in the nozzle throat has been named as *SP#0*. Moving the spindle towards the nozzle exit, the nozzle throat area is reduced, and the area ratio is increased. The effects of the *SP* have been analysed in the range 0-8 mm, with a discretization of 1 mm each position, whose code name will be referred as *SP#0.8*, depending on the spindle position. It is worth noting that 8.0 mm corresponds to the unit minimum capacity (and *VGE*) for operation. The prototype device was manufactured, assembled, and tested with a high-pressure using nitrogen at 60 bar by TBT Technology in Gliwice, Poland. Figure 4 presents the manufactured ejector with the inlet/outlet ports, inner pressure drop ports, and the micrometer used to control the spindle position.

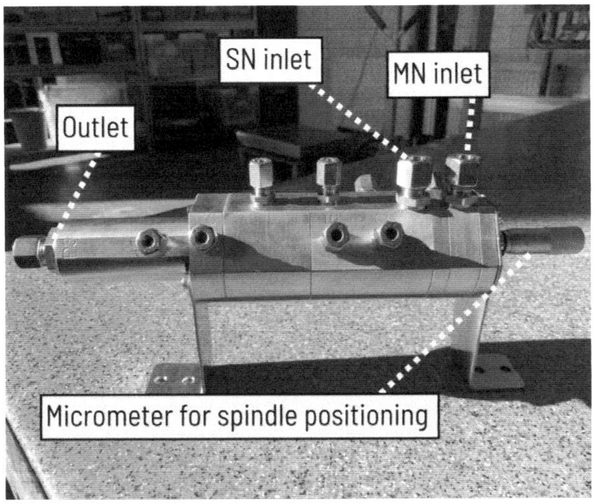

Figure 4. Manufactured R290 variable geometry ejector with spindle.

2.2. Test rig and instrumentation

Figure 5 presents the schematic diagram of the test rig used in this project and Table 1 lists the used instrumentation. All the pipes were covered with thermal polyethylene insulation that withstood the operating temperatures. The ejector inlet and outlet ports were connected with 10 mm copper pipes using Viton straight connectors. According to the goals of the present study, the detailed analysis of the performance of the *VGE* required precise instrumentation, as listed in Table 2. The temperature at the inlets of the ejector and the outlet was measured with calibrated T-type thermocouples. These sensors were attached to the copper walls of the inlet and outlet connectors. The thermal conductive paste was used to ensure proper contact of the sensors with the piping surface. Moreover, thermal insulation was installed at positions of thermocouples to minimize the influence of the ambient temperature on the thermocouple reading. Considering that one of the project aims was related to the definition of the pressure profile inside the ejector, the additional absolute pressure sensor was installed at the motive nozzle inlet. Furthermore, five differential pressure transducers were installed for measuring the pressure profile along the ejector. The mentioned differential pressure sensors marked as DP_1 to DP_5 are presented in Figure 6. It should be noted that the first of these sensors (DP_1) was used to measure the pressure drop inside the motive nozzle. Therefore, the measurement range of that pressure transducer was in the range of 0.0 to 25.0 bar, while the measurement range of DP_2 to DP_5 was in the range of 0.0 to 2.0 bar. The differential pressure sensors were connected with the prototype ejector in positions A-H presented in Figure 7. In the final setup of the system, positions B and G were closed. The high-pressure side for all the differential pressure transducers was defined with respect to the port located closer to the motive nozzle, e.g., in position A for A-C ports for DP_1. All the ports presented in Figure 8 are connected to differential pressure sensors as follows: position A-C - differential pressure sensor DP_1; position C-D - differential pressure sensor DP_2; position D-E - differential pressure sensor DP_3 :position E-F - differential pressure sensor DP_4; position F-H - differential pressure sensor DP_5.

Table 1. Description of the measurement instruments of the SUT R290 heat pump test rig.

Parameter	Sensor type	Model	Range and accuracy
Pressure	Gauge pressure, HP	Danfoss AKS32R	-1 ÷ 32 bar, 0,8% · x
	Gauge pressure, LP	Danfoss AKS32R	-1 ÷ 12 bar, 0,8%·*x*
Temperature	PT1000 RTD	Danfoss AKS11	- 40 ÷ 100 °C, ±0,3◦C
Mass flow rate	Coriollis type	E+H Cubemass C300	0 ÷ 90 kg/h, 0,1% *x*

Table 2. Description of the measurement instruments for the multi-scale study of the *VGE* ejector.

Parameter	Sensor type	Model	Range and accuracy
	Absolute pressure, MN	E+H Cerabar PMP51	0 ÷ 25 bar, 0,1% x
Pressure	Differential pressure, DP_1	E+H Deltabar PMD55	± 0 ÷ 25 bar, 0,1%·*x*
	Differential pressure, DP_{2-5}	E+H Deltabar PMD55	± 0 ÷ 2 bar, 0,1%·*x*
Temperature	T-type thermocouple	Termoaparatura TTP	-40 ÷ 350◦C, ±0,5◦C

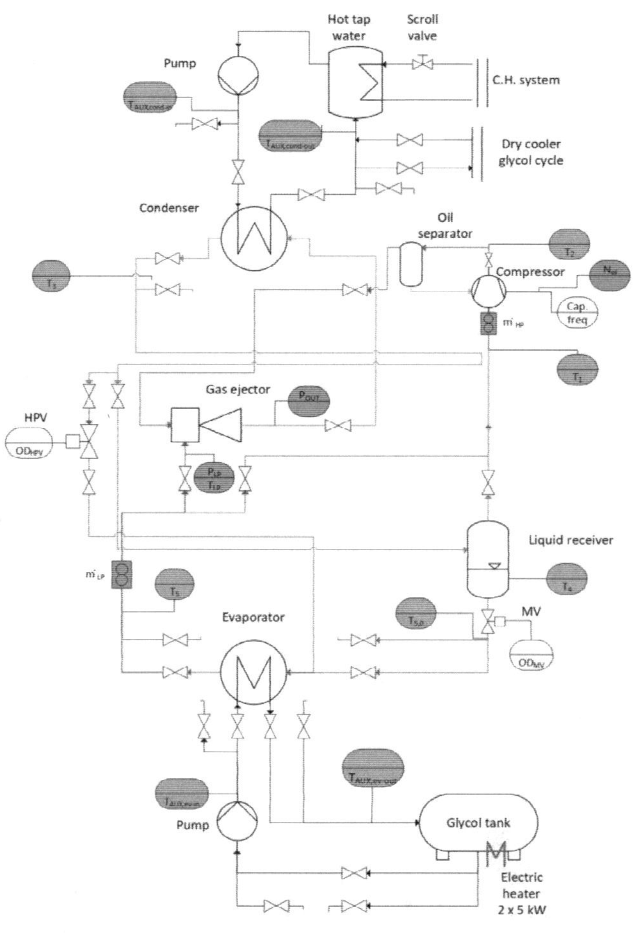

Figure 5. The PID scheme of the SUT R290 heat pump system.

Figure 6. The board for the pressure sensors and the ejector.

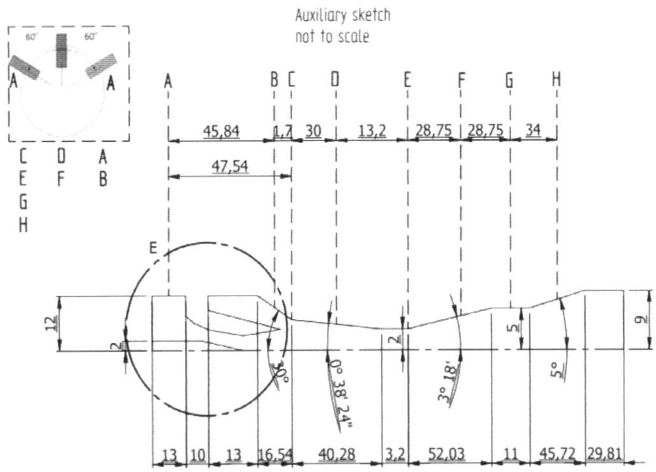

Figure 7. Cross-section of the VGE presenting the measurement ports positioning (A-H).

2.3. Data acquisition system

To ensure the required accuracy of the pressure and temperature measurements, the external data acquisition system was used. All the pressure and temperature sensors described in the previous sections were connected to the National Instruments (NI) cDAQ-9189. For the temperature measurements, a dedicated insulated thermocouple module with built-in cold junction temperature compensation was used (NI-9212). The analogue signals from the pressure transducers were captured with a current input module with an internal current loop (NI-9207). The linear functions defined for the sensors measurement range were used to convert the current signal to the pressure signal. To visualize and record all these signals, the NI LabView application was developed. The frequency of the data recording was set to 5 Hz for all of the investigated cases.

3. Results

The experimental results of these tests are presented in Table 3 and in Figure 8 (ω is the entrainment ratio and P_{lift} is the ratio between the outlet pressure and the secondary flow pressure). For each case, the results were recorded during a period of 5 minutes of the heat pump operation after achieving steady-state conditions. The recorded time period was chosen based on the experimental investigation performed for R744 two-phase ejector. The data sampling frequency was 5 Hz for the motive nozzle pressure, pressure drops and temperatures. For the mass flow rates, the suction nozzle pressure and the outlet pressure, the sampling frequency was 0.2 Hz. All data were averaged, and type-A uncertainty was used to represent the deviation of the results.

Table 3. Experimental results for spindle positions *SP#1-SP#8*.

ID		P_1	T_1	$\dot{m}_1$	P_2	T_2	$\dot{m}_2$	P_3	T_3	$T_{3'}$	P_{lift}	ω
-		$[bar]$	$[°C]$	$[kg/h]$	$[bar]$	$[°C]$	$[kg/h]$	$[bar]$	$[°C]$	$[°C]$	$[bar]$	$[-]$
SP#1_OC1	Av.	17.08	77.61	34.50	4.67	24.57	4.30	5.30	53.35	52.26	0.63	0.12
	Dev.	0.09	2.28	0.00	0.05	0.63	0.00	0.05	2.99	2.98		
SP#1_OC2	Av.	17.75	88.83	35.04	4.99	27.32	4.28	5.60	66.84	65.62	0.61	0.12
	Dev.	0.07	1.35	0.18	0.05	0.29	0.09	0.01	1.45	1.43		
SP#1_OC3	Av.	18.56	96.30	36.01	5.36	28.43	4.30	5.95	74.20	72.88	0.59	0.12
	Dev.	0.08	1.23	0.06	0.05	0.20	0.00	0.05	1.25	1.23		
SP#2_OC1	Av.	18.40	84.55	36.18	5.09	25.63	4.30	5.69	58.93	57.66	0.60	0.12
	Dev.	0.01	2.67	0.27	0.03	0.70	0.00	0.02	3.52	3.57		
SP#2_OC2	Av.	18.77	95.30	35.70	5.24	28.50	4.30	5.86	72.67	71.38	0.61	0.12
	Dev.	0.16	1.27	0.30	0.06	0.21	0.00	0.06	1.30	1.28		
SP#3_OC1	Av.	18.43	90.82	35.50	5.30	27.50	4.60	5.90	67.52	66.27	0.60	0.13
	Dev.	0.03	1.41	0.00	0.00	0.38	0.00	0.00	1.80	1.79		
SP#4_OC1	Av.	19.16	101.14	33.55	5.13	32.17	4.10	5.77	75.66	74.24	0.63	0.12
	Dev.	3.12	1.23	0.15	0.05	8.13	0.00	0.05	1.70	1.70		
SP#5_OC1	Av.	15.46	95.45	33.36	5.11	31.45	4.08	5.74	72.46	70.98	0.63	0.12
	Dev.	0.06	0.63	1.38	0.21	0.29	0.13	0.23	0.98	0.98		
SP#5_OC2	Av.	17.20	87.59	25.88	3.90	26.27	3.30	4.47	62.63	61.35	0.57	0.13
	Dev.	0.03	0.95	0.56	0.00	0.39	0.00	0.05	1.42	1.40		
SP#6_OC1	Av.	18.37	87.21	22.87	3.50	24.54	3.20	4.00	59.45	58.26	0.50	0.14
	Dev.	0.07	1.02	0.18	0.00	0.31	0.00	0.00	1.23	1.23		
SP#7_OC1	Av.	17.77	76.80	17.06	3.20	23.51	1.86	3.50	47.41	46.85	0.30	0.11
	Dev.	0.06	0.50	0.14	0.00	0.29	0.89	0.00	0.94	0.91		
SP#8_OC1	Av.	19.17	76.73	12.50	2.76	26.11	0.00	2.95	47.32	46.21	0.19	0.00
	Dev.	0.04	0.15	1.24	0.20	0.02	0.00	0.20	0.09	0.07		

The motive nozzle pressure and temperature varied from approximately 15.5 bar to 19.2 bar and from approximately 76.7°C to 101.1°C, respectively. Furthermore, the suction nozzle pressure was from approximately 2.75 bar at SP#8_OC1 to 5.4 bar at SP#1_OC3 and the suction nozzle temperature differed from 24.5°C at SP#1_OC3 to 32.2°C at SP#4_OC1. Finally, the outlet pressure was reached in the range from approximately 3.0 bar at SP#8_OC1 to 6.0 bar at SP#1_OC3 and the outlet temperature was from approximately 47.0°C at SP#8_OC1 to 76.0°C at SP#4_OC1. The different operating conditions at both nozzles and the outlet together with the different spindle position caused the motive nozzle mass flow rate reduction. Hence, the motive nozzle mass flow rates presented in Table 4 decreased from approximately 36.0 kg/h at SP#1_OC3 to 12.5 kg/h at SP#8_OC1, which was approximately 27.0% of the highest mass flow rate obtained at SP#0_OC3. In addition, the pressure lift varied from 0.19 bar at SP#8_OC1 to 0.63 bar at SP#1_OC1 and ω was in the range from 0.0 at SP#8_OC1 to 0.14 at SP#6_OC1. Finally, the suction nozzle mass flow rate varied from 0.0 kg/h at SP#8_OC1 to 4.6 kg/h at SP#3_OC1 due to the reached ω at the specific pressure lift and motive nozzle conditions. Pressure drops DP_1-DP_5 shown in Figure 8 demonstrated that the highest pressure drop was reached for DP_1 from approximately 12.0 bar at SP#5_OC1 to 16.5 bar at SP#8_OC1. Then, the pressure increment was obtained for DP_2 from -0.25 bar at SP#8_OC1 to approximately -0.75 bar at SP#3_OC1. Furthermore, the lowest pressure increment was observed for highest mass flow rate reduction at SP#7 and SP#8. A minor pressure drop in the range from 0.04 bar to approximately 0.2 bar was reached for DP_3 due to the mixing process inside the mixing section. The pressure increments at the first part of the diffuser from -0.08 bar at

SP#8_OC1 to -0.36 bar at SP#1_OC3 was reached for DP_4. Finally, the negligible pressure increments of maximum -0.03 bar was obtained for DP_5.

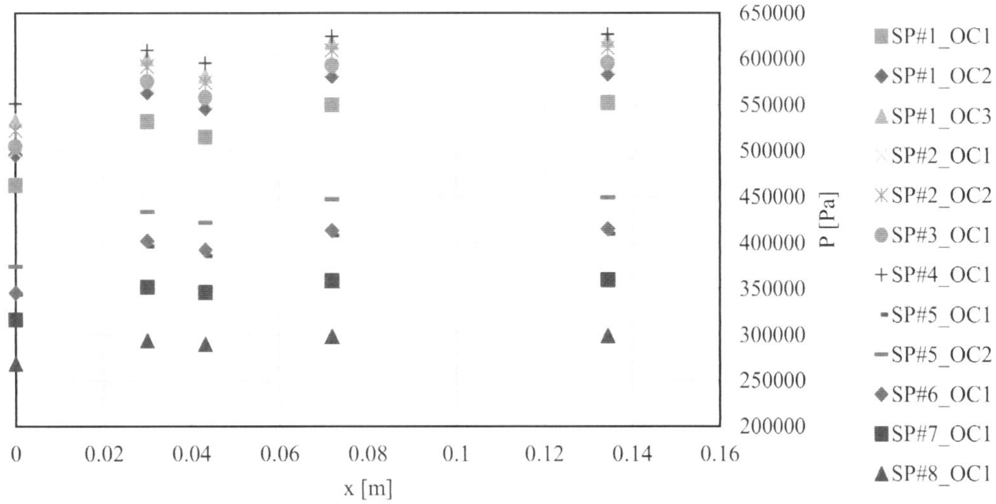

Figure 8. Local pressure drop measurements inside the ejector.

4. Conclusions

The aim of this work was to perform the experimental test campaign of the *VGE* prototype gas ejector using propane as the working fluid. The test campaign considered global parameter measurements and the local pressure drop measurements inside the ejector. The experimental data were gathered for different spindle positions starting from fully open position SP#0 to the spindle position limited by the mass flow rate inside the test rig with the step of 1.0 mm. It was found that the highest pressure drop was reached between the motive nozzle inlet and the half part of the pre-mixer due to the expansion process in the converging-diverging nozzle and the entrainment process in the pre-mixer. The pressure increment between the pre-mixer and the half-part of the converged mixing section was observed due to the sonic and expansion waves appeared near the inner ejector walls. In addition, the off-design conditions of the prototype ejector resulted in a low mass entrainment ratio below 0.2 for all the gathered experimental points. Future works will be devoted to extending this experimental study.

5. Acknowledgements

This work has been financed by the Research Fund for the Italian Electrical System in compliance with the Decree of Minister of Economic Development April 16, 2018.

6. References

[1] Aidoun Z, Ameur K, Falsafioon M and Badache M 2019 *Inventions* **4** 15
[2] Besagni G, Mereu R and Inzoli F 2016 *Renew. Sust. Energ. Rev.* **53** 373-407
[3] Knobloch F, Pollitt H, Chewpreecha U, Daioglou V and Mercure J F 2019 *Energ. Effic.* **12**(2) 521-550
[4] Besagni G and Cristiani N 2021 *Appl. Therm. Eng.* **188** 116612
[5] Cavallini A 2020 *J. Phys. Conf. Ser.* **1599** 012001
[6] Jeon Y, Kim S, Lee S H, Chung H J and Kim Y 2020 *Appl. Energ.* **278** 115715
[7] Zhao H, Yuan T, Gao J, Wang X and Yan J 2019 *Energy* **166** 845-861
[8] Chen Q, Hwang Y, Yan G and Yu J *Appl. Therm. Eng.* 2020 **164** 114456
[9] Del Valle J G, Jabardo J S, Ruiz F C and Alonso J S J 2014. *Int. J. Refrig.* **46** 105-113

Lattice Boltzmann model of a square natural circulation loop with small inner diameter: working fluid effects

Johan Augusto Bocanegra and Mario Misale
DIME - Department of Mechanical, Energy, Management and Transportation Engineering, University of Genoa, Via All'Opera Pia 15-A, (I) 16145, Italy.

augusto.bocanegra@edu.unige.it

Abstract. Natural circulation loops are systems used to transport heat by the free convection of a working fluid in a closed thermo-hydraulic circuit. Many examples of the Lattice Boltzmann method applied to free convection can be found in the literature, but most of them are limited to model simple enclosures. This work is one of the first approaches to an engineering system as the Natural Circulation Loops using this numerical method. This study describes the implementation and validation of a numerical model of a rectangular loop characterized by a small inner diameter. This research is focused on the thermohydraulic response of the NCL varying the working fluid. We have obtained satisfactory results demonstrating that accepted analytical correlations for the steady-state behavior are reproduced by the thermal Lattice Boltzmann Method (double distribution function). Overall, these results suggest that the temperature and velocity oscillations during the transient increase directly with the fluid Prandtl number; instead, the temperature difference between the vertical legs is inversely correlated and decreases.

1. Introduction

Natural circulation loops (NCL) are thermal systems that transport heat from a heat source to a heat sink by the movement of a working fluid into a closed pipes system. The fluid flow is induced by natural convection determined by thermal expansion and gravity [1]. These engineering systems are very interesting in geothermic, solar energy, nuclear plants, or electronic cooling systems, mainly by the absence of an active pump system [2]. The common shape of an NCL is rectangular and oriented vertically. Some analytical relationships can predict the flow characteristics of the loops under selected operational parameters such as geometric dimension, fluid properties, source, and sink characteristics for laminar regime [3,4] and transition or turbulent regime [5]. Those relationships have been validated by comparison with experimental data from laboratories worldwide. In recent years experimental works validated the applicability of the analytical model to NCL with a small inner diameter connected in parallel [7], [8].

Previous works evaluate the performance of rectangular-shaped single-phase NCL with heat end exchangers mainly by developing analytical models, simulations, or experiments. Early computational works use a one-dimensional model [9] solved using finite differences [10]. Yadav et al. in 2014 [11] performed a simulation with different tilt angles and found that the tilt angle increases the mass flow rate and decreases the heat exchange. Moreover, they also found that higher operating pressure reduces the time to reach a steady state. They corroborated the Swapnalee and Vijayan empirical correlation for the turbulent regime [5]. Cheng et al. 2018 [12], using a 3D CFD model, simulate the transient response of a single-phase NCL evaluating mass flow rate and energy generation. They use

Vijayan's correlation [5] for the laminar regime (originally proposed for imposed heat flux at the heater) with a good fitting for the tested Reynolds numbers (lower than 100). Recently a study on the jump of heat transfer coefficient by tilting the square NCL (single and coupled) was developed using 2D and 3D numerical models [13].

In the above papers, the numerical simulation employed the classic techniques of CFD. The Lattice Boltzmann Method (LBM) was adopted in the present work. LBM is a numerical method characterized by a mesoscopic approach based on the solution of the discretized Boltzmann transport equation [14,15]. The LBM differs from the macroscopic approach adopted in the Finite Element Methods, even if, in LBM, some numerical schemes are designed to solve the macroscopic Navier-Stokes equations. Additionally, the LBM is linked to statistical mechanics but differs from the microscopic approach, and no information can be traced back to Molecular Dynamics. This method is highly parallelizable by its local algorithmic rules. It can handle multi-physics by considering different lattices simultaneously, i.e., one lattice for the fluid density and momentum, a second one for the temperature field (thermal double distribution function or DDF), and even additional lattices if particle or chemical species transport are of interest [16]. Different upgrades of the LBM have been developed to account for thermal effects [17]. Examples of studies using LBM to model natural convection show the interest and applicability of the method in this topic: Rayleigh-Bernard [18], rectangular cavities [19], 2D channel [20], nanofluids into a curved boundaries cavity [21], ferrofluids in a linear heated cavity [22], channel heated by active blocks [23], turbulent Rayleigh-Bernard in a channel [24], or a cavity with low Prandtl number fluid [25]. If even the method was successfully applied to natural circulation in cavities and channels, applications to NCL are scarce: our previous work shows the effect of the heater-cooler orientation, and we found an influence on the time to start the natural circulation and on the steady-state converge time [26,27]. This study explores, for the first time, the application of the LBM to NCL under a fixed heater-cooler orientation and different parameter configurations (this is the first approach using LBM to study the effects of changing the working fluid in simple NCL circuits and varying the modified Grashof numbers).

This work aims to present a single-phase NCL characterized by a small diameter, simulated by LBM, and study the thermo-hydrodynamic behavior operating with different working fluids (characterized by different Prandtl numbers). The model was validated by comparison with the analytical model proposed by Cheng et al. [4] for the temperature distribution along the loop and the relationship between the Reynolds number at steady-state and the modified Grashof number. We found a significative effect of the Prandtl number in the thermal oscillations during the transient and on the steady-state Reynolds number. This study provides important insights into the validity of a 2D LBM model to study the steady-state thermal flow in the laminar regime of an NCL.

2. Methods

A two-dimensional model of a squared NCL was developed [26,27]. The loop is characterized by two horizontal heat end exchangers (modeled as fixed hot and cold temperatures at the tube walls) and two vertical legs that connect the heater and the sink (adiabatic), see figure 1. Fluids characterized by different Prandtl numbers ($Pr = 0.1, 1.0, 7.0$) were tested. Thermophysical properties of the working fluid were fixed during the simulations, and a single-phase condition was assumed.

The model was implemented using C++ and PALABOS (Parallel Lattice Boltzmann Solver library), [28]. The LBM double distribution function (DDF) approach was used, the first distribution function was used to solve the fluid dynamic field, and the second one was used to simulate the thermal field. Each distribution function is implemented in a separate lattice; D2Q9 for the fluid dynamic field and D2Q5 for the thermal field, see figure 2. The two lattices are overlapped and coupled through the buoyancy force term. $Pr = 1.0$ in DDF-LBM means that both lattices (D2Q5 & D2Q9) have the same relaxation time (physically have the same diffusivity ratio, i.e., viscosity and thermal conductivity).

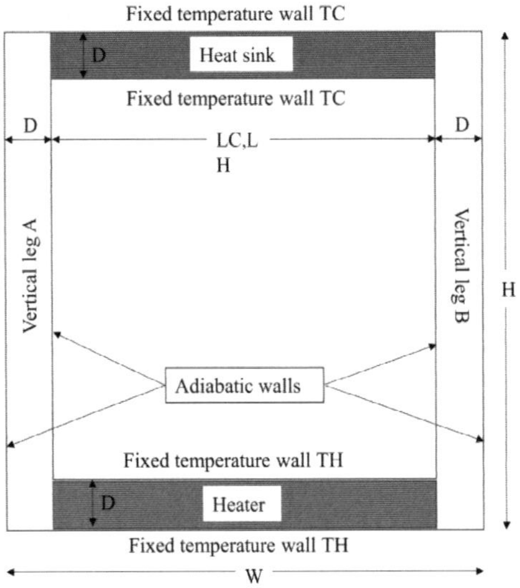

Figure 1. Sketch of the rectangular NCL, *horizontal heater horizontal cooler* configuration.

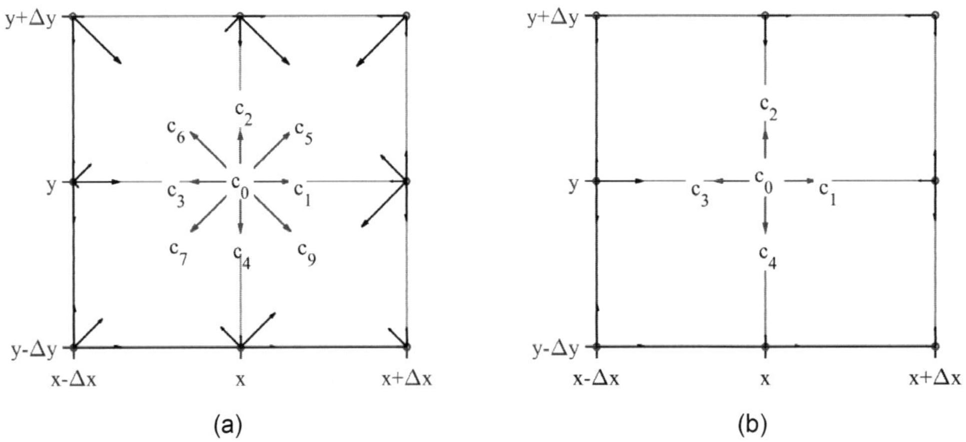

Figure 2. Thermal LBM model using two lattices, every single node has some vectors representing the two discretized probability density functions. (a) D2Q9 hydrodynamic lattice, (b) D2Q5 thermal lattice.

2.1. Stability and convergence

The convergence of the system to the steady-state was determined by calculating the relative standard deviation of the Nusselt number, and if it reaches a given threshold (10^{-16}), convergence is assumed, and the simulation is finished. Additionally, the convergence of the temperature difference between the vertical legs (ΔT_{avg}) and the Reynolds number was studied. In the laminar regime, both quantities converge. The simulations do not present computational instabilities or divergencies in the tested Rayleigh number range (input parameter). The lattice spatial resolution effect on flow conditions was investigated by calculating the Reynolds number at steady-state, changing the number of nodes per unit length. The following spatial resolution values were tested: 500(diverge), 1000, 2000, 4000, 8000, and 10000 grid points per unit length. Results are presented in figure 3. It was noted that the Reynolds number converges for a resolution up to 4000; thus, 40 mesh points per pipe diameter were used in the simulations, i.e., 1000x1000 lattice.

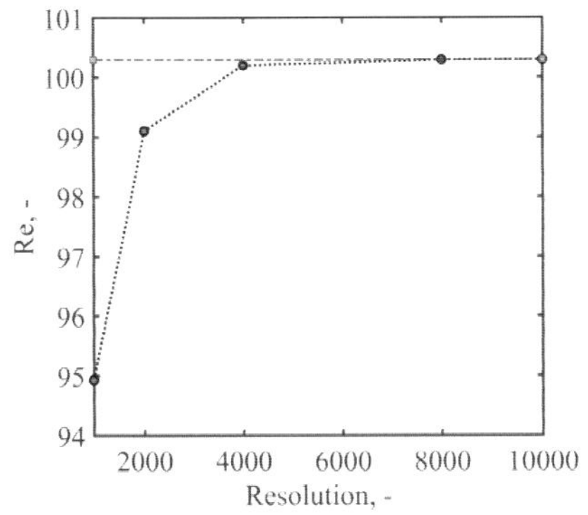

Figure 3. Grid convergence study in laminar regime, Re_{ss} vs. resolution. Pr=1.0.

3. Results

3.1. Steady-state behavior

The one-dimensional model proposed in the literature [3–5] was used to validate the LBM simulation. The effects of the input thermal energy (modified Grashof number, Gr_m, Eq. 1) on the mass flow rate (Reynolds number at steady-state, Re_{ss}) were studied. The analytical model is resumed in the Eq. 2. It is possible to show that this model (used for imposed temperature conditions at the heater and heat sink) is equivalent to the model proposed in [3,5] for the laminar and turbulent region in an NCL (with a heat flux boundary condition at the heater).

$$Gr_m = \frac{g\beta\Delta T_{avg}D^2H}{\upsilon^2} \tag{1}$$

$$Re_{ss} = \sqrt{\frac{Gr_m}{N_G}} \tag{2}$$

The geometrical parameter, N_G, described by Eq. 3, accounts for local and concentrated friction losses, considering the frictional coefficient, f, and the local resistance coefficient, K.

$$N_G = \frac{1}{2}\Sigma_i(f\frac{L_i}{D} + K) \tag{3}$$

The analytical correlation (Eq. 2) between Re_{ss} and Gr_m/N_G was verified regardless of the fluid properties, as presented in figure 4. When the value of the Prandtl number was changed to a higher value, lower values of the modified Grashof number were reached with the same initial setup (global Rayleigh number), and subsequently, lower values of the *steady-state Reynolds number* were obtained. However, all the points fall in the same trend described in Eq. 2, see figure 4. The LBM results for $Pr = 7.0$ show a small deviation for the points corresponding to high Ra numbers even when the flow regime remains laminar.

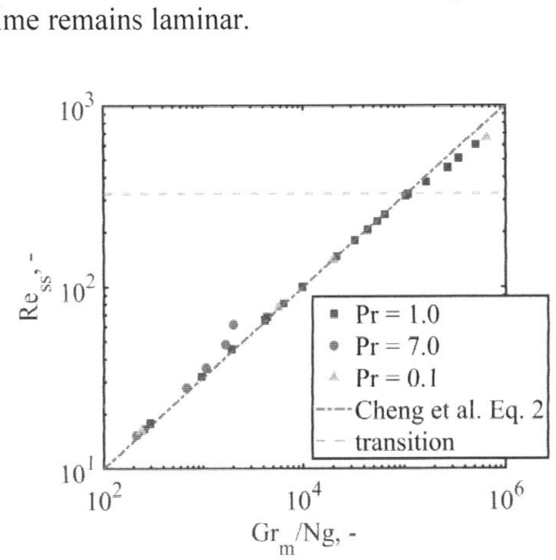

Figure 4. Thermo-hydraulic behavior of the NCL described by the Eq. 2, [3–5].

A change in the slope in this graphic is noticed over $Re_{ss} = 320$. The velocity field, over this transition limit, shows vortex formation near the heater and heat sink walls, see figure 5.

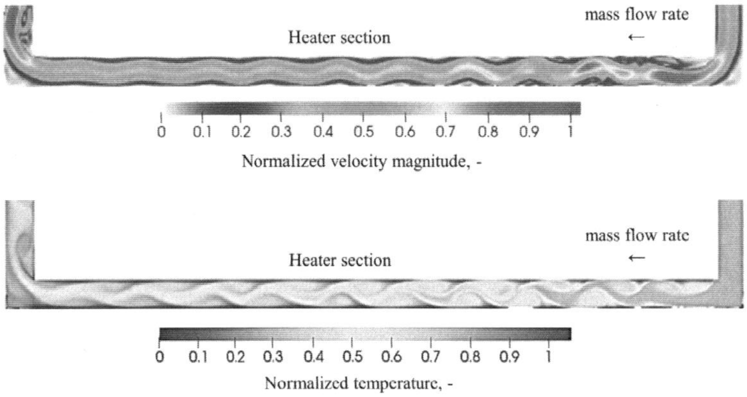

Figure 5. Detail of the heater section. Traveling eddies at the transition regime, $Re_{ss} = 510, Pr = 1.0$.

Resuming, two flow regimes were found, i.e., a laminar regime and a transient regime. The *laminar regime* is characterized by a parabolic velocity distribution in the pipes. The *transition regime* is characterized by traveling eddies near the walls of the heater and cooler (during this regime, the Nusselt number increases).

The temperature distribution predicted by the one-dimensional model is simple and regular because it does not consider the cross-sectional distribution of thermal flow and predicts only the behavior of the *laminar regime.* The vorticity in the transition regime highly influences the temperature distribution along the circuit near the heater and heat sink walls and in the first section of the vertical legs, figure 6(a). The oscillations in the average temperature profile decrease with a lower Prandtl number, as shown in Figure 6(b). The shape of the temperature profile in Figure 6(a) (predicted by the one-dimensional model) has a poor similarity to the LBM results. The difference in the predictions could be linked to resolution and average errors; the average over the tube cross-section shows more variations in the case of higher Pr. In the *transition regime,* the timescale of the flow field variations could be under-resolved in the high Ra number zone, depending on the Pr value (e.g., lower Pr values show more regular cross-sectional profiles).

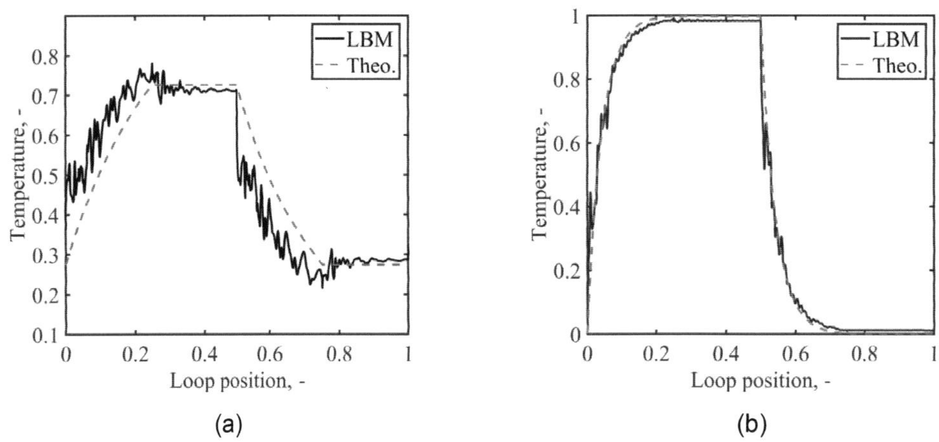

Figure 6. Dimensionless temperature distribution (average over the cross-sections) along the NCL. (a) $Re_{ss} = 510, Pr = 1.0$. (b) $Re_{ss} = 660, Pr = 0.1$.

3.2. *Transient behavior*

Differences in the transient response can be noted; lower Pr fluid presents more dampened oscillations and faster convergence to the steady-state, figure 7(a). A higher temperature difference between the adiabatic legs is noted for the lower Pr fluid (at the same Re_{ss}). This trend is also visible in the velocity oscillations presented in the phase diagrams, figure 7(b). The increment of the Pr implies a dominance of momentum transport over thermal conduction, which means that the fluid transports the heat mainly by convection. The fluid increase or decreases (locally) its temperature with more difficulty, and of course, this difficult the convection because the body force term is considered linear and dependent on this temperature change. In contrast, a fluid with a low Pr is heated easily; therefore, the density changes are quickly induced with a subsequent higher flow rate.

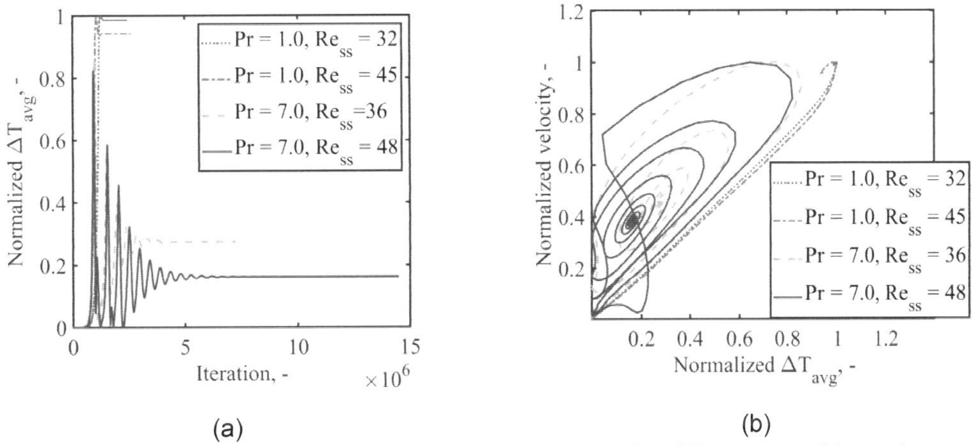

Figure 7. Transient response of the NCL simulated by LBM for different Reynolds numbers and Prandtl numbers. (a) ΔT_{avg} vs. time, (b) phase diagram: normalized velocity vs. normalized ΔT_r.

The main limitation of this work is the scalability of the results to a 3D model. Although the results must be taken with care regarding the limit between the *laminar* and *transition regime* and the complex flow pattern presented in the *transition regime*, our results support further research implementing a full 3D LBM model of an NCL, with relatively small modifications of the tested 2D LBM model. The present results support the applicability of the 2D model to low Re_{ss}, thanks to the positive validation with the analytical model of Eq. 2, this analytical model used in the validation has been verified experimentally in diverse NCL circuits and is a reference correlation in the NCL field.

4. Conclusions

This study shows that a numerical model using the thermal (double distribution function) Lattice Boltzmann Method can reproduce the transient and steady-state behavior of a square single-phase NCL. The results of this study underline that the numerical model reproduces accepted analytical correlations for steady-state behavior.

The main effects of the working fluid were studied by changing the Prandtl number. The observed effects of increasing the Prandtl number are incrementing the temperature and velocity oscillations during the transient response and decreasing the temperature difference between the adiabatic legs at a steady state.

Acknowledgments

This research was funded by Ministero dell'Istruzione, dell'Università edella Ricerca (MIUR, Italy), grant number PRIN-2017F7KZWS.

References

[1] M. Misale, Overview on single-phase natural circulation loops, in: International Conference on Advances in Mechanical & Automation Engineering, Rome (Italy), 2014: p. 13.
[2] D.N. Basu, S. Bhattacharyya, P.K. Das, A review of modern advances in analyses and applications of single-phase natural circulation loop in nuclear thermal hydraulics, Nuclear Engineering and Design. 280 (2014) 326–348. https://doi.org/10.1016/j.nucengdes.2014.09.011.
[3] P.K. Vijayan, Experimental observations on the general trends of the steady state and stability behaviour of single-phase natural circulation loops, Nuclear Engineering and Design. 215 (2002) 139–152. https://doi.org/10.1016/S0029-5493(02)00047-X.
[4] H. Cheng, H. Lei, L. Zeng, C. Dai, Theoretical and experimental studies of heat transfer characteristics of a single-phase natural circulation mini-loop with end heat exchangers, International Journal of Heat and Mass Transfer. 128 (2019) 208–216. https://doi.org/10.1016/j.ijheatmasstransfer.2018.08.136.
[5] B.T. Swapnalee, P.K. Vijayan, A generalized flow equation for single phase natural circulation loops obeying multiple friction laws, International Journal of Heat and Mass Transfer. 54 (2011) 2618–2629. https://doi.org/10.1016/j.ijheatmasstransfer.2011.01.023.
[6] M. Misale, M. Frogheri, Influence of pressure drops on the behavior of a single-phase natural circulation loop: Preliminary results, International Communications in Heat and Mass Transfer. 26 (1999) 597–606. https://doi.org/10.1016/S0735-1933(99)00046-9.
[7] M. Misale, J.A. Bocanegra, D. Borelli, A. Marchitto, Experimental analysis of four parallel single-phase natural circulation loops with small inner diameter, Applied Thermal Engineering. 180 (2020) 115739. https://doi.org/10.1016/j.applthermaleng.2020.115739.
[8] M. Misale, J.A. Bocanegra, A. Marchitto, Thermo-hydraulic performance of connected single-phase natural circulation loops characterized by two different inner diameters, International Communications in Heat and Mass Transfer. 125 (2021) 105309. https://doi.org/10.1016/j.icheatmasstransfer.2021.105309.
[9] N.M. Rao, M. Mishra, B. Maiti, P.K. Das, Effect of end heat exchanger parameters on the performance of a Natural Circulation Loop, International Communications in Heat and Mass Transfer. 29 (2002) 509–518.
[10] K. Kiran Kumar, M. Ram Gopal, Steady-state analysis of CO2 based natural circulation loops with end heat exchangers, Applied Thermal Engineering. 29 (2009) 1893–1903. https://doi.org/10.1016/j.applthermaleng.2008.08.002.
[11] A.K. Yadav, M. Ram Gopal, S. Bhattacharyya, Transient analysis of subcritical/supercritical carbon dioxide based natural circulation loops with end heat exchangers: Numerical studies, International Journal of Heat and Mass Transfer. 79 (2014) 24–33. https://doi.org/10.1016/j.ijheatmasstransfer.2014.07.068.
[12] H. Cheng, H. Lei, C. Dai, Thermo-hydraulic characteristics and second-law analysis of a single-phase natural circulation loop with end heat exchangers, International Journal of Thermal Sciences. 129 (2018) 375–384. https://doi.org/10.1016/j.ijthermalsci.2018.03.026.
[13] A. Dass, S. Gedupudi, Numerical investigation on the heat transfer coefficient jump in tilted single-phase natural circulation loop and coupled natural circulation loop, International Communications in Heat and Mass Transfer. 120 (2021) 104920. https://doi.org/10.1016/j.icheatmasstransfer.2020.104920.
[14] S. Succi, The Lattice Boltzmann Equation for Fluid Dynamics and Beyond, Clarendon Press, Oxford, England, 2001.
[15] S. Succi, R. Benzi, F. Massaioli, A Review of the Lattice Boltzmann Method, International Journal of Modern Physics C. 4 (1993) 409–415.
[16] Y. Wang, M. Xie, Y. Ma, Analysis of the multi-physics approach using the unified lattice Boltzmann framework, Annals of Nuclear Energy. 143 (2020) 107500–107513. https://doi.org/10.1016/j.anucene.2020.107500.

[17] K.V. Sharma, R. Straka, F.W. Tavares, Current status of Lattice Boltzmann Methods applied to aerodynamic, aeroacoustic, and thermal flows, Progress in Aerospace Sciences. 115 (2020) 100616. https://doi.org/10.1016/j.paerosci.2020.100616.
[18] X. Shan, Simulation of Rayleigh-Bénard convection using a lattice Boltzmann method, Phys. Rev. E. 55 (1997) 2780–2788. https://doi.org/10.1103/PhysRevE.55.2780.
[19] Z. Guo, B. Shi, C. Zheng, A coupled lattice BGK model for the Boussinesq equations, Int. J. Numer. Meth. Fluids. 39 (2002) 325–342. https://doi.org/10.1002/fld.337.
[20] A. D'Orazio, S. Succi, Simulating two-dimensional thermal channel flows by means of a lattice Boltzmann method with new boundary conditions, Future Generation Computer Systems. 20 (2004) 935–944. https://doi.org/10.1016/j.future.2003.12.005.
[21] M. Sheikholeslami, M. Gorji-Bandpy, S.M. Seyyedi, D.D. Ganji, H.B. Rokni, S. Soleimani, Application of LBM in simulation of natural convection in a nanofluid filled square cavity with curve boundaries, Powder Technology. 247(2013) 87–94. doi.org/10.1016/j.powtec.2013.06.008.
[22] G.H.R. Kefayati, Natural convection of ferrofluid in a linearly heated cavity utilizing LBM, Journal of Molecular Liquids. 191 (2014) 1–9. https://doi.org/10.1016/j.molliq.2013.11.021.
[23] T. Naffouti, University of Tunis, J. Zinoubi, El Manar, C.S. Nor Azwadi, Malysia, R. Ben Maad, El Manar, Applied Thermal Lattice Boltzmann Model for Fluid Flow of Free Convection in 2-D Enclosure with Localized Two Active Blocks: Heat Transfer Optimization, JAFM. 9 (2016) 419–430. https://doi.org/10.18869/acadpub.jafm.68.224.24198.
[24] Y. Wei, H. Dou, Turbulent Rayleigh–Bénard convection scaling in a vertical channel using the lattice Boltzmann method, in: Ancona, Italy, 2016: pp. 45–52. https://doi.org/10.2495/HT160051.
[25] M.A. Wahba, A.A. Abdelrahman, N.M. El-Sahlamy, M.F.A. Rabbo, Double SRT Thermal Lattice Boltzmann Method for Simulating Natural Convection of Low Prandtl Number Fluids, IJES. 6 (2017) 21–35. https://doi.org/10.9790/1813-0612022135.
[26] J.A. Bocanegra, A. Marchitto, M. Misale, Thermal performance investigation of a mini natural circulation loop for solar PV panel or electronic cooling simulated by lattice Boltzmann method, Int. J. EQ. 5 (2022) 1–12. https://doi.org/10.2495/EQ-V7-N1-1-12.
[27] J.A. Bocanegra, Lattice Boltzmann Method: applications to thermal fluid dynamics and energy systems, Ph.D. Thesis, Genoa University, 2021.
[28] J. Latt, O. Malaspinas, D. Kontaxakis, A. Parmigiani, et al., Palabos: Parallel Lattice Boltzmann Solver, Computers & Mathematics with Applications. (2020) S0898122120301267. https://doi.org/10.1016/j.camwa.2020.03.022.

On the Solved Turbulent Scales in Turbulent Plume Fires

F.S. Ciani[1], P. Bonfiglio[2] and S. Piva[1]

[1] Department of Engineering DE, Università di Ferrara, Via Saragat 1, 44122 Ferrara
[2] Materiacustica Srl, Via Saragat 1, 44122 Ferrara, Italy

pvs@unife.it

Abstract. Plume fires are characterized by a turbulent nature with a large number of different scales. LES is often used to solve the largest structures and to model the smallest ones. Grid size and time steps become decisive to place the limit between resolved and modelled turbulence. Significant information on this limit and its placement can be obtained with spectral analyses of the specific turbulent kinetic energy. While frequency analysis is relatively easy, an analysis in the wavenumber domain is more challenging. The IWC method, typically used in structures and acoustics, is used here for this purpose. IWC method allows to obtain wavenumber spectra with a better resolution than those obtained with a direct approach. Furthermore, in this paper the IWC method is also used in its reverse form to obtain frequency spectra. Although rather dense grids have been chosen, the number of nodes along the plume and their spacing is not such as to guarantee detailed wavenumber spectra with the direct approach and consequently with the reverse IWC. On the contrary, the IWC method provides wavenumber spectra in agreement with those obtained directly, but of much higher quality.

1. Introduction

The freeware CFD code “Fire Dynamics Simulator” (FDS) is an LES CFD code capable of capturing the time evolution of a fire and its turbulent nature. The LES approach allows to resolve the largest eddies and to model the smallest ones. The resolved scale is generally proportional to the size of the grid cells and of the time intervals: the smaller they are, the smaller the resolved scale. This means that a large number of cells and time steps are required to resolve highly turbulent flows and capture a significant number of turbulent structures. It is therefore important to understand where the bound between resolved and modelled turbulence is placed. Significant information on this limit and its location can be obtained with spectral analyses, in particular of the specific turbulent kinetic energy (*tke* in the following). These spectra can be studied both in time and in space. In the time domain a frequency analysis is required, while in the space domain a wavenumber analysis is required. If a frequency analysis simply involves the Fourier transform on time-varying data sets, usually largely available, in the wavenumber domain the methodology for obtaining the spectra is more challenging.

The main method for the computation of spectra in the wavenumber domain is the direct approach, which goes through “the spatial correlation function” between the fluctuations of velocity at a certain instant in a point and in the surroundings. However, in using the direct approach, a difficulty stems from the need of data in a large number of closely spaced points, usually not available. As an alternative to the direct approach, the spectra in the wavenumber domain were obtained using the Taylor’s Hypothesis (TH), relating spatial and temporal characteristics of turbulence [1]. Beside TH, in the literature several approaches were proposed to analyse the spectra in the wavenumber domain, such as different corrections to TH [2], and the models derived from the random sweeping hypothesis [3,4]

A new investigation approach for spectra in the wavenumber domain can be followed using the "Inhomogeneous wave correlation" (IWC) method [5]. The IWC method, starting from the frequency spectra calculated in a certain number of points of the domain, allows to obtain the so-called correlation-fit, i.e. a correlation between frequencies and wavenumbers valid along the path where the frequency spectra are made available. In [5], presumably the first application of the IWC method to turbulence, for a test case of turbulent reacting plumes, *tke* is analysed both in frequencies and wavenumbers and a correlation is found between these parameters.

In the present paper we continue the research started in [5]. A refined grid allows to broaden the range of frequencies and wavenumbers under examination. The intention is to highlight the correlations between grid spacing, time steps and resolved turbulence scales, by means of the *tke* spectra along the centreline of the fire plume. Particular attention is given to the grid spacing. As a starting point, the rule of thumb for the grid recommended by the FDS User Guide [6] is used, consisting in a suggested ratio between the characteristic diameter of the fire and the grid size. Furthermore, it is verified whether the above-mentioned energy spectra exhibit the scaling laws in the inertial subrange and follow the Kolmogorov theory. Furthermore, a study is performed to evaluate the effect of the grid spacing on the *tke* spectra. In addition, we reverse the procedure to examine the results obtained by starting from spectra in the wavenumber domain to obtain the corresponding in the frequency domain.

2. Case Study and Methodology

The plume of a turbulent fire in a cubic domain (10 m side) is considered as the case study (Fig. 1). Top and bottom surfaces are solid walls. The side faces are free to communicate with the external environment. A square burner (L = 1.5 m) is hosted in the centre of the lower plane on a 0.5 m high support. Propane, released at such a flow rate as to achieve an assigned heat release rate (*HRR*), is used as the fuel for the fire. The *HRR* curve reaches, with a linear growth, a maximum power of 5 MW after 10 s (Fig. 2).

The fire problem is addressed numerically using the CFD code FDS release 6.7.1. Compared to what presented in [5], the grid has been modified using the multi-block option with refinements in the areas of interest (plume, ceiling jet), to further halve the finest mesh size previously analysed. The case study is then studied for two different meshes (finer grid size: 0.0625 and 0.03125 m, respectively; larger grid size 0.25 m). Overall, grids of 912 10^3 and 6940 10^3 nodes are generated. The choice of the larger grid (0.25 m) is based on the rule of thumb proposed in the FDS User Guide [6]. For all the runs a constant time step of 10^{-3} s is chosen, having always checked that the stability constraints are satisfied. No slip conditions are assigned to the bottom and top walls. The remaining faces are modelled as openings where the fluid can enter or exit. On these open boundaries, conditions of constant total pressure are assumed. The wind velocity in the far field is zero. It is assumed that the bottom and top walls and the side surfaces of the burner are inert. The LES approach with the Deardorff sub-grid model is used to perform turbulence computations.

Figure 1: Computational grid.

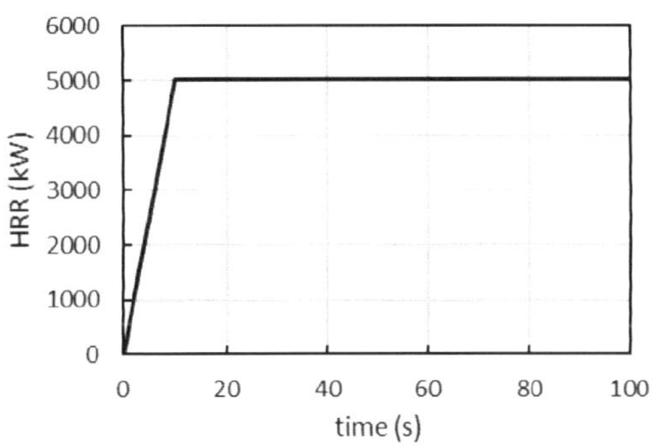

Figure 2: HRR imposed to the burner.

A detailed description of the IWC method is given in [5]. The IWC method is commonly applied to obtain *tke* spectra in the wavenumber domain. To do this, frequency spectra for the points along the centreline are produced as the first step. Next, the IWC method is applied to obtain a correlation between frequencies and wavenumbers. Finally, the wavenumber spectra are obtained by applying the correlation to the frequency spectra.

To deepen the knowledge about the IWC method here we use it in a reverse way, to obtain the *tke* spectra in the frequency domain. In more details, while in [5] the IWC method was used following the standard approach (it means from frequencies to wavenumbers), here we reverse the approach starting from wavenumber spectra calculated through FFTs of the spatial distribution of *tke* at different times. Afterwards, the correlation between frequencies and wavenumbers is obtained with the IWC method. Finally, the frequency spectra are obtained by applying the correlation to the wavenumber spectra.

3. Results and Discussion

The velocity components $\tilde{u}_i\,(\bar{x},t)$ are recorded at each node along the vertical centreline (x_3) of the plume, starting from 1 m up to 9 m (1 m from the ceiling). The Reynolds decomposition is applied to the velocity components $\tilde{u}_i\,(\bar{x},t)$. For example, in the time interval $20 \leq t \leq 100$ s, time-averaged velocities, $U_i\,(\bar{x})$, and velocity fluctuations, $u_i\,(\bar{x},t)$, are separated. In this interval the time average of the velocity fluctuations tends to 0. Next, from the velocity fluctuations, the history of $tke(\bar{x},t)$ along the centreline of the domain is calculated.

3.1. Spectra in the wavenumber domain

The data of *tke* at a point $\bar{x}_0$, $tke(\bar{x}_0,t)$, are the basic elements for the search of the frequency spectra, $TKE(\bar{x},f)$. These are computed with an FFT algorithm. Since the time step is 10^{-3} s, the corresponding Nyquist frequency is 500 Hz.

In Figure 3 the numerical results for the two grids are presented in terms of frequency spectra of *tke* at a point. These spectra refer to a centreline cell at a height of 5 m ($x_3/D = 3$, where for the square burner $D = L$). In these *tke* spectra, the -2/3 power law of the Kolmogorov decay (-5/3 is the exponent of the corresponding Kolmogorov power spectra), shown by a reference line in Fig. 3, is a dominant feature. The inertial sub-range is captured for both the grids.

In FDS an implicit grid filtering is used. The filter size is the mesh spacing, uniform in this case study. The cutoff frequency, *i.e.* the frequency threshold between resolved and modelled turbulence, is placed where the sharp decrease declines. Necessarily this cutoff moves forward as the grid spacing decreases. In particular, the cutoff frequency is identified at approximately 100 Hz for $\Delta x = 0.0625$ m, and at 220 Hz for $\Delta x = 0.03125$ m.

The correlation between frequencies and streamwise wavenumbers, k_{p3}, is then investigated using

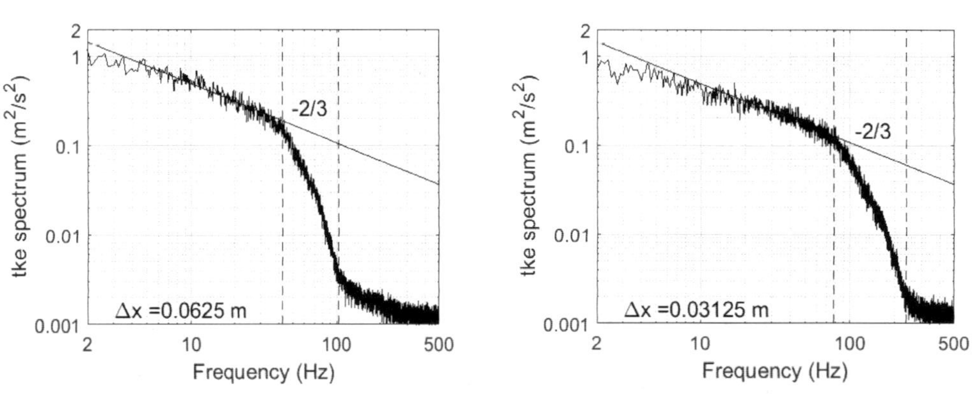

Figure 3: Frequency spectra of *tke* at 5 m of height.

the IWC method [1]. The frequency spectra are then reduced to obtain a correlation between f and k_{p3} (correlation-fit in the following). The results of this data reduction are presented in Fig. 4. The graph shows an area characterized by a high correlation, the "propagation area", and an area of low correlation, the "non-propagation area". In Figure 5, the data are presented for the two different grids with the fitting of the data set pertaining the finest grid. A linear fitting does not fit the two propagation areas. This means that in this case of reactive turbulent plume, Taylor's hypothesis cannot be applied. Conversely, a ratio of polynomials fits well the data in the propagation area. For frequencies tending to 0, the wavenumber values tend to 0. Finally, in Figure 5 the correlation-fit of the finer grid shows a good agreement also with the data set pertaining the coarser grid.

In Figure 6 the wavenumber spectra of *tke* produced by using the correlation-fit of Fig. 5 are presented. Qualitatively the spectra are very similar to those presented in Fig. 3. In both graphs a cutoff wavenumber is well evident, for increasing values when the grid reduces. These cutoff wavenumbers are in accordance with the Nyquist wavenumber, $k_C = \pi/\Delta x$, where Δx is the grid spacing. As for the frequency spectra of *tke* (Fig. 3), the −2/3 power law of the Kolmogorov decay, shown in Fig. 6 by a reference line, is a dominant feature and the inertial sub-range is captured for both grids.

3.2. Spectra in the frequency domain (reverse IWC method)

The IWC method can also be applied to obtain the *tke* spectra in the frequency domain from spectra in the wavenumber domain. This approach will be labelled "reverse IWC method".

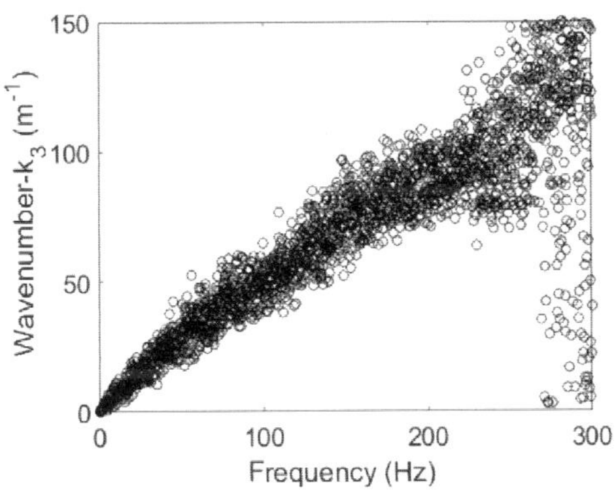

Figure 4: Data distribution (k_{p3}, f) for $\Delta x = 0.03125$ m.

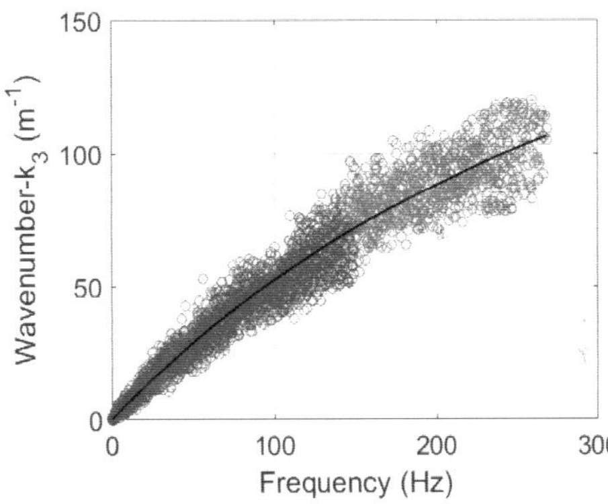

Figure 5: The correlation-fit on the two different grids.

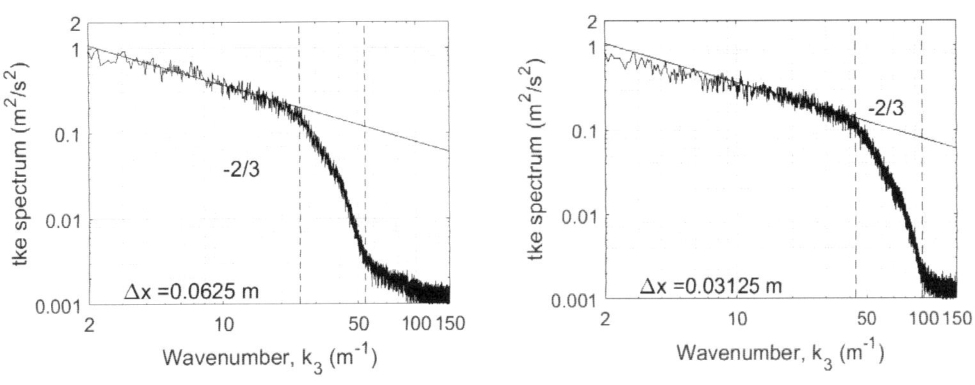

Figure 6: Wavenumber spectra of *tke* along the centreline.

The path for the application of the reverse IWC method is $20 \leq t \leq 100$ s. Since the grid steps are 0.0625 and 0.03125 m and the number of points along the path 128 and 256, respectively, the corresponding Nyquist wavenumbers are 50 and 100 m^{-1}.

The data of *tke* at an instant t_k, $tke(x_3,t_k)$, are the starting elements for the search for the wavenumber spectra, $TKE(k_{p3},t_k)$. To do it, an FFT in the space domain is performed:

$$TKE(k_{p3}, t_k) = \text{FFT}(tke(x_3, t_k)) \tag{1}$$

The FFT is performed on the data available along the centre-line of the plume from 1 m up to 9 m (1 m from the ceiling).

The IWC method consists in evaluating the degree of consistency between the signal under examination, $TKE(k_{p3}, t_k)$, and an inhomogeneous wave (plane in this case):

$$\alpha(f_j, t_k) = \text{A} \exp(-i\, 2\pi f_j\, t_k) \tag{2}$$

Dissipation is assumed to be negligible. The comparison is performed by calculating the indicator IWC for each element, $k_{p3,n}$, of the vector $TKE(k_{p3}, t_k)$:

$$IWC(f_j, k_{p3,n}) = \frac{\left| \sum_{k=1}^{K} \text{A}\, e^{i\,(2\pi f_j t_k)} \cdot TKE(k_{p3,n}, t_k) \right|}{\sqrt{\sum_{k=1}^{K} \left| \text{A}\, e^{i\,(-2\pi f_j t_k)} \right|^2 \sum_{k=1}^{K} \left| TKE(k_{p3,n}, t_k) \right|^2}} \tag{3}$$

Now, we can identify the stream-wise propagative wavenumbers by looking for the maximum of the IWC parameter among those calculated with Eq. (3) for each $k_{p3,n}$ as f_j varies. A dataset that relates k_{p3} with f allows us to obtain a correlation between these parameters.

In Figure 7 the results for the two grids are presented in terms of wavenumber spectra of *tke* at an instant (60 s). In these *tke* spectra, the inertial sub-range is captured and the -2/3 power law of the Kolmogorov decay (-5/3 is the exponent of the Kolmogorov power spectra), shown by a reference line in Fig. 7, is a dominant feature. The quality of the spectra is very low due to the limited number of grid points (128 and 256) available for the computations. The implicit grid filtering used by FDS in its LES implementation is barely visible. On the contrary, in both graphs the cutoff wavenumber, *i.e.* the threshold between resolved and modelled turbulence, is clearly visible. The spectra stop at the Nyquist wavenumbers (50 and 100 m^{-1}) given by $k_C = \pi/\Delta x$, where Δx is the grid spacing. Necessarily, these cutoff wavenumbers advance when the grid becomes finer.

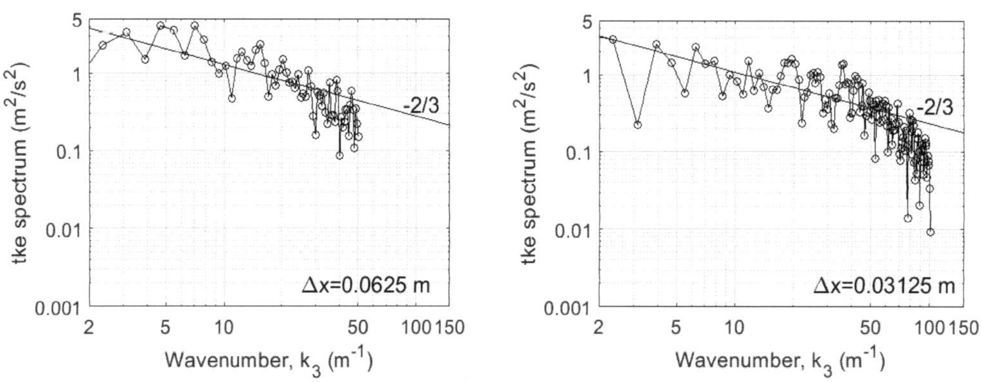

Figure 7: Wavenumber spectra of *tke* at 60 s.

The wavenumber spectra are reduced with the procedure given be Eqs. (1-3) to obtain the correlation-fit. The results of this data reduction are presented in Fig. 8. As expected, the data distribution becomes wider as the resolution of the grid increases. The data of both the distributions show a sharp discontinuity near half of the Nyquist wavenumber (25 and 50 m^{-1}). Both the distributions of data are overlapping up to half of the Nyquist wavenumber of the coarser grid.

In Figure 9 the correlation-fit pertinent to the finest grid is presented. Since the data show a sharp discontinuity near half of the Nyquist wavenumber, we interpolated up to this limit with a function consisting in a ratio of polynomials, as already done in [5]. Justifications for the extrapolation after this limit (dashed line) are addressed in Par. 3.3.

Finally, the correlation-fit is used to produce the frequency spectra of *tke*. In Figure 10 these spectra are presented for the two grids. Qualitatively the spectra are similar to those presented in Fig. 7. In both graphs a cutoff frequency is evident, for increasing values when the grid reduces. As for the wavenumber spectra of *tke* (Fig. 7), the −2/3 power law of the Kolmogorov decay, shown in Fig. 10 by a reference line, is a dominant feature and the inertial sub-range is captured for both grids. The attenuated region is barely visible.

Since the *tke* spectra (Figs. 7 and 10) are characterized by annoying fluctuations, which make the *tke* spectra difficult to read, an ensemble average of the wavenumber spectra at 50, 60, 70, 80 and 90 s is done. In Figure 11 these averaged spectra both in the wavenumber and in frequency domain are shown. Now the attenuated region becomes clearly visible. The lower bound of the attenuated region seems to

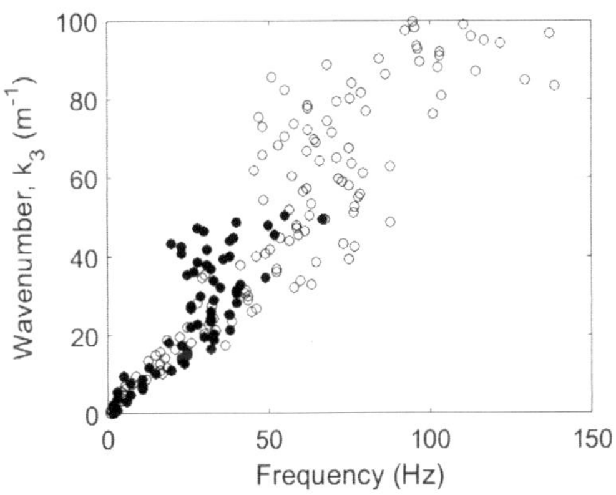

Figure 8: (k_{p3},f) data distributions for the grids.

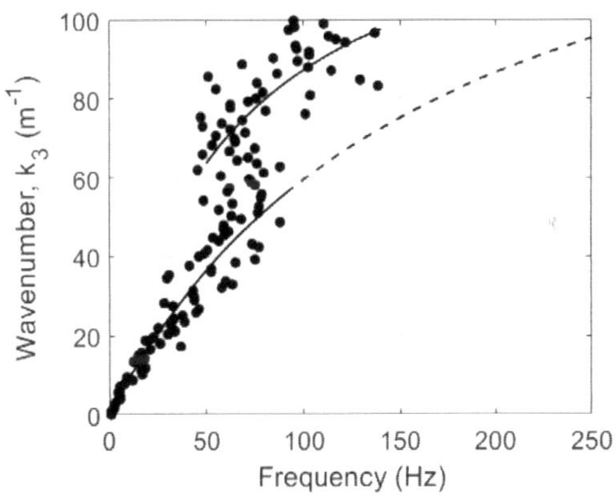

Figure 9: Correlation-fit for the finest grid.

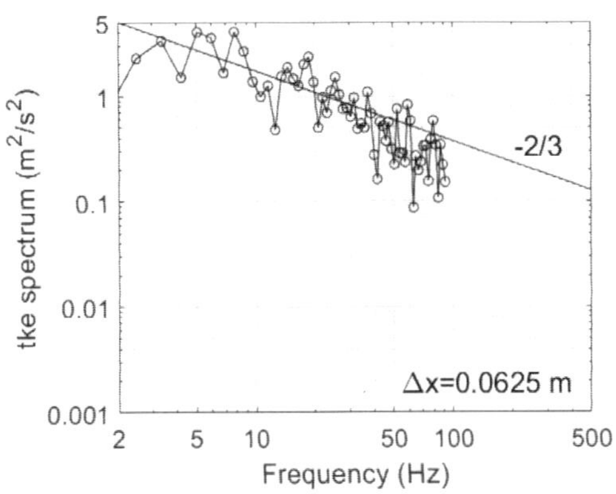

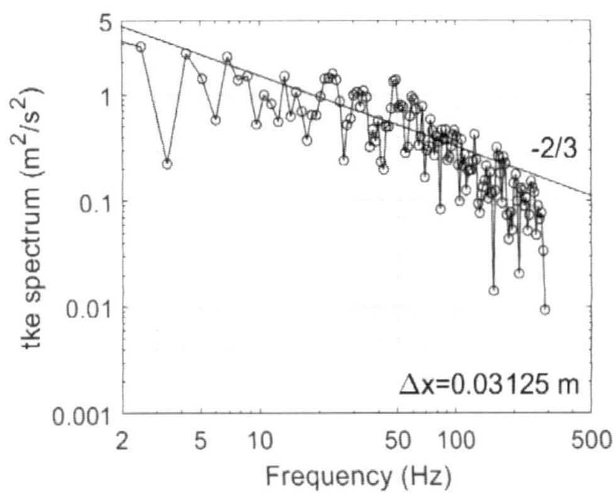

Figure 10: Frequency spectra of *tke* at 60 s.

coincide with half of the Nyquist wavenumber (25 and 50 m^{-1}). The upper bound of the attenuated region here coincides with Nyquist limits (50 and 100 m^{-1}) where the spectra stop. These cut-offs move forward when the grid becomes finer.

3.3. Procedures comparison

Here for the finest grid the results of the two procedures are compared and discussed. In Figure 12 the IWC parameter as a function of the frequency, for k_{p3} = 70 m^{-1}, is shown. These data are interpolated with a summation of Gaussian functions. This fitting allows to detect a second peak with a doubled mean, but with a lower scaling factor. The second Gaussian curve seems to detect the second harmonic of the inhomogeneous wave. In Figure 13, after the sharp discontinuity in frequency clearly visible in Fig. 9, the second harmonic of the IWC function is shown. In the same Figure, the correlation-fit obtained for the data of Fig. 9 is shown. This correlation fits well also the data after the half of the Nyquist number. This justifies our assumption of extrapolating the correlation-fit for the whole range of data of Fig. 13.

For these grids, when starting the procedure with the frequency spectra, the domain of validity of the correlation-fit is significantly larger. This is due to the limited number of points useful for computing the wavenumber spectra which limits the domain of validity of these spectra.

In Figure 14 the data obtained with the IWC method (Fig. 4) and with its reverse (Fig. 13) are compared. The number of data obtained by using the IWC method starting from the frequency spectra

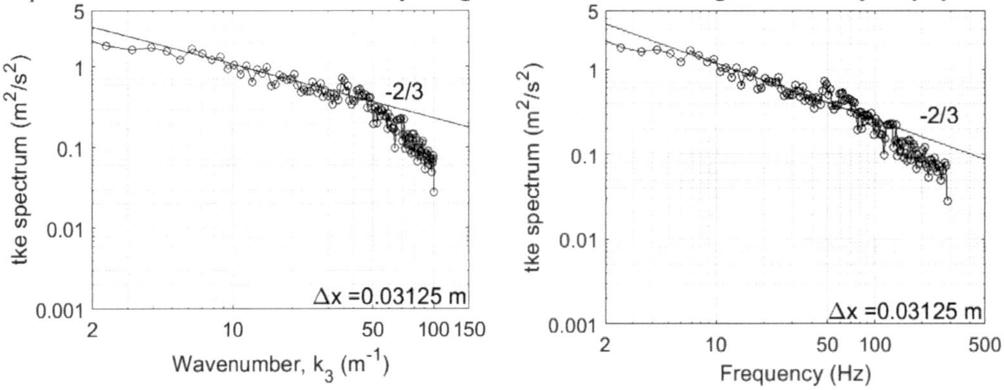

Figure 11 – Ensemble average of frequency and wavenumber spectra.

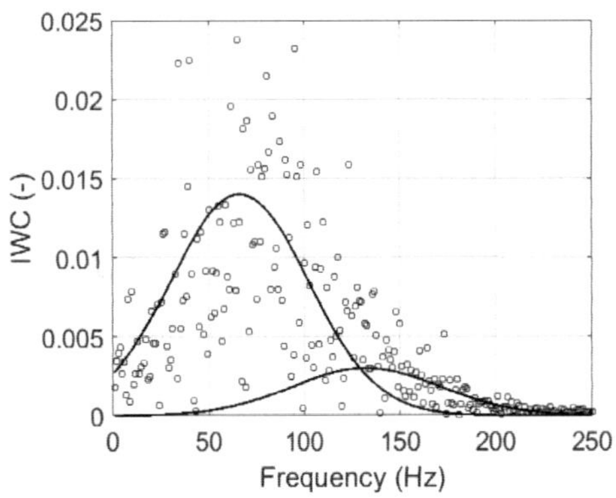

Figure 12 - IWC parameter at k_{p3} = 70 m^{-1}.

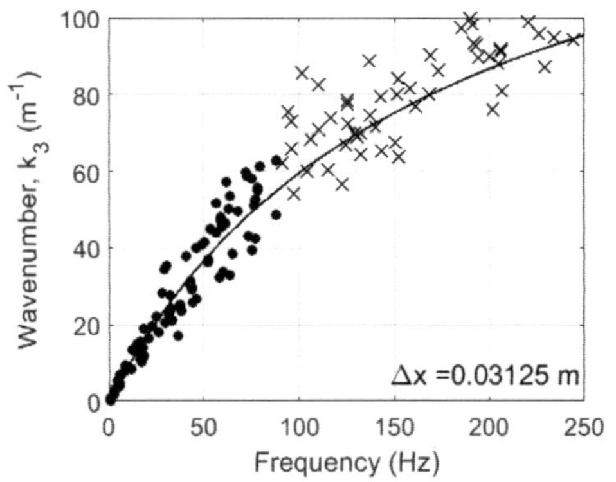

Figure 13 - Reverse IWC data: original (•) and shifted values (x)

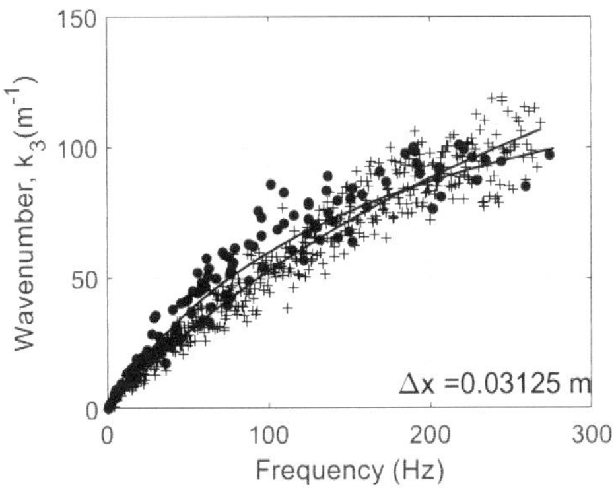

Figure 14 - Propagative dispersion graphs (wavenumber- (●) and frequency-spectra (+)).

is significantly larger than that obtained with the reverse IWC method. This is due to the limited number of points useful for computing the wavenumber spectra. Both the data sets are interpolated with the function $k = af / (f + b)$. The interpolation curves are quite similar. The one computed from the wavenumber spectra bends more sharply. Since the two data distributions of Fig. 14 shows an agreement, this can be a further justification for the data shift after the discontinuity shown in Fig. 8 and finally for the extrapolation.

Despite the large number of grid points (6940 10^3) the quality of the wavenumber spectra obtained with the direct approach (Fig. 7) or after an ensemble average (Fig. 11) is evidently lower than that given by the IWC method (Fig. 6).

4. Concluding Remarks

For a reacting turbulent flow, the IWC method is used to correlate the *tke* spectra in frequency and wavenumber domains. The analysis is carried out for two different grids. The Kolmogorov inertial subrange is always reached. The spectra drop sharply up to the frequency/wavenumber cutoff through an attenuated region, the latter due to the implicit filtering of the LES code.

With the IWC method, spectral analyses in the wavenumber domain become easy. Furthermore, the IWC method shows the interesting advantage of avoiding any physical model.

Even for a very fine grid like that used in this paper (6940 10^3), the number of nodes along the plume and their spacing are not enough to allow the reverse of the IWC method to produce information comparable to those produced with the standard IWC method.

References

[1] Taylor G I 1938 The spectrum of turbulence *Proc. R. Soc.* A **164** 476–490 doi: 10.1098/rspa.1938.0032
[2] Geng C, He G, Wang Y, Xu C, Lozano-Durán A and Wallace J M. 2015 Taylor's hypothesis in turbulent channel flow considered using a transport equation analysis. *Phys. Fluids* **27** 025111 doi: 10.1063/1.4908070
[3] Wilczek M and Narita Y 2012 Wave-number–frequency spectrum for turbulence from a random sweeping hypothesis with mean flow *Phys. Rev.* E, **86** 066308 doi: 10.1103/physreve.86.066308
[4] He G W and Zhang J B 2006 Elliptic model for space-time correlations in turbulent shear flows *Phys. Rev.* E **73** 055303(R) doi: 10.1103/physreve.73.055303
[5] Ciani F S, Bonfiglio P and Piva S 2021 IWC analysis of turbulent plume fires *TECNICA ITALIANA-Italian Journal of Engineering Science* **65** 196-200. doi: 10.18280/ti-ijes.652-408
NIST 2018 *Fire Dynamics Simulator User's Guide* Special Publication 1019 (Gaithersburg, MD, NIST)

Accuracy in evaluating heat transfer coefficient by RANS CFD simulations in a rectangular channel with high aspect ratio – Part 1: benchmark on a channel with plane walls

Maria CORTI, Pasqualino GRAMAZIO, Damiano FUSTINONI, Luigi VITALI and Alfonso NIRO*

Politecnico di Milano, Department of Energy
via Lambruschini 4, 20156 Milano, Italy

* Corresponding author e-mail: alfonso.niro@polimi.it

Abstract. Computational Fluid-Dynamics (CFD) simulations are widely used for designing components and devices for heat transfer enhancement, and to this end the Reynolds Averaged Navier-Stokes (RANS) equations are often chosen, as they are computationally efficient. In this paper, several numerical simulations have been carried out on convective heat transfer of an air flow through a rectangular channel of 1:10 aspect ratio, 120 mm wide, 840 mm long. Numerical results have been compared to analytical values and experimental data. The configuration of the described numerical model will be used as starting point for detailed investigations of fluid-dynamic and thermal performances of ribbed channels in further analysis.

Key Words: *forced convection, heat exchanger, rectangular channel, flat surfaces, experimental tests, CFD validation, RANS*

1. Introduction

Heat transfer enhancement in air forced convection is a contemporary technological issue, despite its decades of life. Especially associated to forced convection inside rectangular channels with various ribbed surfaces, it sees a wide range of applicability in engineering applications [1,2]. A cutting-edge solution is working with ribbed high aspect ratio channels at low Reynolds flow regimes. Channel with high aspect ratio, in addition to fulfil the industrial interest for narrow channels, ensures a wide central area undisturbed by edge losses that can be dedicated to heat transfer. At the ThermALab laboratory at Politecnico di Milano, authors have been and are currently involved in experimental analysis on fluid-dynamics and heat transfer characteristics of forced convection inside a channel with an aspect ratio of 1:10 [3-6]. To complement the experimental campaign with a CFD model, this work is carried out. It will be followed by an analysis regarding the accuracy in evaluating the global heat transfer coefficient by RANS CFD simulations in a high aspect ratio channel with and without ribs. As reported in [7], an appropriate tool is the use of finite volume CFD code. Knowing problems about the computational effort required to solve Navier-Stokes equations, with the present work authors are looking for a CFD model based on RANS that can be used as a starting point for future studies. In this paper, results on velocity profiles, pressure drops, and convective heat transfer coefficient are presented along with a discussion on the comparison between analytical, experimental, and numerical data.

2. Problem configuration and set-up

2.1. Experimental set-up
The experimental set-up is described in detail in [3], briefly it consists in a rectangular channel of 1:10 aspect ratio. The main geometric parameters of the section dedicated to measurements, are summarized in Table 1. In this paper, all the walls of the channel are flat. The channel used in the experimental set-up operates with lower and upper walls maintained at fixed temperature, while side walls are adiabatic. The air flows at Reynolds numbers ranging from 700 to 8000.

2.2. Numerical procedure
Numerical simulations have been carried out using the software Ansys Fluent 19.1 based on finite volume method. The solution domain is a 3D channel with a rectangular cross section as described previously (Figure 1). The working fluid is air, incompressible, with constant specific heat coefficient. To obtain a grid independent solution, a sensitivity analysis has been carried out. Since it is a three-dimensional case, given the three directions x, y, z where x is the direction of the channel width, y the direction of the channel height and z the streamwise direction, a sensitivity analysis has been carried out for each dimension.
Working at different Reynolds numbers [3-4], different flow regimes (laminar or turbulent) are considered so grids have been constructed respecting the needs of each case analysed.
Per example for laminar Reynolds, the number of elements for the whole channel has been around $5{\cdot}10^6$. For turbulent flow with Reynolds higher than 2000 up to 7500, the number of elements for the whole channel has been ranging from $6{\cdot}10^6$ to $40{\cdot}10^6$ ensuring a $y^+=1$ at the wall to solve the laminar sublayer.

2.3. Boundary conditions
Since the flow in the experimental set-up, at the inlet of the measurement section, can be considered hydrodynamically fully developed, this condition is required at the inlet of the present computational domain. For this purpose, some preliminary CFD simulations have been carried out. Considering an adiabatic channel with the same rectangular cross section, with plane walls and enough long in the streamwise direction to guarantee the hydrodynamic development both in laminar and turbulent case (for laminar flow regime: $x_{fd,h} \geq 0.05ReD_h$, for turbulent flow regime: $x_{fd,h} \geq 10D_h$), it has been used to allow the full development of the airflow. The velocity profile (its three components), and, if needed, the corresponding turbulence parameters (Turbulent Kinetic Energy, k [m^2/s^2] and Specific Dissipation Rate, ω [1/s]) obtained at the outlet of this preliminary channel have been used as inlet condition for the analysed cases.

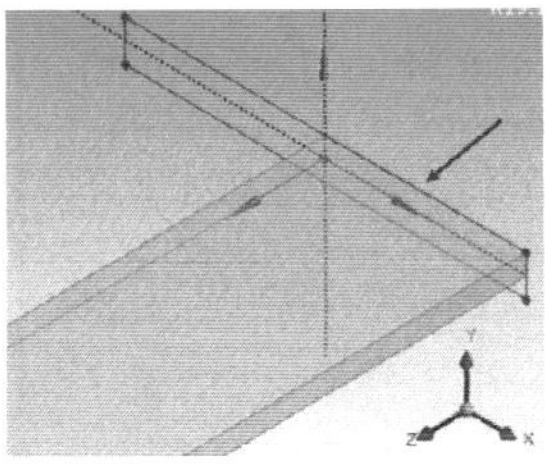

Figure 1. 3D domain.

Table 1. Main channel parameters.

Main channel parameters	
Channel height, 2b	12.0 mm
Channel width, 2a	120.0 mm
Channel length, L	840.0 mm
Hydraulic diameter, D_h	21.82 mm

2.3.1. Preliminary analysis on thermal boundary conditions

A preliminary analysis on thermal boundary conditions has been necessary for finding the most suitable configuration for boundary conditions in the CFD model.

For this reason, a simple case has been studied: a rectangular channel with plane walls and Reynolds number of 1083 has been modelled with CFD.

Boundary conditions of this preliminary case are summarized as follows:

- Inlet: velocity inlet (fully developed velocity profile, previously generated) based on the selected Reynolds number, in this case Re=1083. The working fluid is air, enters the channel at uniform temperature, $T_{in} = 296$ K.
- Outlet: pressure outlet $P_{out} = 0$ Pa.
- Walls: no-slip velocity condition is applied for all the walls, and the thermal condition has been benchmarked.

Coupled algorithm is used for pressure-velocity coupling. The presented laminar flow has been solved with Laminar model. Computational time has been restrained, 4-5 hours per each single simulation using 15 cores Intel(R) Xeon(R) Gold 6148 CPU @ 2.40GHz. Laminar simulations stop at the level-out point of the continuity residual.

The outlet velocity profile obtained numerically has been compared to the analytical velocity profile for a rectangular channel [8] obtaining the expected accurate prediction. This comparison can be graphically seen in Figure 2. Since the case considered is a fully developed velocity profile in a rectangular duct, the equation used for the comparison is the following, equation (1). Notice that n and m are dimensionless coefficients proposed by Purday [8] to reduce the complexity of the equation of the fully developed velocity profile for rectangular ducts. According to that, since the aspect ratio (named α* in the considered equations) is equal to 0.1, values of n=2 and m=13.6 have been used.

$$u_{\mathrm{m}} = -\frac{16}{\pi^3}\left(\frac{d\mathrm{P}}{dw}\right)\frac{a^2}{\mu}\sum_{n=1,3,\ldots}^{\infty}\frac{(-1)^{(n-1)/2}}{n^3}\left(1-\frac{\cosh(n\pi y/2a)}{\cosh(n\pi b/2a)}\right)\cos\left(\frac{n\pi x}{2a}\right) \tag{1}$$

$$u_{\mathrm{m}} = -\frac{1}{3}\left(\frac{d\mathrm{P}}{dw}\right)\frac{a^2}{\mu}\left[1-\frac{192}{\pi^5}\left(\frac{a}{b}\right)\sum_{n=1,3,\ldots}^{\infty}\frac{1}{n^5}\tanh\left(\frac{n\pi b}{2a}\right)\right] \tag{2}$$

$$\frac{u_{\max}}{u_{\mathrm{m}}} = \left(\frac{\mathrm{m}+1}{\mathrm{m}}\right)\left(\frac{\mathrm{n}+1}{\mathrm{n}}\right) \tag{3}$$

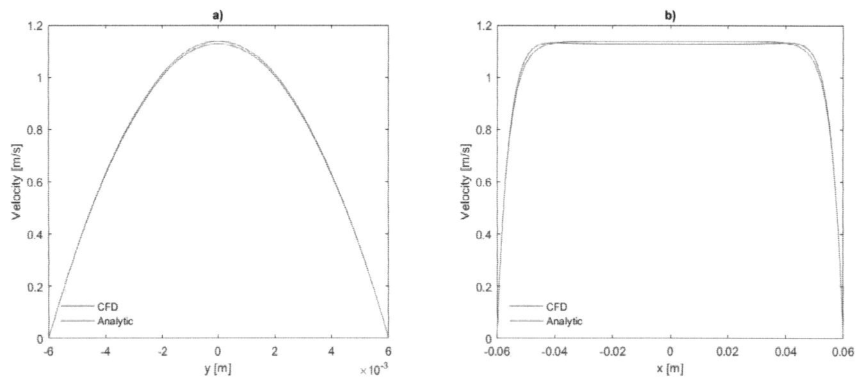

Figure 2. Comparison between numerical and analytic profiles of dimensionless velocity in laminar flow regime at Re=1083. The velocity profile is represented (a) versus y and (b) versus x, respectively height and spanwise directions.

Subsequently, Darcy friction factor obtained from numerical results on the channel at Re=1083 has been compared with the value obtained analytically for a laminar flow, fully developed, inside a rectangular channel adopting equation (4) proposed by Shah and London [9], written below.
Table 2 shows this comparison, it's clear the accurate agreement between the two values considered.

$$f_{Darcy} = 96\left(1 - 1.3553\alpha^* + 1.9467(\alpha^{*2}) - 1.7012(\alpha^{*3}) + 0.9564(\alpha^{*4}) - 0.2537(\alpha^{*5})\right)/Re \quad (4)$$

At this point, considering heat transfer, some boundary conditions for walls have been tested to find the most suitable boundary condition for the studied channel to apply to the CFD model:

- case 1: upper and lower walls heated at constant temperature, imposing $T_w = 323.15K$, while lateral walls considered adiabatic, imposing $\dot{q} = 0\ W/m^2$
- case 2: upper, lower, and lateral walls heated at constant temperature, imposing T_w
- case 3: upper and lower walls heated at constant temperature, imposing T_w, while at lateral walls a boundary condition of the third type has been considered, imposing a convective heat transfer coefficient $h = 4.5\ W/m^2/K$ and a free stream temperature $T_\infty = 293K$
- case 4: upper, lower, and lateral walls heated at constant temperature, imposing $\dot{q}$.

The thermal boundary condition that fits better the experimental set-up is case 2, in fact, the choice of setting all the walls of the channel as heated walls while in the experimental setup lateral walls are considered adiabatic returns an error on the experimental values of the Nusselt number lower than 7-10%. Probably, given the high aspect ratio of the considered channel, in the experimental measurements, the tiny lateral walls tend to be heated from the upper and the lower walls due to conduction phenomena. To better understand why this approximation can be accepted is recommended to deepen the working conditions of the experimental setup, accurately described in [3]. This approximation applied to the numerical boundary conditions to better represent the experimental configuration is applied to all the following CFD simulations carried out.

2.3.2. Adopted boundary conditions

- Inlet: velocity inlet (fully developed velocity profile, previously generated) basing on the selected Reynolds number. The working fluid is air, enters the channel at uniform temperature, $T_{in} = 296$ K. The fluid flow is not thermally developed yet, so the temperature will develop inside the considered channel.
- Outlet: pressure outlet $P_{out} = 0Pa$.
- Wall: no-slip velocity condition is applied for all the walls, moreover the walls (lower, upper, and lateral) are heated at constant temperature, imposing $T_w = 323.15K$.
- Symmetry: since it can be considered a symmetric case in the x-direction and in the y-direction (spanwise), symmetry has been applied at the wall identified from the centrelines to save computational time.

2.4. Solver setting and model

The Semi-Implicit Method for Pressure Linked Equation (SIMPLE) algorithm is used for pressure-velocity coupling. The presented turbulent flows have been solved with RANS, particularly choosing the k-ω-SST model. It uses as default the Enhanced Wall Treatment, a wall function for the near-wall region i.e., viscous sublayer, buffer region and fully turbulent outer region.

Table 2. Comparison between numerical and analytic value of Darcy friction factor in fully developed flow inside rectangular channel at Re=1083.

Re	f_{Darcy}	f_{CFD}	e%
1083	0.07821	0.07823	0.03%

Computational time has been considerable with respect to laminar case; up to 60 hours of calculations were needed per each single case for turbulent flows using 15-30 cores Intel(R) Xeon(R) Gold 6148 CPU @ 2.40GHz. RANS simulations stop when the continuity residual is lower than $1\cdot10^{-6}$.

3. Validation of results

The numbers of Reynolds considered in this study have been precisely: 2168, 3249, 5399 and 7552. Per each Reynolds number a CFD simulation has been carried out analysing in particular the agreement of numerical results with analytical and experimental data. The experimental data considered can be found in [3], in which values of the average Nusselt number inside flat rectangular channel for the range of Reynolds tested experimentally are presented in Figure 3, please refer to the case reported as "Flat T" in the legend.

3.1. Pressure drops

Friction factors obtained from pressure drops deriving from numerical results through equation (5) have been compared with analytical values.

$$f_{Darcy}=\left(\Delta P_0/0.5\rho u^2\right)\left(D_h/L\right) \tag{5}$$

For turbulent flow in a rectangular channel, the correlations used have been the one of Kakac-Shah [8], based on Techo correlation [10] and adjusted for rectangular ducts, and PKN correlation [11-13]. Kakac-Shah correlation, equation (6), is valid for $0 \leq \alpha^* \leq 1$ and, preferentially, $5000 \leq \mathrm{Re} \leq 10^7$, it has the following structure:

$$f=\left(1.0875\text{-}0.1125\alpha^*\right)f_{Techo} \tag{6}$$

PKN correlation is valid for all arbitrarily large Reynolds number as reported in [8]. Errors between those values are shown in Table 3.

For Reynolds number higher than 5000 the agreement with analytical results is quite good, meaning errors lower than 12%, while for Reynolds number lower than 5000 the agreement is lower meaning errors up to 27%.

3.2. Convective heat transfer coefficient

Bulk temperature T_b and heat flux $\dot{q}$ have been calculated as mean section variables, from which can be calculated the local convective heat transfer coefficient $h_z=\dot{q}_z/(T_w\text{-}T_b)$ and consequently the related Nusselt number at any location along the channel $Nu_z = h_z D/k$. This Nusselt number obtained is considered a local Nusselt number at a specified location, in this case the section of interest is the outlet of the channel.

From computational results, since the flow is hydrodynamically and thermally developed (for turbulent flow regime a $x_{fd,th} \geq 10D_h$ can be assumed [14]), also a mean Nusselt number can be calculated from an energy balance.

The mean convective heat transfer coefficient is computed as $h_m=\dot{m}c_p(T_{out}\text{-}T_{in})/A/\Delta T_{m,ln}$ and the associated Nusselt number is $Nu_m = h_m D/k$.

To compare Nusselt numbers obtained from CFD results with Nusselt number obtained by analytical correlation, correlation by Gnielinski [14] for forced convection in turbulent pipe flow has been used, considering the effect related to the entrance to compare also the mean Nusselt number.

For low Reynolds numbers, errors on Nusselt number have been considerable, while good agreement has been obtained for Reynolds numbers higher than 5000, obtaining errors lower than 15-10%.

Values of Nusselt number calculated analytically and from CFD results are presented in Table 4, per each value the associated error is shown, as the error with respect to Nusselt numbers measured experimentally.

3.2.1. Analysis for the choice of RANS model for low Reynolds numbers

Basing on these results, some RANS turbulence models have been investigated to find more reliable results even for low Reynolds number. Considering Reynolds 3249, the following RANS models have been tested:

- k-ω-SST with a mesh that ensures $y^+=1$
- k-ω-SST with a mesh that ensures $y^+=0.8$
- k-ω-SST with a mesh that ensures $y^+=0.2$
- k-ω-SST with a mesh that ensures $y^+=1$ and Low Reynolds option
- k-ε RNG with a mesh that ensures $y^+=0.8$
- k-ε Realizable a mesh that ensures $y^+=0.8$

In Table 5 results are presented. As can be observed in Table 5, changing the RANS model, there was no improvement in the prediction of the Nusselt number, so the k-ω-SST with a mesh that ensures $y^+=1$ has been chosen as future RANS model to apply when Reynolds numbers are higher than 5000, knowing that if Reynolds are lower than 5000, agreement with analytical and experimental results is poor.

Table 3. Comparison between pressure drops. Values obtained from numerical results have been compared to values obtained analytically by means of Kakac-Shah's and PKN correlation.

Re	e% f_{CFD} vs $f_{Kakac-Shah}$	e% f_{CFD} vs f_{PKN}
2168	19%	27%
3249	11%	19%
5399	3%	11%
7552	1%	8%

Table 4. Comparison between Nusselt numbers. Values obtained from numerical results have been compared to values obtained analytically by means of Gnielinski correlation.

Re	e% $Nu_{m,CFD}$ vs $Nu_{Experimental}$	$Nu_{l,Gn}$	$Nu_{l,CFD}$	e% $Nu_{l,CFD}$ vs $Nu_{l,Gn}$	$Nu_{m,Gn}$	$Nu_{m,CFD}$	e% $Nu_{m,CFD}$ vs $Nu_{m,Gn}$
2168	32%	6.67	10.83	62%	7.24	11.53	59%
3249	41%	10.97	14.17	29%	11.90	14.94	26%
5399	7%	17.88	20.26	13%	19.40	21.12	9%
7552	8%	23.81	25.91	9%	25.84	26.80	4%

4. Conclusion

In this paper the CFD model built for a high aspect ratio rectangular channel is presented. Agreement with analytic solution is good both at fluid-dynamic and thermal level with increasing error at low Reynolds numbers regarding turbulent flows.

Boundary conditions suitable to represent numerically the experimental set-up have been found. Regarding the most suitable RANS model, the k-ω-SST has been chosen as future RANS model to apply when Reynolds numbers are higher than 5000 to obtain acceptable numerical results.

The mesh needs to ensure $y^+=1$ near walls and no further thickness of the mesh is required. For low Reynolds number, there is thought to be no room for improvement using RANS models.

This model has been built as a starting point to be able to subsequently expand the thermo-fluid dynamics study of high aspect ratio channels even in the presence of ribbed surfaces considering flows operating at Reynolds number higher than 5000.

Table 5. Comparison between Nusselt numbers obtained from different CFD simulations with analytical and experimental values of Nusselt numbers. Reynolds number considered for this comparison has been equal to 3249.

RANS model tested	e% $Nu_{l,CFD}$ vs $Nu_{l,Gn}$	e% $Nu_{m,CFD}$ vs $Nu_{m,Gn}$	e% $Nu_{m,CFD}$ vs $Nu_{Experimental}$
k-ω-SST with $y^+=1$	29%	26%	41%
k-ω-SST with $y^+=0.8$	30%	26%	41%
k-ω-SST with $y^+=0.2$	30%	26%	42%
k-ω-SST with $y^+=1$ and Low Reynolds	56%	48%	66%
k-ε RNG with $y^+=0.8$	52%	45%	63%
k-ε Realizable $y^+=0.8$	43%	37%	54%

Nomenclature

Symbol	Quantity	SI Unit
A	Area	m^2
a	Channel half width	m
b	Channel half height	m
D_h	Equivalent diameter	m
f	Friction factor	-
h	Convective heat transfer coefficient	W/m^2K
k	Thermal conductivity	W/m/K
k	Turbulent kinetic energy	m^2/s^2
L	Channel length	m
$\dot{m}$	Mass flow rate	kg/s
Nu	Nusselt number	-
P	Pressure	Pa
Re	Reynolds number	-
T_b	Bulk temperature	K
T	Temperature	K
u_m	Bulk velocity (average)	m/s
$\dot{q}$	Wall heat flux	W/m^2
$x_{fd,h}$	Hydrodynamic entry length	m
$x_{fd,th}$	Thermal entry length	m
y+	Dimensionless wall distance $y^+ = \Delta_{y_P}/v\sqrt{\tau_w/\rho}$	-
α^*	Aspect ratio	-
ν	Kinematic viscosity	m^2/s

ρ	Density	kg/m³
τ_w	Wall shear stress	Pa
ω	Specific dissipation rate	1/s

Subscripts

Symbol	Quantity
in	Inlet
l	Local, at a defined position
ln	Logarithmic
m	Mean
out	Outlet
w	Wall
z	Local, at a position that varies with z
∞	Free stream, at ∞ location

References

[1] Webb R L, and Kim N H, (2004). Principles of Enhanced Heat Transfer (2nd ed.). Garland Science. https://doi.org/10.1201/b12413
[2] Nourin F N, and Amano R S, Review of Gas Turbine Internal Cooling ImprovementTechnology, J. Energy Resour. Technol. Trans. ASME, vol. 143, no. 8, pp. 1–8, 2021, 10.1115/1.4048865
[3] D Fustinoni et al 2012 J. Phys.: Conf. Ser. 395 012042, 10.1088/1742-6596/395/1/012042
[4] D Fustinoni et al 2015 J. Phys.: Conf. Ser. 655 012060, 10.1088/1742-6596/655/1/012060
[5] D Fustinoni et al 2017 J. Phys.: Conf. Ser. 796 012015, 10.1088/1742-6596/796/1/012015
[6] Niro A, Gramazio P, Vitali L and Fustinoni D, Local and global heat transfer characteristics inside a rectangular channel of 1:10 aspect-ratio with ribbed surfaces, International Heat Transfer Conference, pp. 5509-5516, 2018, 10.1615/IHTC16.hte.023149
[7] Sharma S K and Kalamkar V R, Computational Fluid Dynamics approach in thermo-hydraulic analysis of flow in ducts with rib roughened walls - A review, vol. 55. Elsevier, 2016, 10.1016/j.rser.2015.10.160
[8] Kakaç S, Shah R K and Aung W. Handbook of Single-Phase Convective Heat Transfer. New York: Wiley, 1987. Print. (Ch. 3-4)
[9] Shah R K and London A L, Laminar Flow Forced Convection in Ducts: A Source Book for Compact Heat Exchanger Analytical Data, vol. Suppl. 1. Academic Press, 1978
[10] Techo R, Tickner R R and James E, (June 1, 1965). An Accurate Equation for the Computation of the Friction Factor for Smooth Pipes From the Reynolds Number. ASME. J. Appl. Mech. June 1965; 32(2): 443, 10.1115/1.3625826
[11] Prandtl L, Führer durch die Strömungslehre, Braunschweig (1944). English Version Transl., Blackie, London 1952
[12] T. von Karman, Turbulence and Skin Friction, J. Aerops. Sci., Vol. 7, pp. 1-20, 1934, https://doi.org/10.2514/8.5
[13] Nikuradse J, Laws of Turbulent Pipe Flow in Smooth Pipes (1933), English transl., NACA TT F-10, 359, 1966,
[14] Bergman T L, Lavine A S, Incropera F P, and Dewitt D P, Fundamentals of Heat and Mass Transfer. John Wiley & Sons, 2011. (Ch. 8-11)

Accuracy in evaluating heat transfer coefficient by RANS CFD simulations in a rectangular channel with high aspect ratio – Part 2: channel with ribbed walls

Maria CORTI, Pasqualino GRAMAZIO, Damiano FUSTINONI, Luigi VITALI, and Alfonso NIRO*

Politecnico di Milano, Department of Energy
via Lambruschini 4, 20156 Milano, Italy

* Corresponding author e-mail: alfonso.niro@polimi.it

Abstract. In this paper, a Computational Fluid-Dynamics (CFD) analysis on heat transfer characteristic using Reynolds Averaged Navier-Stokes (RANS) equations on a rectangular channel with ribbed surfaces is presented. Several numerical simulations have been carried out on convective heat transfer of an air flow through a rectangular channel of 1:10 aspect ratio, 120 mm wide, 840 mm long, with 90° ribs. Ribs have a square cross section with 4 mm side and three different values of dimensionless pitch, namely, 10, 20 and 40. Results on convective global heat transfer coefficient, i.e., averaged over the whole channel, have been compared to experimental data obtained in a channel with the same geometry and boundary conditions. Agreement between numerical and experimental data is discussed for the three different pitches considered. As expected, comparisons show a decreasing reliability for increasing complexity. The aim of this work is to find a suitable configuration of a CFD model with RANS that will permit authors to apply it to the range of dimensionless pitches.

Key Words: *forced convection, heat exchanger, rectangular channel, ribbed surfaces, experimental tests, CFD validation, RANS*

1. Introduction

Heat transfer enhancement in air forced convection, specially associated to forced convection inside rectangular channels with various ribbed surfaces, sees a wide range of applicability in engineering applications. Thinking about applicability, in addition to the well-known use for cooling blades gas turbines [1], nuclear reactors and solar collectors [2,3], a novel challenging application is for automotive batteries cooling [4] which requires large contact area and very small height resulting thus in channel with high aspect ratio. What is peculiar in using ribs is they can be used as turbulence promoters to increase heat transfer in ducts while varying their pitch-to-rib-side ratio, cross section, or blockage ratio. At the ThermALab laboratory at Politecnico di Milano, authors have been and are currently involved in experimental analysis on fluid-dynamics and heat transfer characteristic of forced convection inside a channel with an aspect ratio of 1:10 [5-8]. Since it isn't always possible to test new configurations in the experimental apparatus, an ideal alternative is the use of Computational Fluid-Dynamics (CFD) methods applied on a three-dimensional domain. CFD is an effective tool for investigations of fluid flow in ducts with rib roughened walls. As reported in [9], using finite volume CFD code, RANS turbulence models usually show good agreement with experimental results but it's necessary to highlight that the only use of RANS is not exhaustive and could be inaccurate with respect to other more computational-expensive turbulence models. In this work authors are looking for a CFD model using the RANS k-ω-SST model, widely chosen in literature for similar cases [10-12] with respect to k-ω, k-ε RNG, or k-ε Realizable models. In this paper, results on convective global heat transfer coefficient i.e., averaged over the whole channel, are presented discussing the comparison between experimental and numerical data for different configurations considered. The aim of this work is to find a suitable configuration of a CFD model with RANS that will permit authors to apply it to the range of dimensionless pitches.

2. Problem configuration and setup

2.1. Experimental set-up

The experimental set-up is described in detail in [5], briefly it consists in a rectangular channel of 1:10 aspect ratio with 90° staggered ribs on the upper and the lower walls.

A schematic view of the section dedicated to measurements is presented in Figure 1, and in Table 1 the main geometric parameters are summarized.

In this paper, the upper and the lower surfaces of the channel are characterized by the presence of ribs with squared cross section, having a section side s of 4mm. Ribs are 90° tilted with respect to the stream flow. The pitch-to-side ratio p/s considered have been 10, 20 ,40. They are staggered each other in each configuration considered in this study. The experimental channel operates with lower and upper walls maintained at fixed temperature, while side walls are adiabatic. The air flows at Reynolds number ranging from 700 to 8000, in this specific study Reynolds 7552 has been chosen. For simplicity the tree configurations considered will be named p10s4, p20s4, p40s4 as can be seen in Figure 2.

2.2. Numerical procedure

Numerical simulations have been carried out using the software Ansys Fluent 19.1 based on finite volume method. The solution domain is a 3D channel with a rectangular cross section as described previously. The working fluid is air, incompressible, with constant specific heat coefficient.

The solution domain also includes two added domains: one dedicated to the entrance and one dedicated to the outlet. This configuration, especially while dealing with geometries with obstacles, is recommended to avoid problems with reverse flows. Indeed, to ensure the re-arranging of the flow after the obstacle, the length required is typically 8 to 10 times the obstacle height [13].

The fluid flow inside the rectangular channel has been analysed at Re 7552 and under these applied working conditions the regime of the fluid flow is turbulent. For this reason, the mesh adopted is suitable for the analysis of a turbulent flow. It ensures a $y^+ = 1$ at the wall to solve the laminar sublayer.

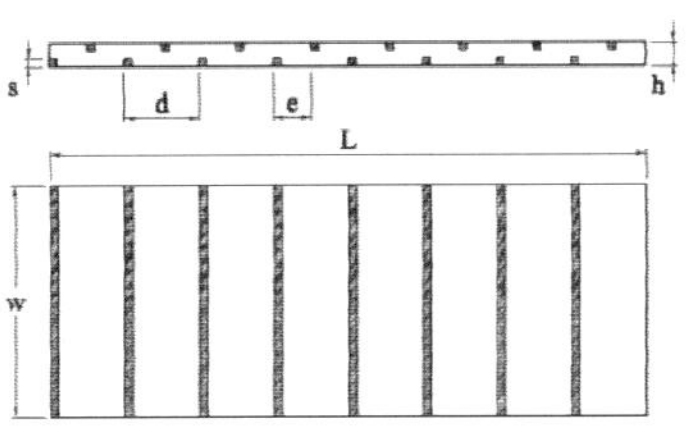

Figure 1. Ribbed channel geometry.

Table 1. Main channel parameters.

Main channel parameters	
Channel height, 2b=h	12.0 mm
Channel width, 2a=w	120.0 mm
Channel length, L	840.0 mm
Hydraulic diameter, D_h	21.82 mm

To obtain a grid independent solution has been carried out a sensitivity analysis. Since it is a three-dimensional case, given the three directions x, y, z where x is the direction of the channel width, y the direction of the channel height and z the streamwise direction, a sensitivity analysis has been carried out for each dimension. It has been observed that the impact of the sensitivity analysis on the solution is higher for the y-direction. Summarizing, various grids with number of elements ranging from $30\cdot10^6$ to $90\cdot10^6$ have been tested and the grids chosen for the discretization of the computational domain have around $65\cdot10^6$ number of elements. An overview of the grid can be seen in Figure 3 while the main quality parameters of the mesh as aspect ratio, orthogonal quality and skewness can be seen in Table 2.

2.3. Boundary conditions

Since the flow in the experimental set-up, at the inlet of the measurement section, can be considered hydrodynamically fully developed, this condition is required at the inlet of the present computational domain. At this scope a preliminary CFD simulation has been used. Considering an adiabatic channel with the same rectangular cross section, without ribs and enough long in the streamwise direction to guarantee the hydrodynamic development ($x_{fd,h}\geq10D_h$), it has been used to allow the fully development of the air flow. The velocity profile (its three components), and the corresponding turbulence parameters (Turbulent Kinetic Energy, k [m^2/s^2] and Specific Dissipation Rate, ω [1/s]) obtained at the outlet of this preliminary channel have been used as inlet condition for the analysed cases.

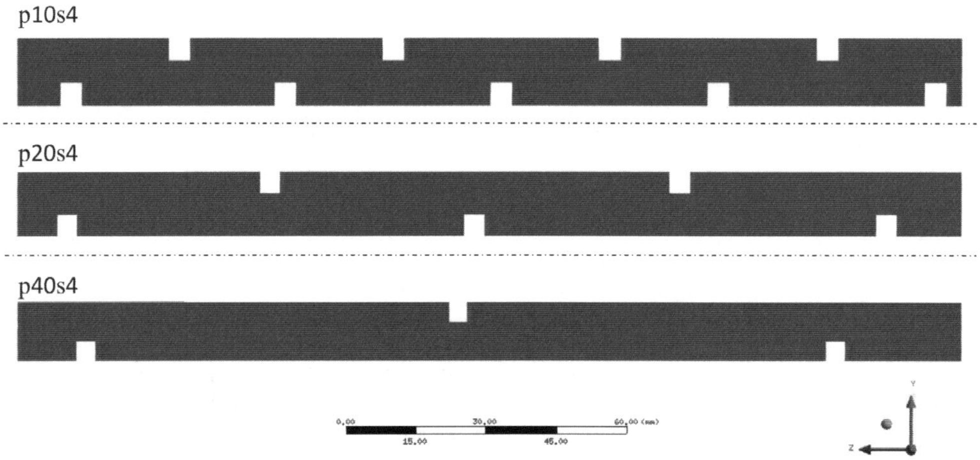

Figure 2. Channel configurations considered.

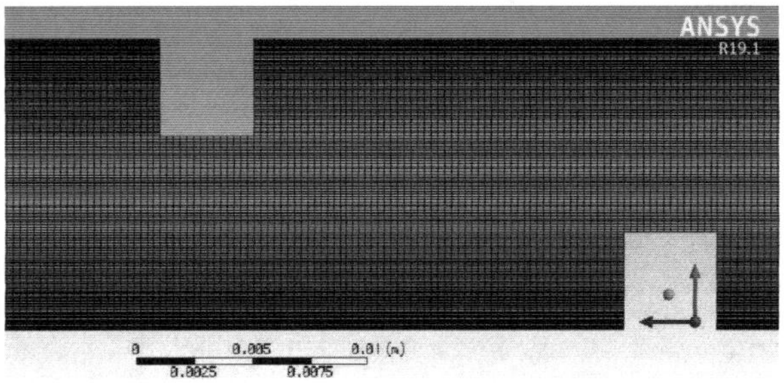

Figure 3. Mesh overview.

Table 2. Mesh parameters.

Mean value of mesh parameters	
Skewness	1.9 e-2
Orthogonal	0.9836
Aspect Ratio	6.9
Element Quality	0.31448

Boundary conditions can be summarized as follow:

- Inlet: velocity inlet (fully developed velocity profile, previously generated) basing on the selected Reynolds number. The working fluid is air, enters the channel at uniform temperature, $T_{in} = 296K$. The fluid flow is not thermally developed yet, so temperature will develop inside the considered channel.
- Outlet: pressure outlet $P_{out} = 0Pa$.
- Wall: no-slip velocity condition is applied for all walls, moreover walls (upper, lower, and lateral) are heated at constant temperature, imposing $T_w = 323.15K$.
- Walls of ribs: no-slip velocity condition is applied, and these walls are considered adiabatic, imposing $\dot{q} = 0\ W/m^2$.
- Symmetry: since it can be considered a symmetric case in the x-direction (spanwise), symmetry has been applied at the wall identified from the centreline.

An additional clarification is needed for motivating the choice of setting all the walls of the channel as heated walls while in the experimental setup the lateral walls are considered adiabatic. Previous simulations [14] have been carried out to choose the most suitable boundary condition for numerical problem able to correctly represent the experimental configuration.

2.4. *Solver setting and models*

Semi-Implicit Method for Pressure Linked Equation-Consistent (SIMPLEC) algorithm is used for pressure-velocity coupling. This choice has been preceded by a comparison of other algorithms, i.e., Coupled and SIMPLE, but they had not managed the simulation to converge, or at least to converge to an unacceptable value of continuity residual in reasonable time.
The presented turbulent flow has been solved with RANS, choosing the k-ω-SST model. This choice has been motivated by evidence in literature and by previous analysis [14].
In literature k-ω-SST and k-ε-RNG models are the mostly used between turbulence models for simulations that required good agreement with experimental results [9]. k-ω-SST uses as default the Enhanced Wall Treatment, a wall function for the near-wall region i.e., viscous sublayer, buffer region and fully turbulent outer region.
Computational time has been considerable, about 96 hours of calculations per each single simulation using 25-30 cores Intel(R) Xeon(R) Gold 6148 CPU @ 2.40GHz.
RANS simulations stop when the continuity residual is lower than $1 \cdot 10^{-4}$.

3. Results

The number of Reynolds considered in this study has been precisely 7552. The choice of this value of Reynolds number is motivated by the fact that according to previous analyses [14] is shown that the accuracy of the solution of a RANS k-ω-SST simulation is better for Reynolds higher than 5000, and the value of 7552 is the highest Reynold number tested experimentally for channels with the three configurations proposed in this paper.
Per each configuration considered (p10s4, p20s4, p40s4) a CFD simulation has been carried out analysing in particular the agreement of heat transfer coefficients of numerical results with experimental data. The experimental data considered can be found in [5]. Values of the average Nusselt number vs Reynolds number for staggered transverse-ribs at different values of dimensionless pitch and height are presented in Figure 7, reference [5]. Values of friction factor vs Reynolds number for transverse-ribs in staggered arrangements at different values of dimensionless pitch and height are presented in Figure 5, reference [5], instead.

3.1. *Convective heat transfer coefficient*

Some sections, equidistant from each other, have been built along the entire channel to calculate the local heat transfer coefficient. For each section, of which an example can be seen in Figure 4, bulk temperature T_b and heat flux $\dot{q}$ have been calculated as mean section variables, from which can be computed the local convective heat transfer coefficient $h_z = \dot{q}_z/(T_w - T_b)$ and consequently the related

Nusselt number at any section $Nu_z = h_z D/k$. Values of the local convective heat transfer coefficients obtained section by section are shown in the graph in Figure 5. The presence of ribs shows a significant change in the trend only for the configuration of p10s4, for other configurations the trend is the one of a channel with plane walls, but with an increase in convective heat transfer coefficient.
By way of example, a velocity and a temperature contour, plotted over a section at z = 0.402 m for channel p20s4, are proposed in Figure 6 and Figure 7 respectively.
Experimentally, the mean convective heat transfer coefficient is calculated as $h_m = \dot{m} c_p (T_{out} - T_{in})/A/\Delta T_{m,ln}$ and the associated Nusselt number is $Nu_m = h_m D/k$ [5]. So, as first comparison, comparing mean convective heat transfer coefficients calculated according to the same formula, errors between experimental measurements and numerical results are around 16% for all the cases considered, showing that the CFD model underestimates the mean convective heat transfer coefficient measured experimentally, probably also due to assumptions made.
But regarding the overall average convective heat transfer coefficient, it can be also computed as follows, applying equation (1), for a more comprehensive evaluation of $\bar{h}$, which considers per example effects of the channel entrance and what happens along the entire length analyzed.

$$\bar{h} = \frac{1}{\sum \Delta A_z} \left(\sum_{z=1}^{n_{sections}} h_z \Delta A_z \right) \quad (1)$$

According to equation (1), each contribution of the local convective heat transfer coefficients h_z has been weighed on the affected heat transfer surface ΔA_z and sum them up. Note that the value of the proposed $\bar{h}$ is independent of the number of sections considered (in this study $n_{sections} = 42$), since once 38 cutting sections are exceeded, the value of the average convective heat transfer coefficient is no longer influenced by the number of cutting planes chosen.

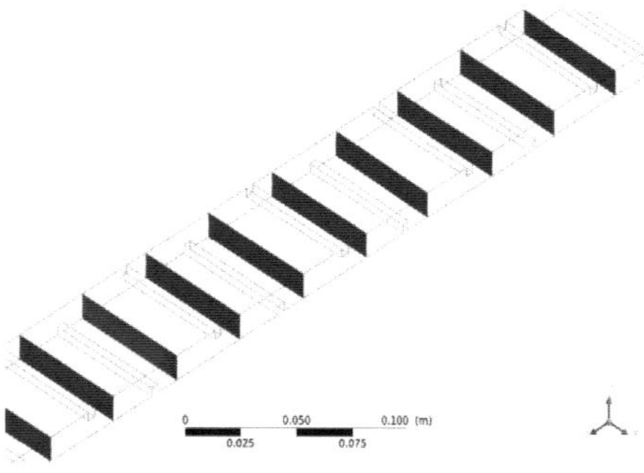

Figure 4. Exemplary section-planes.

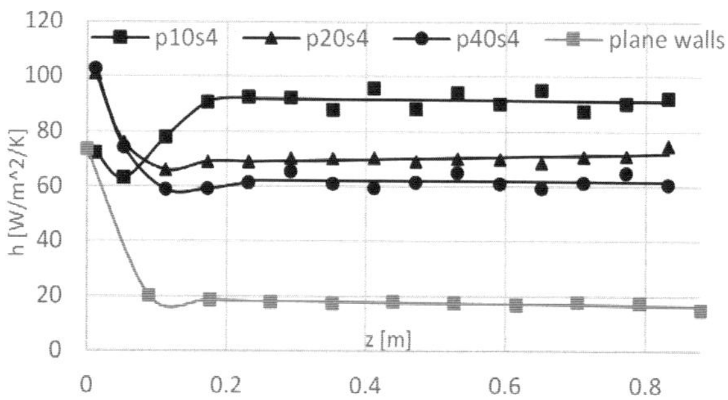

Figure 5. Heat transfer coefficient.

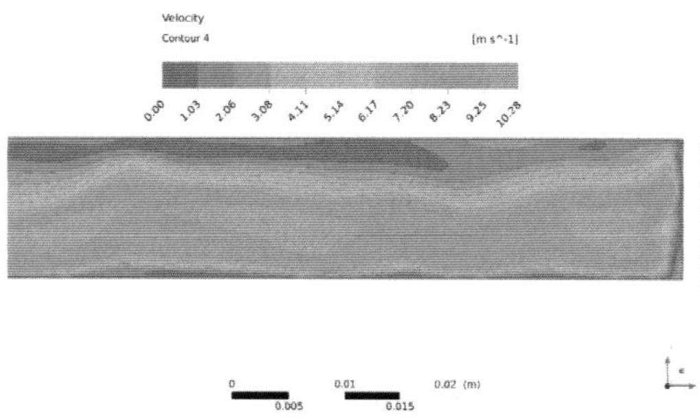

Figure 6. Velocity contour on section plane at z = 0.402 m (p20s4 channel).

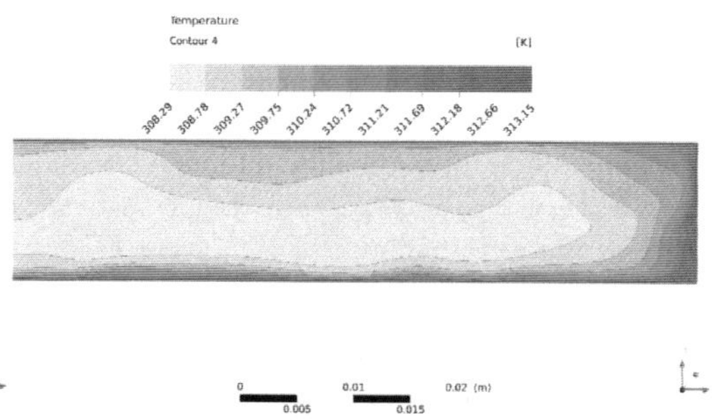

Figure 7. Temperature contour on section plane at z = 0.402 m (p20s4 channel).

Comparing the agreement of $\bar{h}$ with experimental results, the average convective heat transfer coefficient calculated for p40s4 configuration has the best behaviour, with an error of 8.8%. A little less agreement is given by p20s4 configuration with an error of 13.2%, followed by p10s4 configuration with an error of 20.4%. These results are presented graphically in Figure 8 and Figure 9. This can be due to the complexity of the flow since is shown increasing error at lower dimensionless pitch, where the flow is forced to pass in a more complex channel.
Finally, assuming $\bar{h}$, calculated according to eq. (1), as the real value and comparing it with h_m obtained numerically, it could be stated that experimental measurements tend to overestimate the mean convective heat transfer coefficient of the channel by 5-10% depending on the channel configuration.
Looking at the velocity contours in Figure 10, presence of a clockwise recirculating vortex is predicted. These vortexes are present in the low-pressure region between two ribs. Another observable factor is related to blockage effects of ribs in the channels. The flow velocity is higher in the centre of the channel in p10s4 configuration as the available section is reduced.

3.2. Pressure Drops
Friction factors obtained from pressure drops deriving from numerical results through equation (2) have been compared with experimental values, available only for dimensionless pitches of 20 and 40.

$$f_{Darcy}=(\Delta P_0/0.5\rho u^2)(D_h/L) \quad (2)$$

Error between numerical and experimental values [5] has been 11% for a dimensionless pitch of 20 ($f_{Darcy\text{-}CFD}$ = 0.495, $f_{Darcy\text{-}Exp}$ = 0.444) and 9% for a dimensionless pitch of 40 ($f_{Darcy\text{-}CFD}$ = 0.238, $f_{Darcy\text{-}Exp}$ = 0.218).

4. Conclusion
In this paper the CFD model built for a high aspect ratio rectangular channel with 90° staggered ribs on upper and lower walls is presented. k-ω-SST RANS model has been chosen and the agreement with experimental data is good both at fluid-dynamic and thermal level with increasing error at lower dimensionless pitch, where the flow is forced to pass in a more complex channel.

This model has been built as a starting point to be able to expand the thermo-fluid dynamics study of high aspect ratio channels in presence of 90° ribbed surfaces in the range of dimensionless pitches considered without the need of carry out several experimental measurements. CFD could be considered as a helpful tool for the prediction of heat transfer coefficients in rib-roughened channels with 90° ribs, especially during early design-phases of parametric studies.

For more complex configurations which induced more complex flow structures, use of RANS could be not enough accurate, and so further analysis on these configurations will be carried out by exploiting LES to resolve at least larger scale structures.

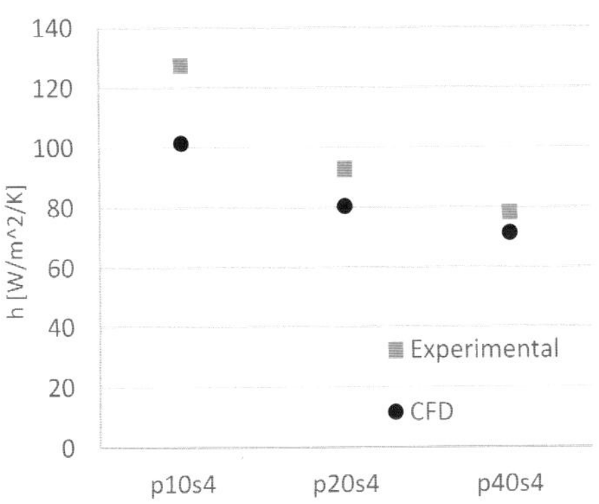

Figure 8. Comparison of global heat transfer coefficient, experimental data versus CFD data.

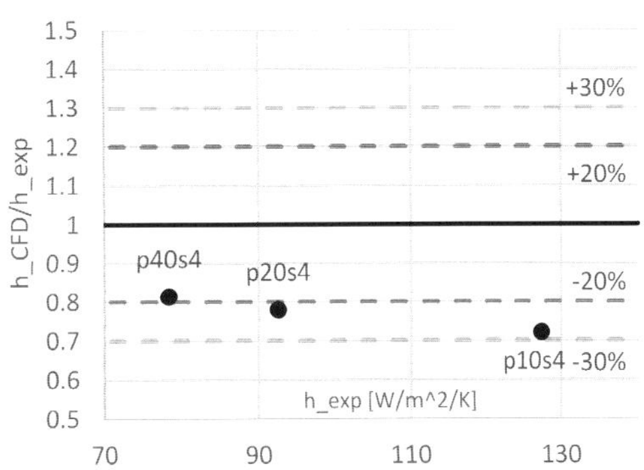

Figure 9. Error between experimental and numerical $\bar{h}$ values.

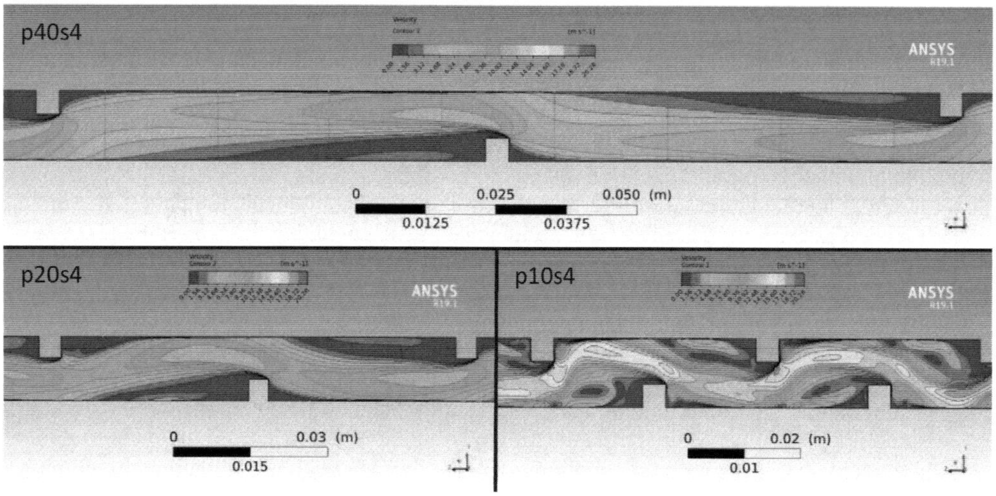

Figure 10. Velocity contours taken on the symmetry plane of the channels considered.

Nomenclature

Symbol	Quantity	SI Unit
A	Area	m^2
a	Channel half width	m
b	Channel half height	m
h	Convective heat transfer coefficient	$W/m^2/K$
k	Thermal conductivity	W/m/K
k	Turbulent kinetic energy	m^2/s^2
L	Channel length	m
$\dot{m}$	Mass flow rate	kg/s
Nu	Nusselt number	-
P	Pressure	Pa
Re	Reynolds number	-
T_b	Bulk temperature	K
T	Temperature	K
u	Bulk velocity (average)	m/s
$\dot{q}$	Wall heat flux	W/m^2
$x_{fd,h}$	Hydrodynamic entry length	m
$x_{fd,th}$	Thermal entry length	m
y+	Dimensionless wall distance $y^+ = \Delta_{y_P}/v\sqrt{\tau_w/\rho}$	-
ν	Kinematic viscosity	m^2/s
ρ	Density	kg/m^3
τ_w	Wall shear stress	Pa
ω	Specific dissipation rate	1/s

Subscripts

Symbol	
in	Inlet
l	Local, at a defined position
ln	Logarithmic
m	Mean
out	Outlet
w	Wall
z	Local, at a position that varies with z

References

[1] Nourin F N and Amano R S, Review of gas turbine internal cooling improvement technology, J. Energy Resour. Technol. Trans. ASME, vol. 143, no. 8, pp. 1–8, 2021, 10.1115/1.4048865
[2] Jain P K, Lanjewar A, Rana K B, and Meena M L, Effect of fabricated V-rib roughness experimentally investigated in a rectangular channel of solar air heater: a comprehensive review, Environ. Sci. Pollut. Res., vol. 28, no. 4, pp. 4019–4055, 2021, 10.1007/s11356-020-11415-6
[3] Jain S K, Das Agrawal G, and Misra R, A detailed review on various V-shaped ribs roughened solar air heater, Heat Mass Transf. und Stoffuebertragung, vol. 55, no. 12, pp. 3369–3412, 2019, 10.1007/s00231-019-02656-4
[4] Huynh D and Mahjoob S, Numerical Analysis of Structured Ribbed Channels for Electronic Cooling Applications, Intersoc. Conf. Therm. Thermomechanical Phenom. Electron. Syst. ITHERM, vol. 2020-July, pp. 766–771, 2020, 10.1109/ITherm45881.2020.9190181
[5] D Fustinoni et al 2012 J. Phys.: Conf. Ser. 395 012042, 10.1088/1742-6596/395/1/012042
[6] D Fustinoni et al 2015 J. Phys.: Conf. Ser. 655 012060, 10.1088/1742-6596/655/1/012060
[7] D Fustinoni et al 2017 J. Phys.: Conf. Ser. 796 012015, 10.1088/1742-6596/796/1/012015
[8] Niro A, Gramazio P, Vitali L and Fustinoni D, Local and global heat transfer characteristics inside a rectangular channel of 1:10 aspect-ratio with ribbed surfaces, International Heat Transfer Conference, pp. 5509-5516, 2018, 10.1615/IHTC16.hte.023149
[9] Sharma S K and Kalamkar V R, Computational Fluid Dynamics approach in thermo-hydraulic analysis of flow in ducts with rib roughened walls - A review, vol. 55. Elsevier, 2016, 10.1016/j.rser.2015.10.160
[10] Liu J, Hussain S, Wang J, Wang L, Xie G and Sundén B (2018). Heat transfer enhancement and turbulent flow in a high aspect ratio channel (4:1) with ribs of various truncation types and arrangements. International Journal of Thermal Sciences, 123, 99–116, 10.1016/j.ijthermalsci.2017.09.013
[11] Saidi A, SundÉn B (2000) Numerical Simulation Of Turbulent Convective Heat Transfer In Square Ribbed Ducts, Numerical Heat Transfer: Part A: Applications, 38:1, 67-88, 10.1080/10407780050134974
[12] Wei Peng, Pei-Xue Jiang, Yang-Ping Wang , Bing-Yuan Wei. Experimental and numerical investigation of convection heat transfer in channels with different types of ribs Applied Thermal Engineering 31 (2011) 2702-2708, 10.1016/j.applthermaleng.2011.04.040
[13] Pope S, (2000). Turbulent Flows. Cambridge: Cambridge University Press. doi:10.1017/CBO9780511840531
[14] Corti M, Gramazio P, Fustinoni D, Vitali L, and Niro A. Accuracy in evaluating heat transfer coefficient by RANS CFD simulations in a rectangular channel with high aspect ratio – Part 1: benchmark on a channel with plane walls, submitted to Journal of Physics Conference Series

Preliminary design and modeling of mini channels to enhance heat transfer in a millimetric catalytic combustor

Caterina LA TERRA[1], Maria CORTI[1], Adelaide NESPOLI[2], Carlo FANCIULLI[2] and Alfonso NIRO[1*]

[1] Politecnico di Milano, Department of Energy,
via Lambruschini 4, 20156 Milano, Italy,
[2] CNR-ICMATE, Via Previati 1/E, 23900 Lecco, Italy
* Corresponding author e-mail: alfonso.niro@polimi.it

Abstract. Mini channel solution is used in devices that require a high density of transmitted thermal power as a very large-scale integration design in computer systems and compact exchangers. Furthermore, the mini channels are extensively investigated in the literature for turbulent and laminar regimes. In this project, different configurations of mini channels have been studied to enhance heat transfer, using simulations with a commercial multi-physics code. Thanks to the results of the models, more promising configurations with 3D printing technique may be built. The project challenge is improving convective thermal power extracted by the exhaust gases of a mini-catalytic combustor. The combustor feeds six modules for thermoelectric power production (TEMs). As the first step, three different mini channel geometries have been chosen; the first one with 19 channels with rectangular cross-section, the second one with 6 channels with a convergent profile, and the latter with 2 channels with a fractal branching geometry. Simulations started from studying fluid dynamic to investigate the velocity field at the exit of the mini channels. The analysis has been extended by adding the conjugate heat exchange between fluid and combustor wall. The results show an increase in heat exchange compared to the base case for all configurations, with a maximum value for the 19 mini channels configuration.

Key Words: *forced convection, thermoelectric generator, mini channels*

1. Introduction

The problem of power supply for decades is no more just about the production and distribution of electricity on a large scale. The development of electronics and the increasing demand of this market, with systems ranging from integrated circuits to microsensors, have opened up the need to power devices with the greatest possible autonomy that requires electric power in small dimensions [1]. The traditional solution to provide electricity to these products is storage through batteries. Batteries need replacement, when after a series of charge and discharge cycles, they are no longer performing. Thermoelectric generators (TEGs) are a remarkable alternative [2]. The thermoelectric phenomenon is based on the passage of charges in two types of semiconductors: type-n and type-p, which lead to the generation of voltage due to an imposed temperature gradient (Seebeck effect) and vice versa (Peltier effect). Since 1959, when General Motors launched the first thermoelectric module (TEM) which

couples the semiconductors thermally in parallel and electrically in series, it has been possible to develop power generators that exploit this phenomenon [3].
Thermoelectric generators have ideally infinite service life, carrying out a direct conversion of thermal power in electrical power, without the aid of moving parts. Being solid state generators, their high reliability and durability, it has made them the focus of a renewed scientific interest. In addition, their modularity allows them to be used for applications of varied dimensional scales without losing in performances. On the other hand, the great limitation of this technology lies in the low efficiency of thermoelectric modules which, for commercial solutions, always ranges between 5 and 9% [4]. For all these features, they are in no way comparable to traditional large engines with high efficiency, but on small dimensional scales - at small powers and temperature differences - they become the most advantageous solution [5].

1.1. TEG prototype
The prototype of a TEG has been studied on a millimetric scale. The device, like those built previously [6], has been designed to replacing common power banks solution for smartphones. The challenge of its design is to create a safe, reliable, and portable system. With the scope of creating a marketable device, it is necessary to increase the initial performances as much as possible. It has been decided to reduce the problems that affect the efficiency of the modules by raising the heat transfer in the hot side system.
The TEG operates via six commercial modules in Bismuth-Tellurium of size 18·18 mm^2. TEG's dimensions are 75 mm in diameter to 21 mm in height and a total weight of 400g. The system is propane fueled. The choice of propane is due to its high calorific value and the commercial availability of small propane tank for refilling lighters. The premixed fuel and air in stoichiometric conditions are injected into the bottom of a hexagonal combustion chamber. The chamber is built exploiting additive manufacturing technique in SS316L (Figure 1). The gas mixture reacts on 52 cylindrical platinum pellets to coat an inert alumina support. The catalysis allows to reach lower combustion temperatures than traditional flame combustion and to obtain a safer process that can be coupled with the TEMs, which have operating temperatures that cannot exceed 300°C. The exhaust gases are piped to the walls where they exchange with the hot face of the modules, providing the thermal power needed for the generator. To manage the temperature difference between the faces of the thermoelectric module, the TEG is equipped with six water cooling circuits made on polymer through the fuse deposition melting technique. For the optimization of the cold side system authors have built serpentine ducts with an unconventional section in previous work [7]. The prototype thus constituted generates a total power of 9 W with an efficiency of 3.5%.

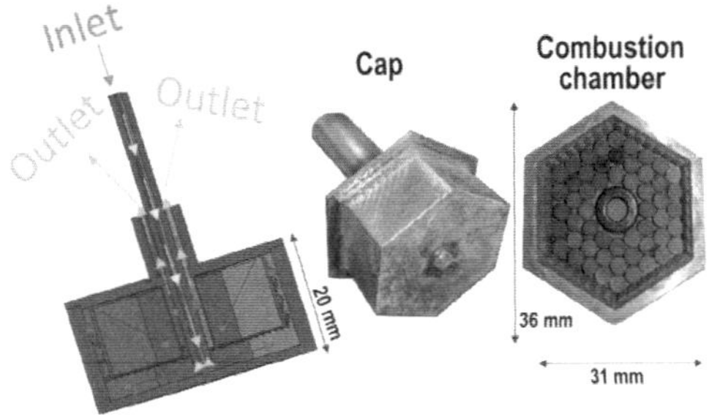

Figure 1. Details of the 3d printed TEG's catalytic combustor: the total drawing with the flow circuit description (on the left), the sealing cap (central) and the arrangement of the pellets in the bottom of the chamber with dimensions(right).

1.2. Aim of the project

To increase the current generator performance, new heat transfer enhancement solutions have been evaluated. Specifically, the focus has been on designing mini channels solutions at the interface between combustor and TEMs. Mini channels are used to increase the power density in compact systems. Tuckerman et al. [8] first used mini channels in highly integrated electronics cooling systems and showed an increase in power density up to 790 W/cm^2 from an initial condition of 20 W/cm^2. In this work, it has been evaluated the feasibility of combining the mini channels into a system for the exploitation of thermal power by exhaust gases, in millimetric scales. Three different mini channel configurations have been selected and then numerically tested. The first one is a flat rectangular cross-section configuration, because in the work of Tuckerman this geometry has reported excellent results. The second one is a convergent configuration, because the Duryodhan's study [9] has shown how on micro size for bioengineering tests, it can break down thermal gradients through the combination of convection and conduction. The last is a branched fractal configuration, this network solution allows the reduction of thermal gradients in conditions of imposed thermal flux [10].

2. Method

The computational model has been developed utilizing the commercial code COMSOL Multiphysics, using CFD and heat transfer packages. In this model, chemical reaction and preheating of inlet gases are not considered, and the composition of inlet mixture and exhaust gases was neglected, i.e., gases have been simply considered as air. This assumption is based on the consideration that propane-air mixture is in a volumetric ratio of 1:25, and therefore it is in high excess of nitrogen which does not participate in chemical reaction.

The project consists of two computational models with two different inlet mass flow rates. The first model considers the overall combustion chamber and is aimed at its validation with experimental data (Figure 2). This model was run with the volume flow rate of the reagent mixture currently used to fuel the TEG, namely 3 Nl/min, which determines a mass flow rate of $1 \cdot 10^{-5}$ kg/s in each of the six chamber's walls (D_h=0,0019m), an input velocity of 0.5 m/s, and a Reynolds number of 64. The second model is used instead to compare mini channel configurations in the best-operating conditions achievable by TEG, i.e., using the input mass flow rate that optimizes the power transmitted to the thermoelectric modules. In this perspective, this model has been reduced to the study of a single wall to reduce computational time, after verifying the uniformity of the velocity field between the six walls with simulations of the overall combustor. The optimal mass flow inlet is equal to $1{,}9 \cdot 10^{-4}$ kg/s for each wall, with a Reynolds number of 1155 and an inlet velocity of 9m/s. This value is the result of an analytical model that combines the project targets to maximize the electrical power generated by the modules, by increasing the temperature difference ΔT between the faces of the module itself.

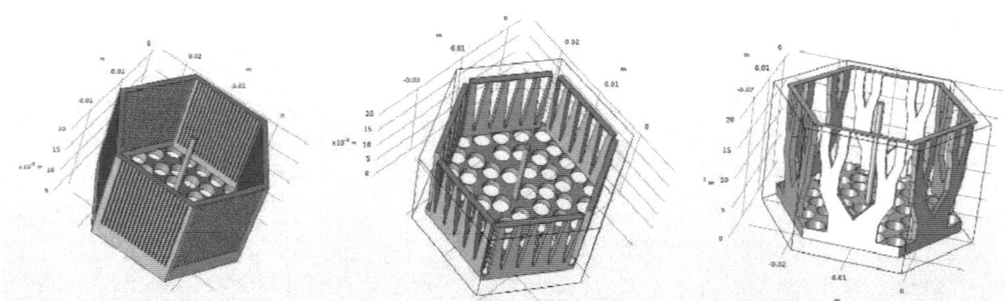

Figure 2. The fluid domain in COMSOL laminar simulations. From left to right: chamber with planar mini channels (configuration1), convergent mini channels (configuration 2), with fractal mini channel (configuration 3).

As the commercial TEMs can support up to 200°C of ΔT whereas the other project constraints are:

- The air-propane mixture has an optimum catalysis temperature of 600 ° C, moreover, the mixture cannot be below 300 ° C, which allows the self-sustaining of the chemical reaction;
- The combustion must be stoichiometric, so the inlet mixture always has an air-propane ratio of 25:1 by volume;
- The need to have 600 ° C in the combustion chamber requires the use of 3D printing steel, which withstands temperatures up to 1400 ° C, but has a low thermal conductivity;
- The operating temperatures of the thermoelectric modules cannot exceed 300 °C;
- Hot gas current must have the most constant thermal profile possible during the passage through the channel-wall in a direction transversal to the thermal flow, since the thermal gradients on the gas side affect the faces of the TEM and reduce its performance;

with an increase in the total volumetric flow rate of the inlet mixture from 3 Nl/min to 27Nl/min, and with a bypass of air in a ratio of 1:1, the transmitted power has been increased from 1.7 to 8.4 W and has been supplied to each wall $1,9\cdot10^{-4}$ kg/s of mass flow inlet.

2.1. Model setup

The geometric parameters chosen to build the mini channels are a compromise between the reduced size of the wall and the need to create as many mini channels as possible in the same section. To allow the comparison, the three configurations have been built with the same length and height of the cross-section. With these two parameters, 19 mini channels have been built with configuration 1 and an aspect ratio of 2.5. To create at least six converging channels, it was necessary to imprint an angle of inclination not greater than 2.95°, that provide an aspect ratio of 0.5 at outlet and 2.5 at inlet. However, for the fractal configuration that requires at least two bifurcations, a single mini channel has been inserted (aspect ratio of 0.667).

Table 1. Mini channels geometric parameters.

	Configuration 1	Configuration 2	Configuration 3
Base of cross-section	0.4 mm	0.4 mm	0.4 mm
Height of cross-section	1 mm	1 mm	1 mm
Length of mini channel	18 mm	18 mm	18 m
Angle of inclination	-	3°	80°
Number of mini channels	19	6	1

For the model of the whole combustor, a laminar study was chosen due to the low inlet velocity.
For the mini channel geometries comparison, given the high value of Re at the entrance region [11], a k-ε turbulence model has been used.
For the thermal analysis, a constant temperature condition has been set on the steel wall corresponding to the interface with thermoelectric modules. The mesh was performed with the default triangular and tetrahedral mesh algorithm available in COMSOL Multiphysics with a minimum mesh size of $6.5\cdot10^{-4}$ mm. The mesh consists of $8.5\cdot10^{5}$ elements, this choice is due to the results of a sensitivity analysis, conducted between $17\cdot10^{5}$ and $4.45\cdot10^{5}$ elements. Simulations were carried out with a multi-grid algebraic solver. The choice of this solver is dictated by the magnitude and complexity of the CFD model to be developed [12]. The solver works in a segregated way, i.e., first flow field is solved and then the temperature field is determined based on velocity results previously found.

Computational time has been about 1 hour and 20 minutes using four cores of 3.60 GHz. Simulations have been stopped when the continuity residual is lower than $1 \cdot 10^{-4}$.

2.2. Thermography analysis

The model validation was carried out by comparing the results of the 3D simulation on the whole combustion chamber with the experimental data. The mass flow rate inlet of $6.275 \cdot 10^{-5}$ kg/s has been used, with a laminar solver. The experimental campaign has been conducted with the use of thermographic images of the side walls of the combustion chamber isolated from thermoelectric modules and exposed to stationary air. The FLIR Termovision A40 IR camera has been set to an emissivity value of 0.24 and used jointly with K-type thermal probes placed on the combustor surface and. Under these operating conditions, thermal profiles in transverse direction have been obtained for the four walls under analysis. All the four thermographs have a decreasing trend, in agreement with physics, as the temperature reduces linearly from the region in contact with the pellets how far the cap is sealed. The thermal gradient is 25°C for all the analyzed surfaces. Validation has been made by imposing a convective heat transfer coefficient of 22 W/m^2/K on all walls for the basic condition without mini channels. The model, in the same operating conditions, shows a thermal trend with a temperature drop of 25 °C that agrees with the measurements, but the temperatures deviate from the measured values of 200 °C (Figure3). This is because the simulations do not consider all the phenomena that occur simultaneously within the combustion chamber, first the preheating of the entering mixture.

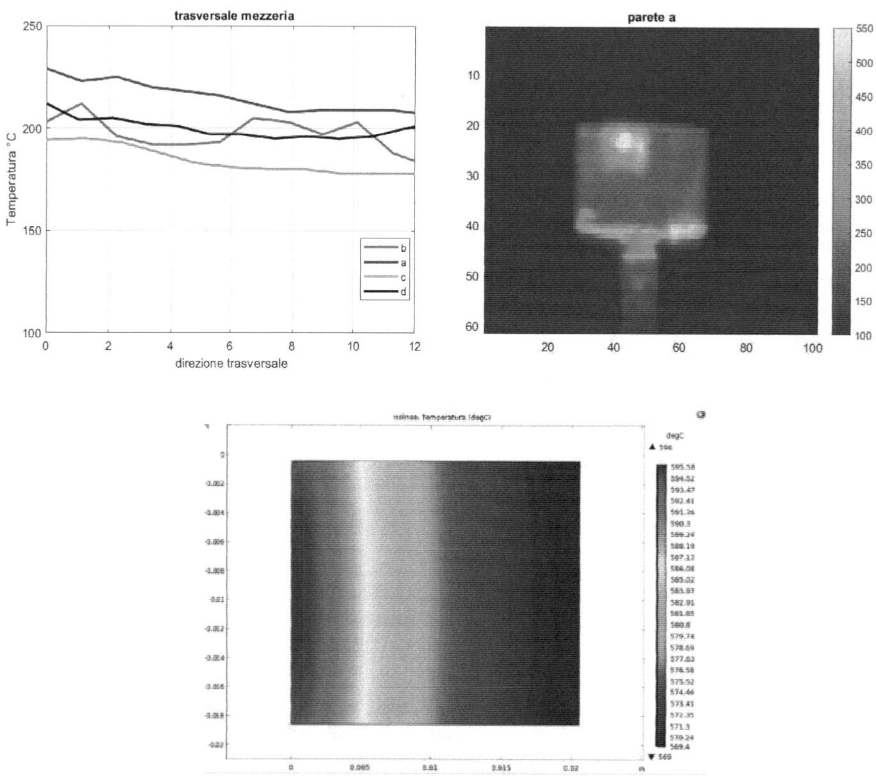

Figure 3. Temperature gradient from experimental data in four chamber's walls (left), IR image (right), the model results at the same operating conditions of the experimental data (bottom).

3. Results

The results, presented in Figure 4, show a section in the x-y plane of the mini channels, in the half of inlet section at z=0mm. All four cases are evaluated at system optimum inlet mass flow rate of $1{,}9\cdot10^{-4}$ kg/s and display the velocity field that is generated inside the mini channels beyond an entrance region of 8 mm length. The basic case highlights how, with an inlet velocity of 9 m/s in the outlet section with a flat duct, it's possible to achieve local velocities no higher than 11m/s. For configuration 1, the velocity profile accelerates in all 19 mini channels, in a uniform trend, up to reach a velocity of 38 m/s in the exit section, i.e., with an increase of 300% with respect to the base case. In configuration 2, convergent channels allow to overcome the results of the straight mini channels achieving local velocities in the outlet section of 75.8 m/s. Configuration 3, with network geometry, attains slightly higher velocities at the outlet of the mini straight mini channels, due to the regions following the bifurcations of the fluid and the fact that all the input mass flow rate is conveyed in a single channel.

Regarding pressure drop, Table 2 shows ΔP for each configuration across inlet and outlet sections. It is evident that the convergent configuration is the one with maximum pressure drop, followed by configuration 1, and then by configuration 3. For heat transfer analysis, the same boundary conditions have been applied to be able to compare the 3 cases. Precisely a wall temperature of 200 °C has been set, to assess the bulk temperature at the fluid outlet sections having imposed an inlet temperature of 20 °C. Numerical results display that the 19-mini channels can heat the gas more effectively, with a temperature increase of 52% compared to the case without mini channels. The converged configuration stops at a 27%, while for the network configuration is the one with less increase in temperature (only 15%). These results highlight that even if the configuration 1 is not where the fluid flow has the maximum velocity, the 19 mini channels solution is the best from the heat transfer point of view. This is due to the fact that the exchange areas between fluid and solid increases as the number of mini channels.

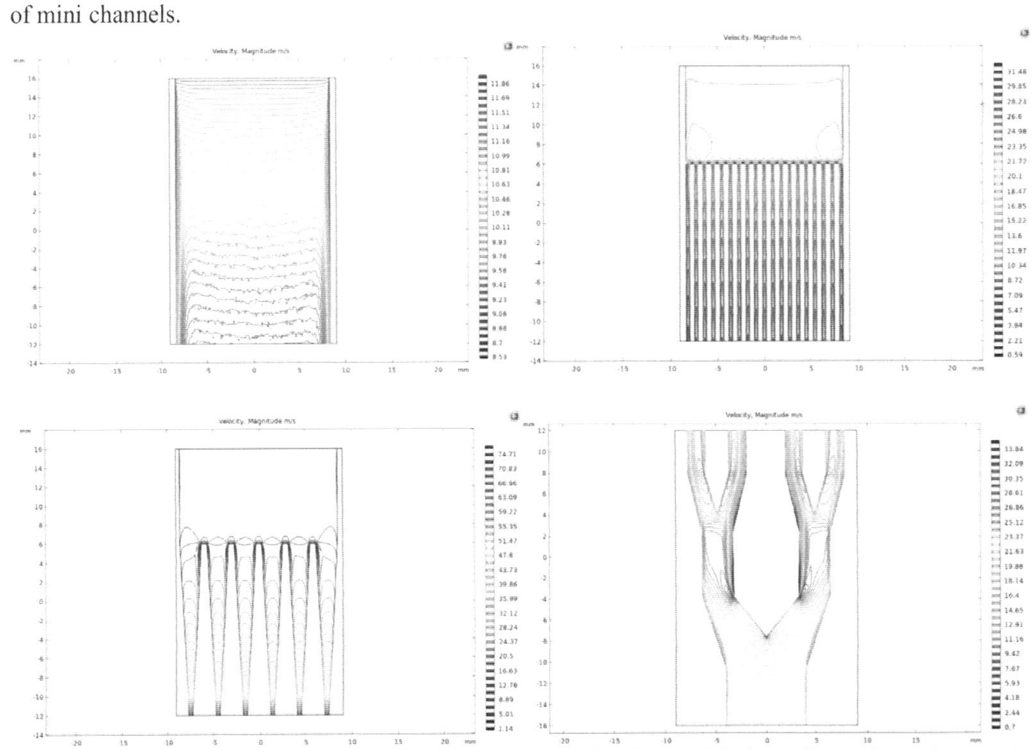

Figure 4. Velocity field contours for the base case and the three mini channels configurations at the same inlet velocity of 9m/s and external wall temperature of 200°C.

Table 2. Pressure drops and outlet bulk temperature results of the computational models.

	ΔP [Pa]	T_{bulk_outlet}[°C]
Base case	71	99
Configuration 1	1146	151
Configuration 2	3020	125
Configuration 3	790	114

4. Conclusion

It has been analysed how it is possible to increase heat transfer between a mini catalytic combustor and a thermoelectric module, using mini channel solutions. The three configurations chosen for the computational simulations show promising results, regarding heat transfer enhancement. Configuration 1, with 19 rectangular section straight mini channels is the best compromise between heat transfer enhancement and pressure drops penalization. In view of new measurements, in force of the promising modelling results, the three mini channels configurations have been printed in SS316L with Selective Laser Melting technique (Figure 5). The design of the combustion chambers is a trade-off between technological feasibility and best numerical results.

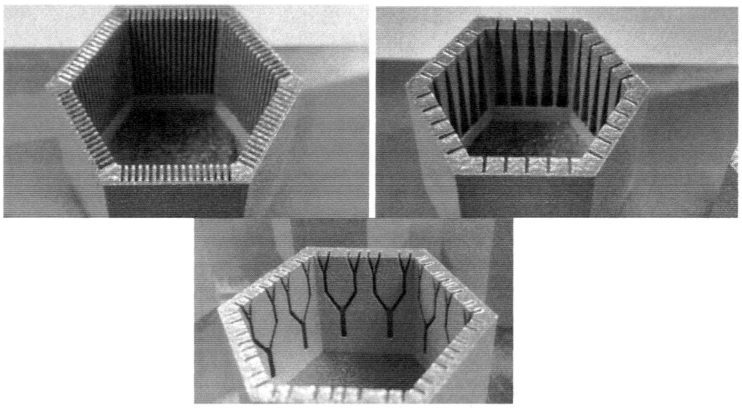

Figure 5. New combustion chamber prototypes with different mini channels configurations.

References

[1] Sundarraj S, Maity D, Roy S S and Taylor R A, Recent advances in thermoelectric materials and solar thermoelectric generators-a critical review, *RSC Adv.*, vol. 4, no. 87, pp. 46860–46874, 2014, doi: 10.1039/c4ra05322b

[2] Jeffrey Snyder G, Small thermoelectric generators, 2002. doi: 10.1109/ICT.2002.1190351.

[3] Jaziri N, Boughamoura A, Müller J, Mezghani B, Tounsi F and Ismail M, A comprehensive review of Thermoelectric Generators: Technologies and common applications, *Energy Reports*, vol. 6, pp. 264–287, 2020, doi: 10.1016/j.egyr.2019.12.011

[4] Tian Z, Lee S and Chen G, Heat transfer in thermoelectric materials and devices, *J. Heat Transfer*, vol. 135, no. 6, p. 061605, May 2013, doi: 10.1115/1.4023585

[5] Vining C B, An inconvenient truth about thermoelectrics, *Nat. Mater.*, vol. 8, no. 2, pp. 83–85, 2009, doi: 10.1038/nmat2361

[6] Abedi H, Migliorini F, Dondè R, De Iuliis S, Passaretti F and Fanciulli C, Small size thermoelectric power supply for battery backup, *Energy*, vol. 188, 2019, doi:

10.1016/j.energy.2019.116061
[7] M Corti et al 2022 J. Phys.: Conf. Ser. 2177 012011
[8] Tuckerman D B and Pease R F W, High-Performance Heat Sinking for VLSI, *IEEE Electron Device Lett.*, vol. EDL-2, no. 5, pp. 126–129, 1981, doi: 10.1109/EDL.1981.25367
[9] Duryodhan V S, Singh S G and Agrawal A, Heat rate distribution in converging and diverging microchannel in presence of conjugate effect, *Int. J. Heat Mass Transf.*, vol. 104, pp. 1022–1033, 2017, doi: 10.1016/j.ijheatmasstransfer.2016.09.014
[10] Lyu Z, Pourfattah F, Arani A A A, Asadi A and Foong L K, On the Thermal Performance of a Fractal Microchannel Subjected to Water and Kerosene Carbon Nanotube Nanofluid, *Sci. Rep.*, vol. 10, no. 1, pp. 1–16, 2020, doi: 10.1038/s41598-020-64142-w
[11] Qu W and Mudawar I, Experimental and numerical study of pressuredrop and heat transfer in a single-phase micro-channel heat sink, *Int. J. Heat Mass Transf.*, vol. 45, no. 12, pp. 2549–2565, 2002, doi: 10.1016/S0017-9310(01)00337-4
[12] William T V, "COMSOL CFD Module User 's Guide," *CFD Modul. User's Guid.*, pp. 1–710, 2017, [Online]. Available: https://doc.comsol.com/5.3/doc/com.comsol.help.cfd/CFDModuleUsersGuide.pdf

Local wall heating effects on aeroacoustic field radiated by an isolated square cylinder in a laminar flow

M. Falone, V. D'Alessandro, L. Giammichele and R. Ricci

Dipartimento di Ingegneria Industriale e Scienze Matematiche
Università Politecnica delle Marche
Via Brecce Bianche, 60131 Ancona (AN), Italy

E-mail: `v.dalessandro@univpm.it`

Abstract. The aim of this work is to gain an insight on the effect of the wall heating on the aeroacoustic sound radiated by bluff bodies in laminar flows. In particular, the local thermal treatment of the wall boundary was investigated as a possible method for active controlling the emitted noise. This technique was studied performing direct numerical simulations of the aeroacoustic noise produced by an isolated square cylinder operating at a Reynolds and Mach numbers equal to 150 and 0.2, respectively. In the considered case, the Karman vortex street deriving by the flow/cylinder interaction, produces a lift and drag pulsation on the body surface, leading to a dipolar-like noise emission. In this context, different local thermal fluxes were applied to the cylinder wall in order to reduce its aerodynamic forces fluctuation and, consequently, the associated pressure disturbance that produces the radiated sound. The computations are performed using an OpenFOAM solver that adopts an explicit Runge-Kutta scheme for time integration and a second-order, energy conserving scheme for the convective part of the Eulerian flux. Moreover, the spurious numerical waves reflections at the far-field boundary are damped adopting a sponge-layer approach.

1. Introduction

The study of the aeroacoustic emission derived from air/body interaction is still an engineering open challenge, with particular emphasis to the self-noise produced by bluff bodies, that is an important issue related to built environment noise pollution, [1]. For the above mentioned application area, the interaction between a body and an incoming flow produces unsteadiness in the wake region resulting in an acoustic perturbation, [1, 2]. In this context, the possibility to control the aeroacoustic field by altering the wake geometry is promising strategy, either using passive techniques or active ones.

Passive control techniques, including the use of additional objects near the reference body to mitigate acoustic disturbance, have been widely studied, [3, 4]. On the other hand, the flow field control through the imposition of line, transverse, or rotary oscillations was studied as active control techniques, [5, 6].

It is worth nothing that heating the body's surface allows to control the wake geometry as experimentally observed in a circular cylinder by Lecordier et al. [7, 8]. The acoustic effects related to wall heating were recently investigated by D'Alessandro et al. [9], who extended their analyses to non-circular geometry sections as well. It was showed that wall temperature increase, reduces the lift and drag pulsation acting on the considered body. The damping of the force pulsation generated by the Karman vortex street, which is shed over bluff bodies in laminar flows, leads to aeroacoustic perturbations decay.

In the present work, we want to deeply investigate the wall heating effects on aeroacoustic field radiated by an isolated square cylinder in a laminar flow. In particular, the consequences on the aeroacoustic sound generation and propagation derived from an applied local thermal flux is studied.

An open-source solver called `caafoam`, developed by the authors themselves, [10, 11], was adopted to perform direct computation of the flow and acoustic fields derived from the aforementioned configuration.
This paper is organized as follows: in Sec. 2 the governing equations are presented, Sec. 3 provide a brief description of the discretization technique, Sec. 4 contains the numerical results and Sec. 5 summarizes the conclusions.

2. Flow model

Considering the set of conservative variables $\boldsymbol{\phi} = (\rho, \rho u_i, \rho E)^T$, the flow is modelled following mass, momentum and energy equations in their perturbed form:

$$\frac{\partial \boldsymbol{\phi}}{\partial t} + \nabla \cdot \mathbf{F}_c(\boldsymbol{\phi}) - \nabla \cdot \mathbf{F}_v(\boldsymbol{\phi}, \nabla\boldsymbol{\phi}) = \sigma\left(\boldsymbol{\phi}_{ref} - \boldsymbol{\phi}\right), \tag{1}$$

$\mathbf{F}_c(\boldsymbol{\phi})$ and $\mathbf{F}_v(\boldsymbol{\phi}, \nabla\boldsymbol{\phi})$ in eq. 1, represent convective and viscous fluxes given by:

$$\mathbf{F_{c,j}} = \begin{pmatrix} \rho u_j \\ \rho u_i u_j + p\delta_{1j} \\ \rho u_j H \end{pmatrix}, \quad \mathbf{F_{v,j}} = \begin{pmatrix} 0 \\ \tau_{ij} \\ \tau_{ji} u_i - q_j \end{pmatrix}. \tag{2}$$

In eq. 2, ρ denotes the density, u_i is the generic Cartesian component of the velocity vector, p is the pressure, and H is the total enthalpy. H is obtained from $H = E + p/\rho$, where E is the total internal energy. Moreover, the viscous stress tensor and heat flux vector components, τ_{ij} and q_i , are respectively computed using the constitutive relation for Newtonian fluids and the Fourier postulate:

$$\tau_{ij} = \mu\left(\frac{\partial u_i}{\partial x_j} + \frac{\partial u_j}{\partial x_i} - \frac{2}{3}\frac{\partial u_k}{\partial x_k}\delta_{ij}\right), \quad q_j = -\lambda\frac{\partial T}{\partial x_j}. \tag{3}$$

In this work, thermal conductivity λ is considered temperature independent, while the Sutherland law is adopted to describe the dynamic viscosity, μ. Moreover, the fluid temperature is evaluated starting from the total internal energy and the pressure is computed following the ideal gas equation of state:

$$\mu = \mu_\infty\left(\frac{T}{T_\infty}\right)^{\frac{3}{2}}\frac{T_\infty + S}{T + S}, \quad c_v T = E - \frac{1}{2}u_j u_j, \quad p = \rho(\gamma - 1)\left(E - \frac{1}{2}u_j u_j\right), \tag{4}$$

in which μ_∞ and T_∞ are the free stream dynamic viscosity and temperature respectively, S indicates the Sutheland's constant, c_v is the specific heat at constant volume and γ denotes the specific heat ratio of the fluid.
The artificial term on the right–hand side of eq. 1 provides to damp the conservative variables to a known solution, $\boldsymbol{\phi}_{ref}$, in proximity of the domain external boundaries: this is the so–called sponge layer treatment [12], that admits to avoid spurious wave reflections in the far–field that may affect the solution. Specifically, the damping function σ depends on the layer width, L_{sp}, and the minimum distance from the inflow/outflow boundaries, d, as follows:

$$\sigma = \sigma_0\left(\frac{L_{sp} - d}{L_{sp}}\right)^n \tag{5}$$

where σ_0 and n are constant parameters controlling the strength and the shape of the sponge profile.
Two sponge layer features need to be properly calibrated in order to ensure its effectiveness: the sponge strength, -η_{target} [dB], and the sponge width. The first one is defined as:

$$-\eta_{target} = 20\frac{2\log_{10} e}{1 - M_\infty^2}\int_{L_{sp}} \sigma d\mathbf{x}, \tag{6}$$

while the sponge width, in its dimensionless expression, must be designed under the following constraint:

$$0.5 \leq \frac{L_{sp} \cdot f}{a_\infty} \leq 2 \tag{7}$$

where f is sound disturbance frequency and a_∞ is the free stream speed of sound.
In this work, the characteristic frequency obtained by the spectral analysis of lift coefficient time–history is considered for the dimensionless width calculation. Moreover, $n = 2$ in eq. 5 was chosen, while -η_{target} was imposed to be 40 dB, according with Mani et al., [12]. Lastly, the dimensionless parameter $(L_{sp} \cdot f)/a_\infty$ was setted to 0.5 in order to limit the computational load.

3. Numerical approximation

A cell–centered finite volume method, available within OpenFOAM library, is adopted to solve the governing equations described in Sec. 2. Convective terms approximation is handled following Pirozzoli's second–order accurate, energy–conserving scheme [13], while standard central schemes for diffusive contributions are adopted.
For time integration a 2N storage explicit Runge–Kutta (ERK) scheme was implemented in order to exploit its good performance for large scale computations. In particular we have considered a five stage, fourth order accurate, ERK scheme available in the open literature, [14]. In our computational experience this scheme is really appealing since it allows to use a maximum Courant number equal to 1. Thus, it allows to exploit good parallel behaviour of the explicit approaches without a severe time–step size restriction.
In this study, a fully structured computational grid was used: the grid cells were clustered near the cylinder walls using a dimensionless first cell height, y_c/D, equal to $5 \cdot 10^{-3}$ (see Fig. 1), and the computational domain was extended up to 200 D from the center of the cylinder as schematised in Fig. 2. The obtained computational mesh has a total cells number, N_c, of $4.4 \cdot 10^6$.

Figure 1. Fully structured mesh, near wall refinement.

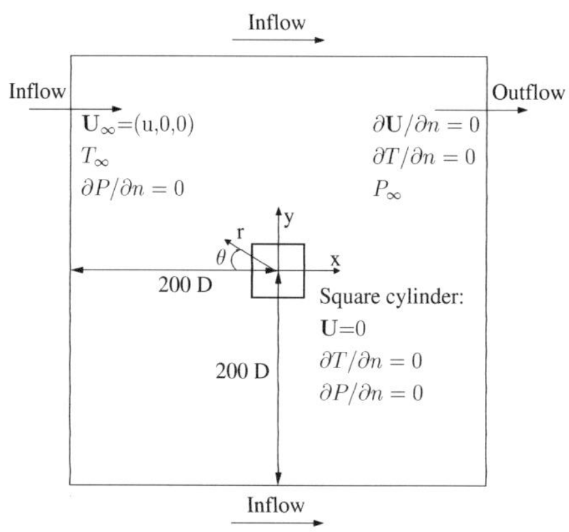

Figure 2. Schematic of square cylinder base configuration.

4. Results

In this section the numerical results of the performed computations are presented. We focused on the analysis of the acoustic noise produced by an isolated square cylinder operating at $Re = 150$, $M_\infty = 0.2$. This configuration was chosen because is a well–established literature benchmark [15]. In all the

following cases the undisturbed flow parameters where setted as in Inoue [15]. In particular we have adopted for the specific heat ratio, the Prandtl number and the dimensionless Sutherland's constant: $\gamma = 1.4$, $\mathrm{Pr} = 0.75$ and $\mathrm{S}/\mathrm{T}_\infty = 0.3855$, respectively.
Standard boundary conditions were adopted both for inflow and outflow and a no-slip, adiabatic wall is employed to model the cylinder surface coupled with a zero normal gradient for the pressure, as highlighted in Fig. 2. The results are presented in terms of aerodynamic coefficients and standard parameters related to acoustic fields: (i) drag and lift coefficients, (ii) Strouhal number and (iii) dimensionless acoustic pressure.
Drag and lift coefficients are given by:

$$C_D = \frac{2D'}{\rho_\infty u_\infty^2 A_{ref}}, \quad C_L = \frac{2L'}{\rho_\infty u_\infty^2 A_{ref}}, \tag{8}$$

where D' and L' denotes drag and lift forces, ρ_∞ and u_∞ represent the density and velocity values at the free stream, and A_{ref} is the reference area.
Being the cases treated in the paper markedly in unsteady regime the solutions are compared on the basis of the statistics of the force coefficients: time–averaged drag coefficient, $\langle C_D \rangle$ and the amplitude of oscillations of the force coefficients over their mean value, $\Delta C_D = (C_{D,max} - C_{D,min})/2$ and $\Delta C_L = (C_{L,max} - C_{L,min})/2$.
The Strouhal number is defined as:

$$St = \frac{f L_{ref}}{u_\infty}. \tag{9}$$

Instead, the acoustic field was evaluated by means the acoustic pressure, defined as the root mean square of the dimensionless fluctuating pressure, p$'$:

$$p' = \frac{p - \langle p \rangle}{\rho a_\infty^2}. \tag{10}$$

where $\langle p \rangle$ denotes average pressure field, resulting from the time averaged pressure. Finally, the dilatation rate field $\partial u_j / \partial x_j$ was also used to visualise the acoustic waves because is strictly related to the change in density and, consequently, to the noise propagation.

4.1. Validation

For the sake of validation, the flow past a square cylinder under an adiabatic wall boundary condition was analysed. The above described flow conditions produces a Karman vortex street, that is shed behind the cylinder as noticeable looking at the vorticity magnitude in Fig. 3. The vortexes detach alternately from top and bottom surfaces of the cylinder, leading to a pulsation in lift and drag forces clearly visible in the time–histories in Fig.4. L' and D' fluctuations act like an acoustic source and, considering the main

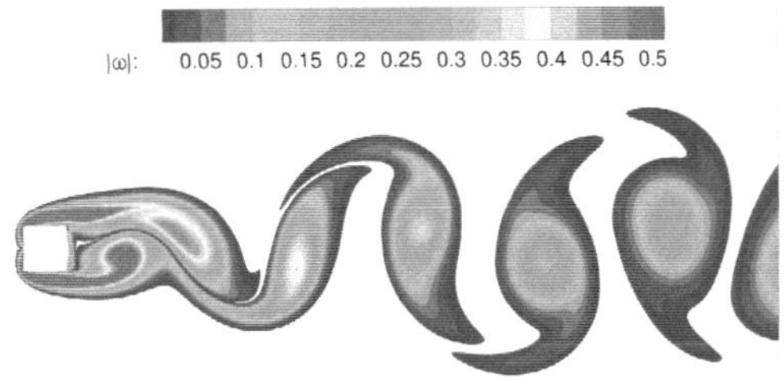

Figure 3. Vorticity magnitude.

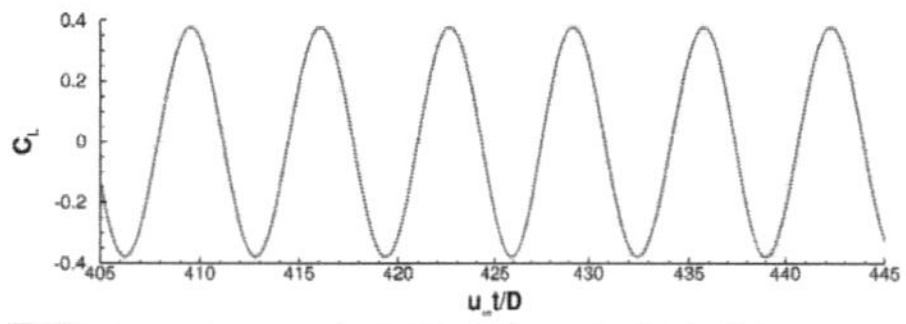

(a) Lift coefficient.

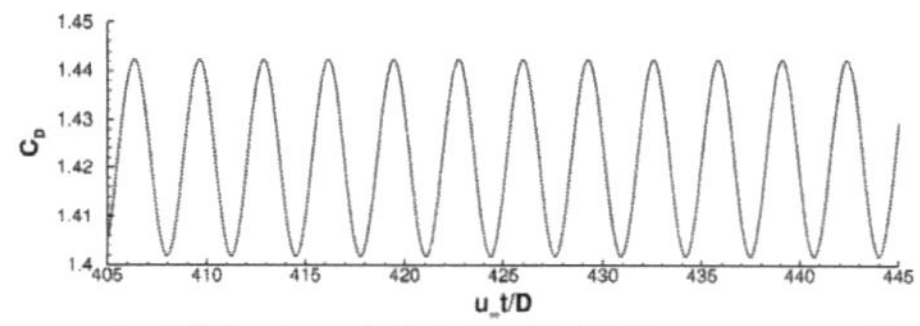

(b) Drag coefficient.

Figure 4. Force coefficients time–histories.

Table 1. Aerodynamic coefficients.

	Inoue et al. [15]	`caafoam`
St	0.151	0.152
ΔC_L	0.386	0.387
$\langle C_D \rangle$	1.414	1.431
ΔC_D	$2.04 \cdot 10^{-2}$	$2.01 \cdot 10^{-2}$

direction involved, an acoustic field having a quadrupole nature is expected. However, the amplitude of lift oscillation is two order of magnitude greater than the drag one, as reported in Tab. 1. For this reason, vertical dominance results in a dipolar–like sound emission as can be observed from the dilatation rate field in Fig. 5.

Furthermore, it is easy to note that the acoustic field is not effected by nonphysical pressure waves, confirming that the adopted numerical approach, including the sponge–layer treatment, is effective. This

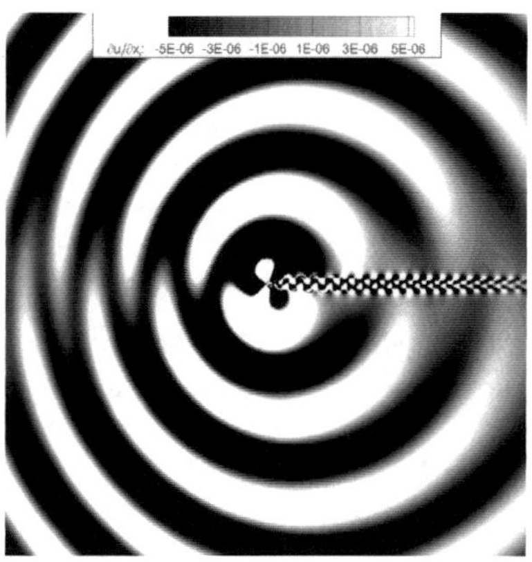

Figure 5. Dilatation rate field.

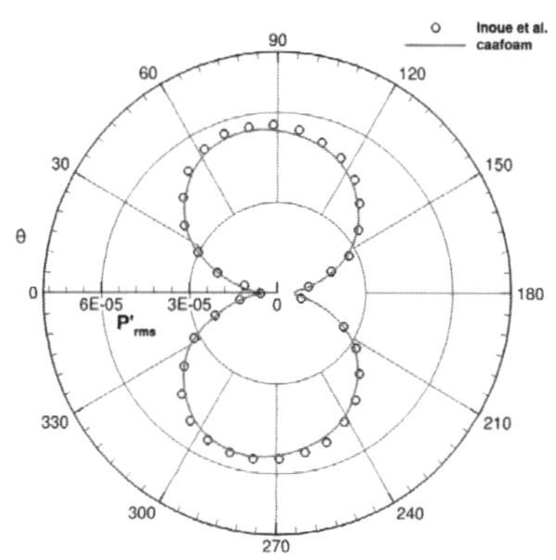

Figure 6. p'_{rms} polar plot, $r/D = 75$.

evidence is also corroborated by the aerodynamic coefficients showed in Tab. 1. The predicted values are in an excellent agreement with Inoue's reference data, [15].

The emitted noise in the considered flow/acoustic problem is evaluated in Fig. 6: it refers to the acoustic pressure extracted on a circumference that originates from the cylinder centre and having a dimensionless radius $r/D = 75$. Also in this case the obtained results are in a very good agreement with literature data, confirming the capabilities of `caafoam` in resolve this kind of problems.

Table 2. Wall thermal effect - Aerodynamic coefficients.

	$\partial T/\partial n\|_w = 0$	$\partial T/\partial n\|_w = 0.5$	$\partial T/\partial n\|_w = 1$	$\partial T/\partial n\|_w = 2$
St	0.152	0.150	0.149	0.145
ΔC_L	0.387	0.363	0.331	0.247
$\langle C_D \rangle$	1.431	1.439	1.446	1.456
ΔC_D	$2.01 \cdot 10^{-2}$	$1.79 \cdot 10^{-2}$	$1.45 \cdot 10^{-2}$	$0.87 \cdot 10^{-2}$

4.2. Local thermal effect on the aeroacoustic field

In this section we want to provide an insight on the alteration of the aeroacoustic field due to the cylinder local wall heating. It was shown that a thermal boundary condition leads in a lift and drag pulsation abatement as well as a decrease in vortexes detachment frequency, [10]. Our intent is to investigate if the same effect could be obtained limiting the thermal treatment to specific zones of the square cylinder. Firstly, a dimensionless temperature gradient, $\partial T/\partial n|_w$, ranging from 0.5 to 2 was imposed next to the wall of the whole cylinder in order to compare the global thermal effect on the aeroacoustic field with the local heating one. Considering that in the adopted approach λ is temperature independent, an imposed temperature gradient next to the wall corresponds to a constant heat flux through the cylinder surface. Looking at Tab. 2 is evident that when a non–adiabatic condition is applied at the wall boundary, the time–averaged drag coefficient increases, while the lift and drag pulsations are damped as already noted experimentally by Lecordier et al. [7, 8] in a similar contexts. Moreover, is important to remark that the damping of the amplitude of forces oscillation increases with the applied temperature gradient rises. These conditions lead to an abatement of the emitted sound from the flow/body interaction as noticeable from the polar plot in Fig. 7. The acoustic pressure was extracted over a circumference having a radius $r/D = 40$ in order to better evaluate the thermal effect on the aeroacoustic field. Also in this case, the applied thermal flux plays a major role on the radiated noise decay. In fact, it is easy to note from Fig. 7 that p'_{rms} decreases when $\partial T/\partial n|_w$ moves to higher values.

Furthermore, it is important to put in evidence that the applied heat flux also slow down the vortexes detachment from the cylinder, resulting in a lower frequency for the radiated noise (see Strouhal number in Tab. 2). Starting from the above described setup, four different configurations of local wall heat flux were analyzed. In the first one, only the top and bottom surfaces (hereinafter TB) of the bluff body were heated, while in the second analysis the heat flux was applied to the front and back (FB) faces of the cylinder as represented in Fig. 8(a). Moreover, the sole front surface (F) has a non–adiabatic boundary condition and, for sake of completeness, also the configuration with a wall normal temperature gradient applied only to the back surface (B) of the square cylinder was investigated (see Fig. 8(b)).

It is important to remark that for all the above mentioned cases, a dimensionless temperature gradient, $\partial T/\partial n|_w = 1$, was employed and the heating strategy was chosen in order to preserve the symmetry of the considered problem. Also for the aforementioned investigation, the effect of the local wall heating on the acoustic emission was evaluated by means of the root mean square of the fluctuating pressure over a circumference having a radius $r/D = 40$.

Looking at Fig. 9, it is interesting to note that the heat flux in TB configuration, slightly increase the radiated sound, but this configuration produced very similar results if compared to a fully adiabatic condition. By the contrast, $\partial T/\partial n|_w = 1$ applied to the surfaces orthogonal to the incoming flow direction, produces an acoustic pressure decay equivalent to the temperature gradient application on the whole body surface.

The observed phenomena were deeply investigated performing the computations for the configurations B and F. In this context, the obtained results show that the square cylinder back surface heating is the only responsible of the sound emission abatement. In fact, the heat flux applied in proximity of the flow separated region provides a reduction of the root mean square of the fluctuating pressure quite similar to the full surface heating, as shown in the polar plot in Fig. 10.

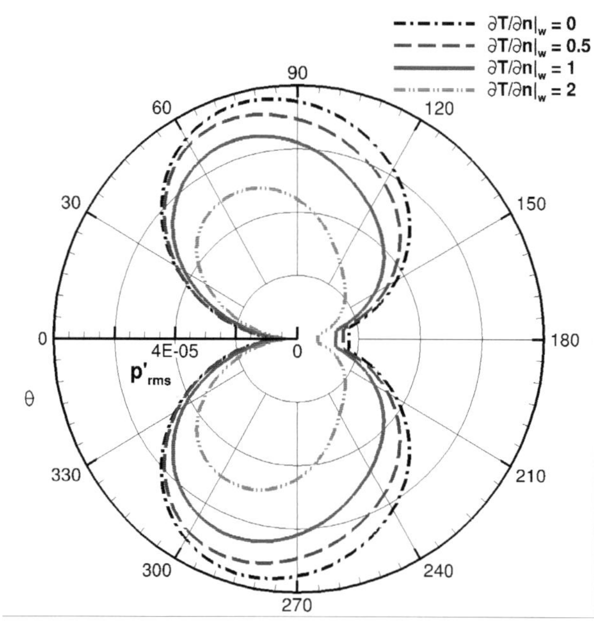

Figure 7. Wall thermal effect - p'_{rms} polar plot, $r/D = 40$.

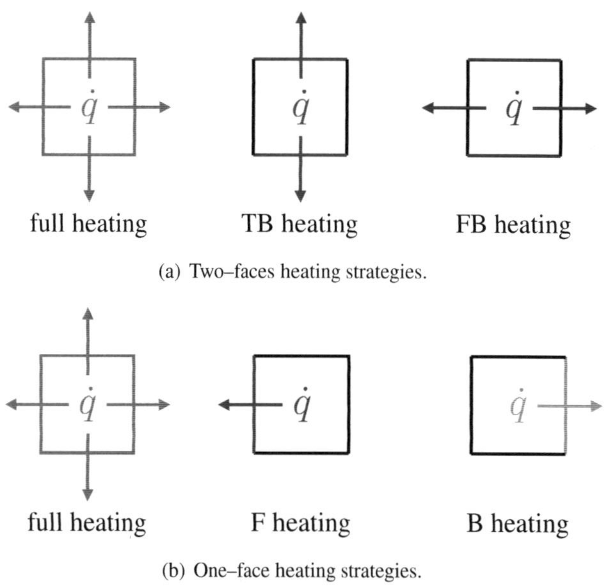

Figure 8. Schematic of local wall heating strategies.

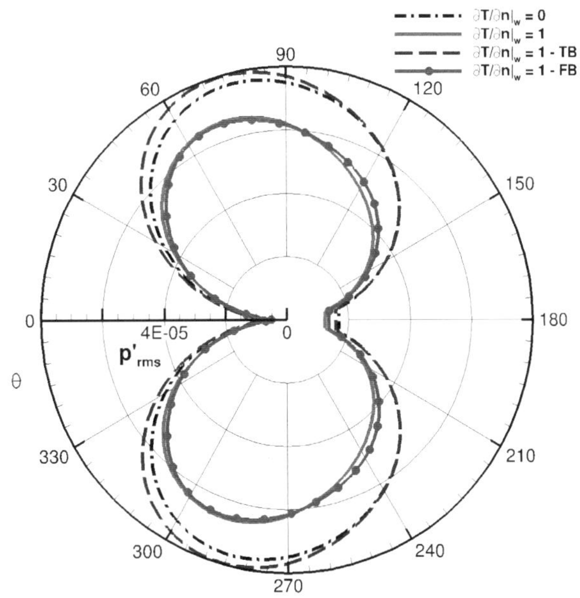

Figure 9. Local thermal effect, TB and FB configurations - p'_{rms} polar plot, $r/D = 40$.

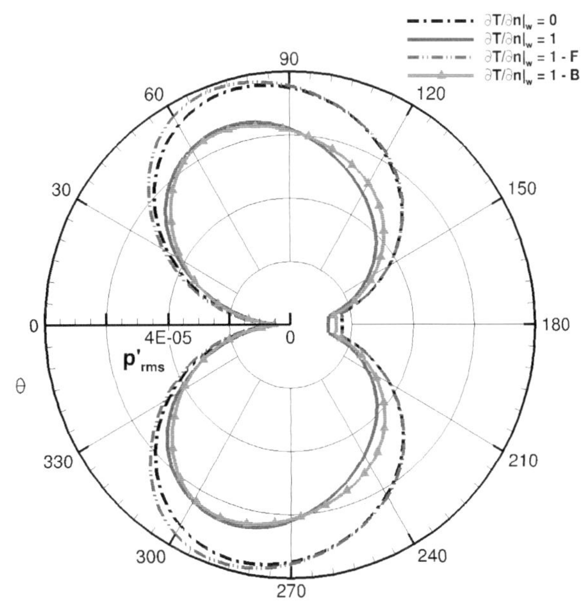

Figure 10. Local thermal effect, F and B configurations - p'_{rms} polar plot, $r/D = 40$.

5. Conclusions

This paper addresses the effectiveness of local wall heating as active control technique for the aeroacoustic emission radiated by a square cylinder in a laminar flow.

We have found that a heat flux applied to the cylinder wall reduces the aerodynamic forces pulsations generated by the Karman vortex street that is shed behind the bluff body, resulting in the far–field sound abatement. This condition holds true for different wall normal temperature gradients.

In particular, the damping of the amplitude of forces oscillation increases with the applied temperature gradient rises, leading to an higher impact on the emitted sound from the flow/body interaction. Moreover, the wall heat flux condition produces to a lower frequency for the radiated noise. More in depth, it was shown that the square cylinder back surface heating provides a sound abatement quite similar to the full surface effect. It is worth also noting that anheat flux focused on separated flow region is the key point to control adequately the aeroacoustic emission.

Finally, the proposed technique can be considered effective for active control the sound/noise produced by a square cylinder in a laminar flow.

References

[1] Lam, B and Gan, WS and Shi, D and Nishimura, M and Elliott, S 2021 *Build. Environ.* **200** 107928 ISSN 0360-1323
[2] Inoue, O and Hatakeyama, N 2002 *J. Fluid Mech.* **471** 285–314
[3] Mahato, B and Ganta, N and Bhumkar, YG 2019 Control of aeroacoustic noise generation during flow past a circular cylinder using splitter plate *INTER-NOISE 2019 MADRID - 48th International Congress and Exhibition on Noise Control Engineering*
[4] Chauhan, MK and Dutta, S and Gandhi, BK 2019 *J. Wind Eng. Ind. Aerodyn.* **184** 342–361 ISSN 0167-6105
[5] Bearman, P W 1984 *Ann. Rev. Fluid Mech.* **16** 195–222
[6] Williamson, CHK and Govardhan, R 2004 *Ann. Rev. Fluid Mech.* **36** 413–455
[7] Lecordier J C, Browne L, Le Masson S, Dumouchel F and Paranthoën P 2000 *Exp. Therm. Fluid Sci.* **21** 227–37
[8] Lecordier J, Hamma L and Paranthoen P 1991 *Exp. Fluids* **10** 224–9
[9] D'Alessandro V, Falone M, Giammichele L and Ricci R 2022 *Int. J. Heat Mass Transfer* **190**
[10] D'Alessandro V, Falone M and Ricci R 2020 *Comput. Fluids* **203** 104517
[11] D'Alessandro V, Falone M, Giammichele L and Montelpare S 2019 A low-storage Runge-Kutta OpenFOAM solver for compressible low-Mach number flows: Aeroacoustic and thermo-fluid dynamic applications *XII International Conference on Computational Heat and Mass Transfer, 2019* vol 128
[12] Mani A 2012 *J. Comput. Phys.* **231** 704–16
[13] Pirozzoli S 2011 *Annu. Rev. Fluid Mech.* **43** 163–94
[14] Kennedy C, Carpenter M and Lewis R 2000 *Appl. Numer. Math.* **35** 177–219
[15] Inoue O, Mori M and Hatakeyama N 2006 *Phys. Fluids* **18**

Optimised Electro-Osmotic Flow in Rectangular Microchannels with Smoothed Corners

Marco Lorenzini[1],**Nicola Suzzi**[2]

[1] Università di Bologna, DIN – Dipartimento di Ingegneria Industriale, Via Fontanelle 40, I-47121, Forlì (FC), Italy

[2] Università di Udine, DPIA - Dipartimento Politecnico di Ingegneria e Architettura, Via delle Scienze 206 - I-33100 Udine (UD), Italy

E-mail: marco.lorenzini@unibo.it

Abstract. Electro-osmotic flows are a means of circulating polar fluids through microchannels without resorting to mechanical pumping. The lack of moving parts, of noise and the ease of integration in silicon chips make them an interesting option for microchip cooling and miniaturized total analysis systems. This paper describes an optimization in terms of first- and second-law analysis of the cross-section of a microchannel subject to electro-osmotic flow. Starting from rectangular cross-sections of different aspect ratios, its corners are progressively smoothed and the resulting Poiseuille and Nusselt numbers computed. Performance evaluation criteria are then used to assess the change in, among others, heat transfer rate, temperature difference between wall and bulk fluid, and equivalent pumping power. The entropy number is also computed and the results commented. It is found that the latter criterion highlights a configuration of minimum entropy generation, whereas the trends of the heat transfer and temperature difference are opposite to that of the equivalent pumping power.

1. Introduction

Practical applications of micro-flow devices (MFDs) are growing in number and the future outlook appears promising, yet MFDs are still the subject of much fundamental research, [1]. Among MFDs, microchannel heat sinks can remove high heat flux densities, but pressure drop across the channels, particularly when liquids are employed, can be a crippling drawback. When charged surfaces are involved, electro-osmosis can be used, since it allows motion of a liquid relative to a charged surface under the action of an electric field, [2], without the need for moving parts. Also, the system lends itself to being realized directly onto electronic chips, [3], and micro total analysis systems (μ- TAS), [4]. Performance of a microchannel heat sink can be improved by modifying the shape of its cross-section, optimising it according to a chosen set of constraints, [5,6]. In spite of their great applicative potential where removal of high heat fluxes is in demand, the related pressure drop may be significant, especially when liquids are employed as coolants, with the reduction in hydraulic diameter of the ducts quickly leading to viscous heating of the fluid circulated, [7,8], which reverses the direction of the heat flow. At small scales, however, flow devices can exploit micro-effects, [7], such as electro-osmosis, which allows the motion of a liquid relative to a charged surface, induced by an applied external potential gradient across a microchannel, [2,9]. If a polar fluid is used in combination with channel walls possessing a net electrical charge, an inhomogeneous charge distribution develops in the liquid, which can be moved by applying electric field over the length of the channel, which acts on the layer of mobile ions close to the walls: the flow which originates is called electro-osmotic (EOF). EOF devices have several advantages: they do not have moving parts, are therefore free of noise and vibrations and no lubrication is needed. Also, liquid reservoirs require tiny volumes, making them ideal for direct connection to the chips. EOFs have been reported to yield experimental Nusselt numbers about 10% larger than pressure-driven flow (PDF) for the same geometry, although Joule heating may become significant when the applied voltage between the electrodes increases beyond a certain threshold, [10]. Voltage signals sent to electrodes make implementation of EOF devices easier than manufacturing and controlling a micro mechanical pump. Among the drawbacks, the chemical composition at the interface can significantly alter the velocity profiles, which are markedly different from those exhibited by pressure-driven flows, and Joule heating in the fluid may partly or completely offset its cooling action, [2]. Electro-osmotic flows in microchannels have been the subject of several fundamental investigations, for a variety of cross-section, ranging from the common circular ducts and parallel plates, [11], to unusual geometries such as polygonal, elliptical and triangular ducts, [3, 12, 13], as has been the use of electro-osmotic pumps (EOPs), [14, 15], sometimes with the specific purpose of electronics cooling, [10, 16, 17]. The use of non-Newtonian fluids, [18, 19], and nanofluids, [20], is also a recent field of investigation.

For the single microchannel, one of the earliest works is that of Rice and Whitehead, [11], who dropped the assumption of negligible EDL thickness as in the classical treatment by Smoluchowski, and predicted an electro-viscous effect, i.e. a maximum in the apparent viscosity of the flow. Maynes and Webb, [21], investigated the convective heat transfer associated to purely electro-osmotic and both pressure and electro-osmotically driven flows for fully developed flow in circular ducts and parallel plate channels, [22], assuming that the Debye length reached to half-width Z of the channel. Their findings demonstrated that the factors with the strongest influence on the temperature profile and Nusselt number are Z, the intensity of the non-dimensional Joule heating and the ratio of the external electric field to pressure gradient.

Vocale et al., [13], focussed their investigations on triangular microchannels both with and without superimposed pressure gradient. The velocity field was found to be strongly influenced by the aspect ratio and the electro-kinetic diameter of the channel. The role of the relative ratio of pressure forces and electrical forces acting on the fluid was also analysed and characteristic curves for EOPs with triangular channels were obtained.

The discussion above reveals how research on electro-osmotic flow in microchannels has been very active in the recent past, touching a variety of subjects such as the fluids employed, the electric field magnitude, the presence of the pressure gradient, and the shape of cross-section which can take several forms (circular, trapezoidal, rectangular, etc.) thanks to the current development of micro-fabrication technologies. Optimisation of the cross-section can lead to improved performance in terms of increased heat transfer and decreased pressure drop and entropy production, and this can be achieved through smoothing of its corners, as demonstrated in [23, 24]. Investigation has been extended to pressure-driven flows in microchannels with and without viscous dissipation, [8], and to electro-osmotic flows in the fully developed and constant heat flux and perimeter temperature case, [25], employing the results to carry out a first-law analysis based on performance evaluation criteria (PECs), a method detailed in e.g. [5], which is based on a first-law approach. This work investigates numerically the performance of a microchannel subject to EOF and with uniform wall temperature when its base, rectangular, cross-section has its corner progressively smoothed, keeping the thermal entry region into account, thus solving a so-called Graetz problem. Optimisation is carried out using PECs and entropy generation minimisation, [6], to account for both the first and second law of thermodynamics, as already proposed by [26]. The investigation is similar to what was done in [25] for uniform heat flux and temperature over the cross-section and fully-developed flow. The factors influencing the thermal behaviour of the microchannel are investigated and an analysis of the influence of Joule heating to heat transfer ratio is also given. The practical uses of the results are highlighted, as their usefulness in conducting PEC-type analyses.

2. Problem modelling and method of solutions

The geometry investigated is plotted in Fig. 1, which shows the cross-section a × b in dimensions, with smoothed corners of radius of curvature r, Fig. 1(a), and the channel of length L, Fig. 1(b), with its axis along the L direction of a reference frame. To obtain the Poiseuille and Nusselt numbers, Po and Nu, needed for optimisation, the electric potential, velocity and temperature fields must be computed. The corresponding governing equations with suitable boundary conditions are written in the following directly in non-dimensional form; for a detailed description of how the equations are obtained using the Gouy-Chapman model, [2], see e.g. [25, 27].

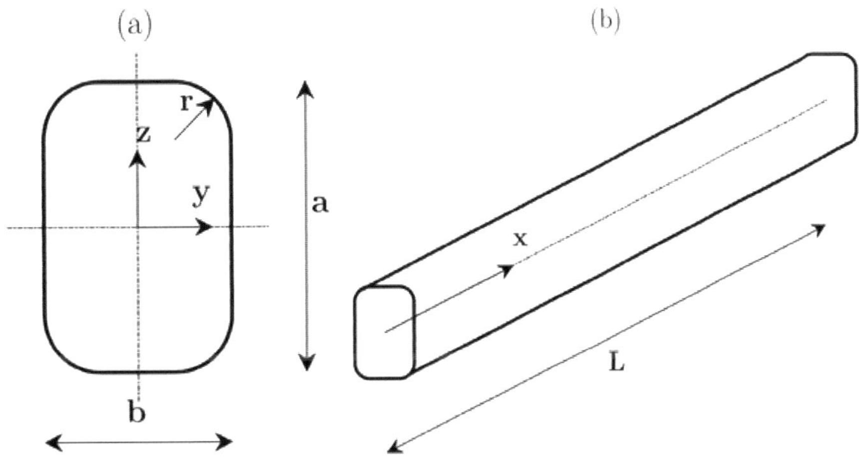

Figure 1: Cross-section (a) and channel (b).

2.1. Electric potential

The non-dimensional formulation of the equation for the electric potential distribution, Ψ is,

$$\frac{\partial^2 \Psi}{\partial Y^2} + \frac{\partial^2 \Psi}{\partial Z^2} = (k_D\, D_h)^2 \sinh \Psi \tag{1}$$

D_h is the hydraulic diameter and k_D is the Debye–Hückel parameter (m^{-1}), [2, 7], and represents the inverse of characteristic length (Debye length) for the electric double layer (EDL). The non-dimensional value, Ψ_0, of the potential at the Stern plane which separates the two layers forming the EDL is used as a boundary condition over the non-dimensional perimeter, $\partial\Omega$:

$$\Psi|_{\partial\Omega} = \Psi_0 \tag{2}$$

2.2. Velocity field

A steady, hydrodynamically developed laminar flow of a Newtonian, incompressible fluid with uniform thermophysical properties through a microchannel of uniform cross section is considered.

The non-dimensional governing equation for the only non-zero component of the velocity U is thus:

$$\frac{\partial^2 U}{\partial Y^2} + \frac{\partial^2 U}{\partial Z^2} = \tilde{E} \sinh \Psi \tag{3}$$

where $\tilde{\mathrm{E}}$ is the non-dimensional external electric field. The no-slip boundary condition is imposed through the solid wall, $\mathrm{U}|_{\partial\Omega} = 0$. Knowledge of the velocity field allows computation of Po, [28].

2.3. Temperature field

The non-dimensional temperature distribution $\Theta(\mathrm{X}, \mathrm{Y}, \mathrm{Z})$ is obtained from the solution of the non-dimensional energy equation below,

$$\mathrm{Re}\,\mathrm{Pr}\,\frac{U}{U_b}\frac{\partial \Theta}{\partial X} = \frac{\partial^2 \Theta}{\partial Y^2} + \frac{\partial^2 \Theta}{\partial Z^2} + M_z \tag{4}$$

Pr is the Prandtl number of the fluid, while $\mathrm{U_b}$ and Re are the non-dimensional bulk velocity and the Reynolds number. The parameter $\mathrm{M_z}$ appearing in Eq. (4) is defined as

$$M_z = \frac{\sigma\,(D_h\,E_x)^2}{\lambda\,(T_w - T_i)} \tag{5}$$

σ and λ are the fluid's electrical resistivity and thermal conductivity, $\mathrm{E_x}$ is the electric field applied at the ends of the channel, and $\mathrm{T_w}$ and $\mathrm{T_i}$ the wall temperature and the fluid temperature at the inlet respectively. $\mathrm{M_z}$ relates Joule heating within the fluid to the heat transfer at the wall, and its role is similar to that of the Brinkman number in pressure-driven flows with significant viscous dissipation, [8, 29] A uniform wall temperature is imposed, and a uniform temperature profile at the channel entrance ($\mathrm{X_0} = 0$) is chosen, so that

$$\Theta|_{\partial\Omega} = 0, \;\; \Theta(Y, Z)|_{X_0} = 1 \tag{6}$$

As discussed in [27], use of a preparation length, [30], to account for Joule heating within the fluid is not necessary in this case.

2.4. Numerical modelling

The numerical solution of Eqs. (1), (3) and (4) was obtained using the open source software `GNU Octave` and a number of additional libraries, as detailed in [27]. The electric potential and velocity fields are solved sequentially over the two-dimensional domain corresponding to the cross-section; this would not appear possible for the energy equation, Eq. (4), yet it has the form of a parabolic partial differential equation, so the three-dimensional physical problem can be reduced to a two-dimensional mathematical problem. Implementing the $\theta-$weighted scheme for marching over channel axis, Eq. (4) is discretised as if it were a two-dimensional, unsteady problem, in the form

$$\mathrm{Re}\,\mathrm{Pr}\,\frac{U}{U_b}\left[\frac{\Theta}{\Delta X}\right]^{n+1} - \theta\left[\frac{\partial^2 \Theta}{\partial Y^2} + \frac{\partial^2 \Theta}{\partial Z^2}\right]^{n+1} = \mathrm{Re}\,\mathrm{Pr}\,\frac{U}{U_b}\left[\frac{\Theta}{\Delta X}\right]^{n} + (1-\theta)\left[\frac{\partial^2 \Theta}{\partial Y^2} + \frac{\partial^2 \Theta}{\partial Z^2}\right]^{n} + M_z \tag{7}$$

where n denotes the integration step along the channel axis.

2.5. Entropy generation

Entropy generation is obtained in non-dimensional form by writing the entropy balance over a length dx of the channel and integrating it between entrance and exit, see e.g. [26]. Knowledge of the temperature and velocity fields allows computation of the entropy generation number N_S, [6], which is the ratio of the total entropy production in a tentative configuration (i.e. for a given smoothing radius) to the reference one (i.e. with sharp corners). N_S can be written as a function of two contributions, one, N_T, due to heat transfer, which includes Joule heating, and one, N_P, due to the driving force, and of their ratio $\Phi_r = N_P/N_T$:

$$N_S = \frac{N_T + \Phi_r N_P}{1 + \Phi_r} \tag{8}$$

The two contributions N_T and N_P can be written as

$$N_T = \frac{\dot{Q}}{\dot{Q}_r}\left[\frac{T_{b_{ir}}}{T_{b_{or}}} + \frac{\dot{Q}/\dot{Q}_r}{\dot{m}/\dot{m}_r}\left(1 - \frac{T_{b_{ir}}}{T_{b_{or}}}\right)\right]^{-1} \frac{(T_{b_i} - T_w)}{(T_{b_i} - T_w)_r} \cdot \exp\left[4 \cdot St_r \frac{L_r}{D_{h_r}}\left(1 - \frac{St}{St_r}\frac{L/L_r}{D_h/D_{h_r}}\right)\right] \tag{9}$$

$$N_P = \frac{P}{P_r} = \frac{\left(\dot{m}/\dot{m}_r\right)^2 Po/Po_r \cdot L/L_r}{\Omega/\Omega_r D_h/D_{h_r}} = P' \tag{10}$$

the subscript r denotes quantities in the reference configuration, i is for 'inlet', o for 'outlet' and b for 'bulk'; the superscript $'$ is used for the ratio of a quantity to its value in the reference configuration. The heat flux is $\dot{Q}$, $\dot{m}$ is the mass flowrate and St is the Stanton number $St = Nu/(Re \cdot Pr)$. Since the fluid properties are assumed to be constant, $St' = Nu'/Re'$.

3. Results and discussion

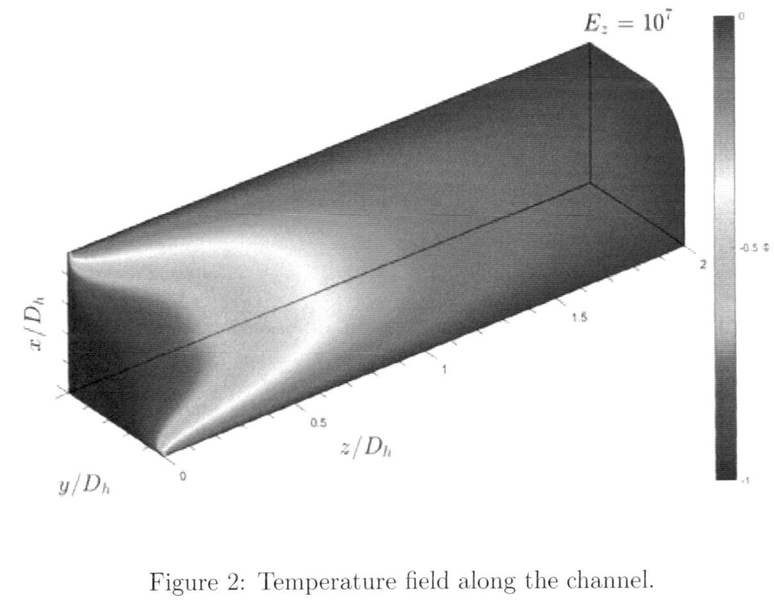

Figure 2: Temperature field along the channel.

The code was verified with the analytical and numerical results in [31] and [25] respectively, and validated with the experimental results of [10], and was then employed to determine the

Journal of Physics: Conference Series **2509** (2023) 012013 doi:10.1088/1742-6596/2509/1/012013

velocity and temperature distribution under the same conditions as in [27] for four values of the aspect ratio, $\beta = 1, 0.50, 0.25, 0.10$ and a non-dimensional radius of curvature $0 \leq \mathrm{R_c} \leq 1$; since one further geometrical constraint must be added to apply PEC optimisation [25], it was chosen to keep the hydraulic diameter unchanged, $\mathrm{D'_h} = 1$. To compute $\mathrm{N_S}$, $\mathrm{N_T}$ and $\mathrm{N_P}$,the relative entropy generation in the reference condition was chosen as $\Phi_\mathrm{r} = 0.1$, a reasonable estimate, see e.g. [23] for details, and $\mathrm{T_{b_{i_r}}}/\mathrm{T_{b_{o_r}}} = 0.95$, the latter corresponding to an inlet temperature $\mathrm{T_{b_{i_r}}} = 298\,\mathrm{K}$ and a temperature increase of $15\,\mathrm{K}$ between inlet and outlet.

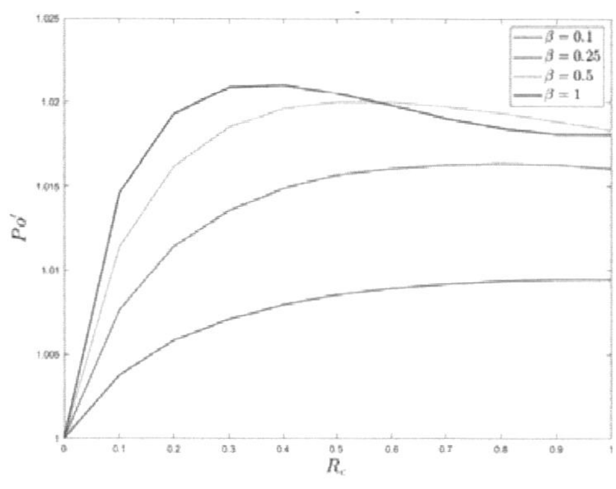

Figure 3: Relative Poiseuille number.

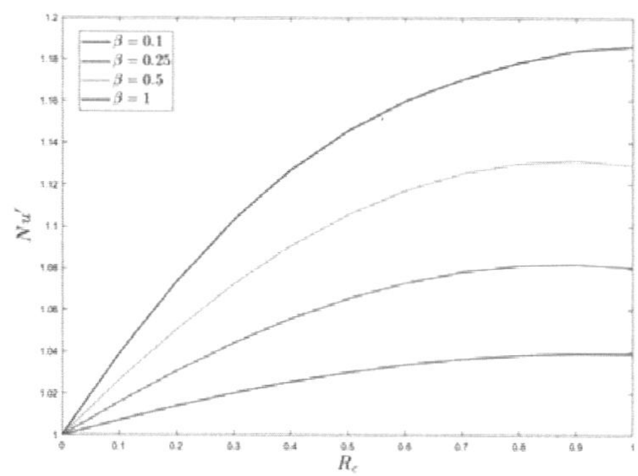

Figure 4: Relative Nusselt number.

A typical temperature profile is shown in Fig. 2, for one quarter of the channel (thanks to the problem's symmetries); the thermal entry region is easily discernible.

From the velocity and temperature distribution, Po and Nu were computed according to [27, 28] and expressed as third order polynomials in the β and $\mathrm{R_c}$, whose coefficients are given in [27]. The relative values of the two quantities Po$'$ and Nu$'$, which are needed for PEC optimisation are easily obtained from the above data and are plotted in Fig. 3 and Fig. 4 respectively. It can be noticed that Nu$'$ increases with $\mathrm{R_c}$ and reaches a rather flat maximum for the intermediate values of β at high radii of curvature, Po$'$ has a maximum for $\beta = 1, 0.5, 0.25$ and a relative minimum for $\beta = 1$, $\mathrm{R_c} = 0.9$. This testifies that smoothing the corners increases the transport phenomena, with benefits to heat transfer and penalties for pumping power, yet, for larger aspect ratios, where the curved stretches take up a significant length of the sides, maximum increase in heat transfer does not necessarily correspond to the maximum frictional losses.

3.1. Performance evaluation criteria and related entropy production: FG1a

Due to space constraints, only maximisation of heat transfer under fixed mass flowrate, channel length and temperature difference between wall and bulk fluid at the inlet is shown and discussed. This PEC is known as FG1a. Under these conditions, the objective function for $\dot{\mathrm{Q}}'$ becomes:

$$\dot{Q}' = \frac{1 - \exp\left(- 4St_r \frac{L_r}{D_h} \left(\frac{St'}{Dh'} \right) \right)}{1 - \exp\left(- 4St_r \frac{L_r}{D_{h_r}} \right)} \tag{11}$$

with $\mathrm{Pr} = 7$, $\mathrm{L_r} = 1\,\mathrm{cm}$, $\mathrm{D_{h_r}} = 3\,\mu\mathrm{m}$ and, as mentioned, $\mathrm{D'_h} = 1$. The trend of $\dot{\mathrm{Q}}'$ is shown in Fig. 5 and is similar to that of Nu$'$.

The pumping power under the FG1a constraints is closely related to M_z, in fact

$$P' = \sqrt{M_z'} = Po'/\Omega' \tag{12}$$

Since M_z is closely related to the external electric field, which is responsible for fluid motion in EOF, it is the trend of M_z' which is reported in Fig. 6. Again, pumping power increases with the radius of curvature, and the change is steeper, the larger the aspect ratio.

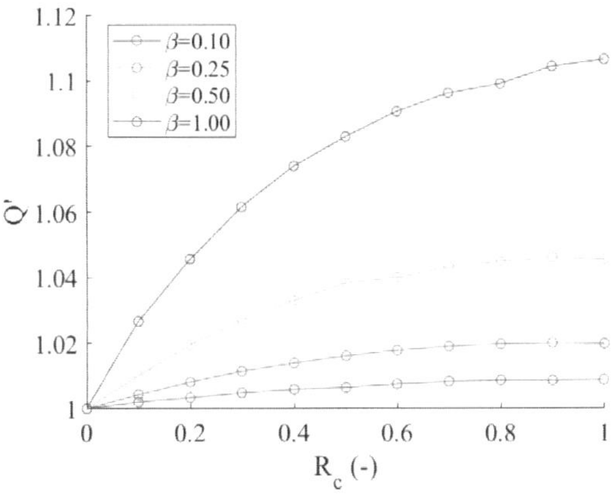

Figure 5: Relative heat flux, criterion FG1a.

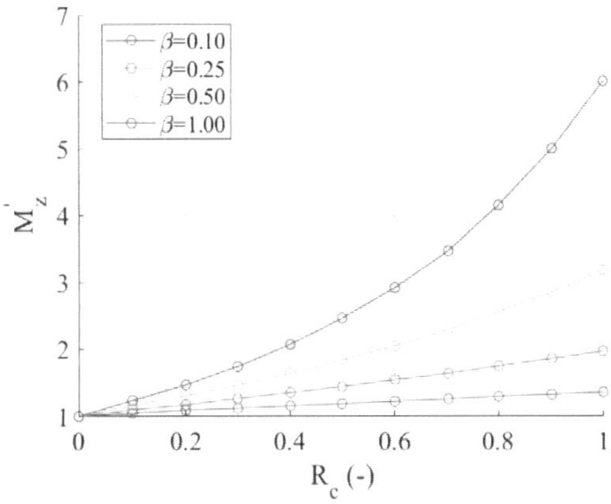

Figure 6: M_z', criterion FG1a.

If the PECs alone are considered, the trends of heat transfer and pumping power are conflicting, so it is not clear which configuration would yield the best results; second-law analysis can supply some further insight. The results for N_T and N_P are plotted in Figs. 7 and 8, and the resulting entropy generation number, N_S in Fig. 9.

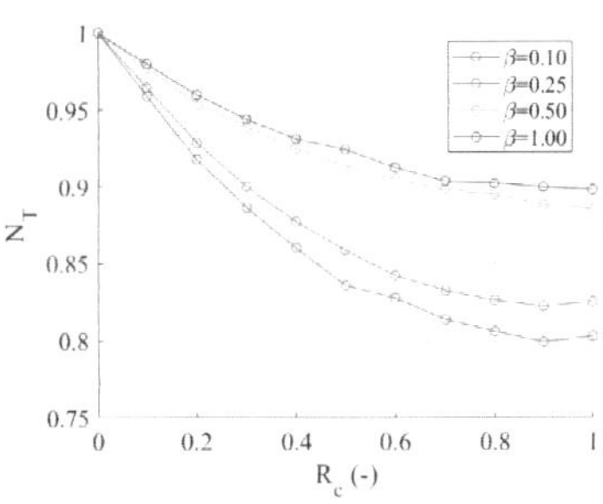

Figure 7: Entropy number for temperature, criterion FG1a.

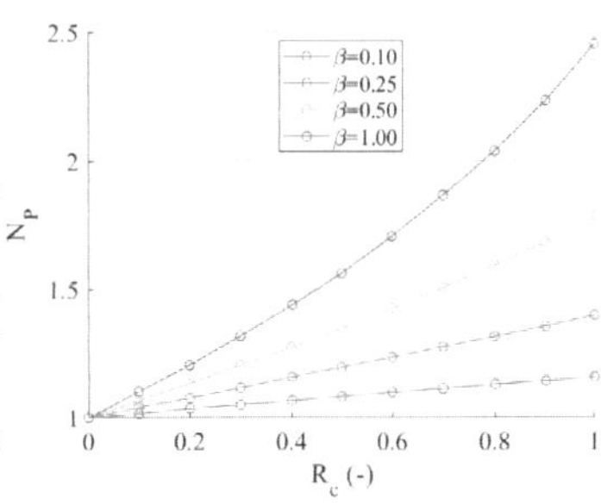

Figure 8: Entropy number for pressure, criterion FG1a.

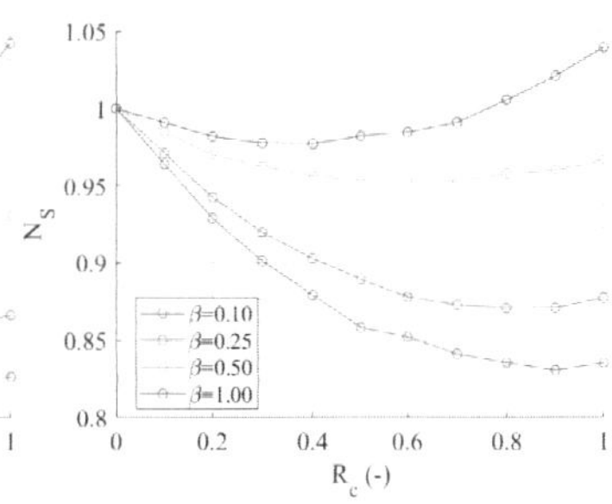

Figure 9: Entropy generation number, criterion FG1a.

It can be seen clearly that entropy generations due to heat transfer and frictional losses have opposing trends: entropy generation is always lower for sections with smoothed corners, as compared to sharp corners, and the smaller aspect ratios also exhibit a minimum, whilst the entropy generation due to pumping losses is always increasing with the radius of curvature and with the aspect ratio. The total entropy generation, which is represented by N_S, exhibits a minimum for all aspect ratios considered, at increasing values of the radius of curvature, the larger β is. Also, for $\beta = 1$ N_S exceeds unity, i.e. is larger than in the reference configuration, for $\beta \geq 0.8$, and is about $N_S \approx 0.97$ at its minimum, hinting at almost no changes in the entropy generated (but with an increase in the heat flux of over 10% at large R_c, see Fig. 5). When the aspect ratio is low, on the contrary, a decrease of almost 20% in entropy production can be achieved, yet a minimal increase at best in the heat transfer can be expected.

4. Conclusions

In this work, the Graetz problem is solved for the electro-osmotic flow in a microchannel with uniform wall temperature, with a rectangular cross-section and rounded corners. Velocity and temperature fields have been used to compute the Poiseuille, Nusselt and the entropy generation numbers for the evaluation of the microchannel performance according to both PEC and entropy minimisation. It is found that the heat transfer increases with the radius of curvature, with a rather flat maximum for high R_c's, whereas the increase in pumping power, as represented by M_z is much steeper, and in both cases the results increase with the aspect ratio. Entropy generation increases steadily with both the radius of curvature and the aspect ratio when friction losses are involved, whilst the heat transfer has a decreasing entropy production with the increase of the radius of curvature and the decrease of the aspect ratio. The entropy generation number shows that there is a minimum in generated entropy for every aspect ratio, and that the decrease is larger for smaller aspect ratios. The method is of general validity and can therefore be applied to other values of the parameter involved and to other PECs.

Nomenclature

D_h	Hydraulic diameter	m
$\tilde{E}$	Nondimensional electric field	-
k_D	Debye–Hückel parameter	m^{-1}
L	Channel length	m
M_z	Joule heating to heat transfer ratio, Eq.(5)	-
N_P	Contribution of pressure drop to entropy generation number, Eq.(10)	-
N_T	Contribution of heat transfer to entropy generation number, Eq.(9)	-
N_S	Entropy generation number, Eq.(8)	-
P	Pumping power	W
Po	Poiseuille number	-
Pr	Prandtl number	-
$\dot{Q}$	Thermal power	W
R	Nondimensional radial coordinate	-
Re	Reynolds number	-
St	Stanton number	-
T	Temperature	K
U	Nondimensional velocity	-
x	Axial coordinate	m
X	Nondimensional x-coordinate	-
Y	Nondimensional y-coordinate	-
Z	Nondimensional z-coordinate	-

Greek letters		
β	Aspect ratio of the channel	-
γ	Nondimensional smoothing radius	-
$\partial\Omega$	Nondimensional perimeter	-
λ	Thermal conductivity	$\mathrm{W{\cdot}m^{-1}{\cdot}K^{-1}}$
μ	Fluid viscosity	Pa·s
Phi	Ratio of N_P to N_T	
Ψ	Nondimensional electric potential	-
Ψ_0	Nondimensional Stern potential	-
σ	Electric conductivity	$\mathrm{S{\cdot}m^{-1}}$
Θ	Nondimensional temperature	K
Subscripts and superscripts		
b	Bulk	
c	Curvature	
i	Inlet	
o	Outlet	
r	Reference	
th	Thermal	
w	Wall	
x	Component along the x-direction	

References

[1] Han Y, Liu Y and Li Mand Huang J 2012 *Energy Procedia* **14** 148–153
[2] Kirby C 2010 *Micro- and Nanoscale Fluid Mechanics* (Cambridge, UK: CUP)
[3] Ohadi M, Choo K, Dessiatoun S and Cetegen E 2013 *Next Generation Microch. Heat Exch.* (NY: Springer)
[4] Tabeling P 2010 *Introduction to microfluidics* (Oxford, UK: Oxford University Press)
[5] R L Webb 1994 *Principles of Enhanced Heat Transfer* (New York: Wiley)
[6] A Bejan 1982 *Entropy generation through heat and fluid flow* (New York: Wiley)
[7] Bruus H 2010 *Theoretical microfluidics* (Oxford: OUP)
[8] Lorenzini M and Morini G 2011 *Heat Transfer Engineering* **32** 1108–1116
[9] Hossan M, Dutta D, Islam N and Dutta P 2018 *Electrophoresis* **39** 702–731
[10] Al-Rjoub M, Roy A, Ganguli S and Banerjee R 2011 *Int. Journal of Heat and Mass Tr.* **54** 4560–4569
[11] Rice C and Whitehead R 1965 *Journal of Physical Chemistry* **69** 4017–4024
[12] Wang C Y and Chang C C 2011 *Electrophoresis* **32** 1268–1272
[13] Vocale P, Geri M, Morini G and Spiga M 2014 *Journal of Physics: Conference Series* **501**
[14] Geri M, Lorenzini M and Morini G 2012 *International Journal of Thermal Sciences* **55** 114–121
[15] Al-Rjoub M, Roy A, Ganguli S and Banerjee R 2015 *Journal of Electronic Packaging, Trans. ASME* **137**
[16] Al-Rjoub M, Roy A, Ganguli S and Banerjee R 2012 *ASME 2012 3rd International Conference on Micro/Nanoscale Heat and Mass Transfer, MNHMT 2012* 829–833
[17] Pramod K and Sen A 2014 *Journal of Electronic Packaging, Trans. ASME* **136** 03101201–03201214
[18] Shamshiri M, Khazaeli R, Ashrafizaadeh M and Mortazavi S 2012 *J. of Non-Newt. Fl. Mech.* **173-174** 1–12
[19] Shit G, Mondal A, Sinha A and Kundu P 2016 *Physica A: Statist. Mech. and its Applications* **462** 1040–1057
[20] Al-Rjoub M, Roy A, Ganguli S and Banerjee R 2015 *Int. Comm. in Heat and Mass Transfer* **68** 228–235
[21] Maynes D and Webb B 2003 *International Journal of Heat and Mass Transfer* **46** 1359–1369
[22] Maynes D and Webb B 2003 *Journal of Heat Transfer* **125** 889–895
[23] Ray S and Misra D 2010 *International Journal of Thermal Sciences* **49** 1763–1775
[24] Chakraborty S and Ray S 2011 *International Journal of Thermal Sciences* **50** 2522–2535
[25] Lorenzini M 2020 *Thermal Science and Engineering Progress* **19**
[26] Zimparov V 2000 *International Journal of Heat and Mass Transfer* **43** 3137–3155
[27] Suzzi N and Lorenzini M 2021 *Fluids* **6**
[28] Morini G, Lorenzini M, Salvigni S and Spiga M 2006 *International Journal of Thermal Sciences* **45** 955
[29] Lorenzini M, Daprá I and Scarpi G 2017 *Applied Thermal Engineering* **122** 118–125
[30] Suzzi N and Lorenzini M 2019 *International Journal of Thermal Sciences* **145** 10603201–10603210
[31] Y Jian, L Yang and Q Liu 2010 *Phys. Fluids* **22** 042001

An exhaustive search optimization of heat transfer and pressure drop in Kelvin's open cell foams

M Iasiello[1,#], G M Mauro[2], N Bianco[1], A Andreozzi[1], W K S Chiu[3] and V Naso[1]
[1] Dipartimento di Ingegneria Industriale - Università degli Studi di Napoli Federico II Piazzale Tecchio 80 - 80125 – Napoli – Italy
[2] Dipartimento di Ingegneria - Università degli Studi del Sannio – Piazza Roma 21 - 82100 – Benevento – Italy
[3] Department of Mechanical Engineering – University of Connecticut – 191 Auditorium Road – 06269 – Storrs (CT) - USA

[#] Email: marcello.iasiello@unina.it

Abstract. Thanks to their high effective thermal conductivity, specific surface area, and tortuosity, open-cell foams are well-known for their capability to enhance heat transfer in applications such as heat exchangers and volumetric solar air receivers. In the very recent years, innovative manufacturing techniques, including 3D designing and printing, have been looked very helpful to find foam morphologies that allow to maximize heat transfer and minimize pressure drop. Optimal foam structures can be obtained by means of pore-scale simulations, employing an exhaustive search with a bearable computational effort. A multi-objective optimization of convective heat transfer and pressure drop in Kelvin's foams with air is presented in this paper. A pore-scale numerical model, with a uniform heat flux at the solid/fluid interface, is used to predict the interfacial convective heat transfer coefficient, h_c, and pressure drop, Δp, in the foam. The cell size, porosity, cell anisotropy stretching factor, as well as the inlet velocity and the direction of the air, are assumed as the design variables for the optimization model, while the interfacial convective heat transfer coefficient and pressure drop are chosen as the objective functions to be maximized and minimized, respectively. Pareto fronts ranging from h = 110 W/m^2 K and Δp = 0.77 Pa to h_c = 460 W/m^2 K and Δp = 51 Pa are predicted, within which the optimum point for the chosen foam morphology, air velocity and direction can be selected, according to the chosen criterion.

Nomenclature

a, b	Length of the ellipsoid semi-axes (m)	u_{in}	Inlet velocity (m/s)
a^*	Cell anisotropy factor (-)	V_c	Cell volume (m^3)
C_p	Heat capacity (J/kg K)	x, y, z	Cartesian coordinates (m)
d_c	Cell size (m)	**Greek symbols**	
h_c	Interfacial heat transfer coefficient (W/m^2 K)	ε	Porosity (-)
k	Thermal conductivity (W/m K)	μ	Dynamic viscosity (Pa s)
p	Pressure (Pa)	ρ	Density (kg/m^3)
q_{in}	Heat flux (W/m^2)	**Subscripts**	
T	Temperature (K)	f	Fluid
$\mathbf{u}$	Velocity vector (m/s)	s	Solid

1. Introduction

Thanks to their high effective thermal conductivity, heat exchange area to volume ratio, and their tortuosity, metal and ceramic open cell foams can be usefully employed in applications such as heat exchangers, heat sinks for thermal management, thermal energy storage systems, volumetric solar air receivers [1].

Closure coefficients, such as permeability, inertial factor, interfacial heat transfer coefficient, must be known a priori, in order to predict thermal performances with a local thermal non-equilibrium volume-averaged approach [2]. The closure coefficients can be evaluated by means of single-phase pore-scale simulations in order to save computational power and costs. Tomography-based geometries with computations based on realistic geometries can also be employed [3]. However, it has been shown through the years that computationally-based geometries can be generated, such avoiding costly tomographic scans of realistic samples. Plateau's [4], Kelvin's [5] and Weaire-Phelan's [6] geometries have been used instead of tomographic scans; Kelvin's foams well replace tomographic foams [7] whereas Kelvin's and Weaire-Phelan's foams provide similar convective heat transfer coefficient and pressure drop especially for higher porosities [8]. Besides, the accuracy of computationally-generated models is strongly affected by how detailed is the foam's design. For instance, since the strut shape is influenced by the foam porosity [9, 10], the accuracy of predictions by the Kelvin's foam model improves if the strut shape is better modeled [10].

Recent additive manufacturing techniques allow to obtain structures at small scales. Therefore, this is really useful when one wants to customly design foams starting from well-known geometrical models, in order to improve their characteristics. Pusterla et al. [11] showed that the radius and tapering of the strut as well as the cell inclination angle of Kelvin's foams strongly affected heat convection and pressure drop. For instance, higher ligament tapering and cell inclination decreased pressure drop and increased heat transfer performances. Kumar and Topin [12] analyzed the impact of the strut shape on pressure drop in Kelvin-based foams with constant ligament cross section. By varying porosities and Reynolds number in the 0.60 - 0.95 and 10^{-6} – 3000 ranges, respectively, they derived correlations for Ergun-like pressure drop closure coefficients accounting for different strut shapes. Ambrosio et al. [10] investigated Kelvin's foams with different strut shapes and analyzed heat convection and pressure drop; they concluded that circular shapes maximized heat transfer and minimized pressure drop. Iasiello et al. [13] induced anisotropy in Kelvin's foams, without varying the cell; they found that convection heat transfer was penalized when the cell was stretched in the flow direction. Elliptical struts with either the same circumference or the same flow cross-section area as those of circular struts were investigated by Moon et al. [14], in order to evaluate the effects on the wake region of the struts. The authors found a 44% reduction in the pressure drop and a slight reduction in the heat transfer performances, because of the reduced free stream velocity around the strut. Different geometrical configurations of cylindrical and elliptical struts as well as four struts orientations in Kelvin's foams were studied by Calati et al. [15]. The performance evaluation criteria (PEC) [16] of foams with elliptical struts were found to be better than those of circular struts, with a 22% maximum difference between them.

In the recent years, mono- and multi-objective optimization techniques have been widely used to find the best mix of design variables which optimizes heat transfer and fluid flow performances, which are widely known to be in contrast, especially with reference to convection and pressure drop. Optimization has been used to investigate heat transfer and pressure drop in helically corrugated tubes [17], heat exchanger tubes with compound insert geometries [18] and plate-fin heat exchangers [19]. With reference to porous materials, optimization has been applied to volumetric solar receivers [20, 21], heat exchangers with foam and nanoparticles [22], finned-foam heat sinks both under forced convection [23] and coupled with phase change materials for thermal management [24]. Very recently, Mauro et al. [25] performed both PEC mono-objective optimization and heat transfer and pressure drop multi-objective optimization in a foam-filled channel with a graded foam, achieving a 1.51 PEC, with a bunch of solutions which maximized heat transfer and pressure drop performances.

The literature survey highlights that customizing pore-scale geometry of a foam allows to increase its heat transfer performances; on the other hand, it is also clear that optimization plays a role in achieving the best heat transfer and pressure drop performances of foam-equipped devices. An exhaustive brute-force search on Kelvin-based foams, in order to evaluate the design variables that maximize the interfacial convective heat transfer coefficient and minimize the pressure drop, is reported in this paper. Inlet velocity, anisotropy elongation ratio, cell size and porosity are assumed as the design variables for both geometrical and CFD models. Governing equations with appropriate boundary conditions are solved numerically; the Pareto front is obtained.

2. Mathematical model

2.1. Geometry and governing equations

The foam geometry is built up starting from the Kelvin's foam model [5], which was proved to predict heat convection and pressure drop in good accordance with Weaire-Phelan's [8] and tomography-based [7] models, with substantial savings in complexity, that make this foam still interesting. The foam generation process, similar to that carried out in [7, 8, 13], is resumed in the following. A so-called wet closed-cell foam is first generated with Surface Evolver [26]: the dry foam is obtained by confining the liquid phase inside both foam struts and junctions at a constrained porosity. Some commands are also performed to reduce the minimal surface energy of the structure; the resulting cell of a 0.90 porosity foam is depicted in the left side of figure 1. The structure is then exported in a different software that builds up a bounding box around it, in order to allow the simulation of the air flow through the foam.

In order to account for the anisotropy effects as an input variable for the foam model, an artificial elongation is given to the foam, as in [13] (see the right side in figure 1). An ellipsoid with equal lengths of the minor semi-axes, b, is assumed to be inscribed inside the foam cell. The volume, V_c, of an ellipsoid having equal axes, i.e., a sphere, is given by the following expression:

$$V_c = \frac{4}{3}\pi\left(\frac{d_c}{2}\right)^3 = \frac{4}{3}\pi\, a\, b^2 \tag{1}$$

with d_c the cell size and a the length of the semi-axis that accounts for anisotropy.

A cell anisotropy factor, a^*, is defined as:

$$a^* = \frac{a}{d_c/2} \tag{2}$$

which is equal to 1 in a spherical ellipsoid and higher than 1 in elongated ellipsoids. From equation (1), the length of the minor axis, b, of cells having the same total volume can be derived:

$$b = \left(\left(\frac{d_c}{2}\right)^3 \Big/ a\right)^{1/2} \tag{3}$$

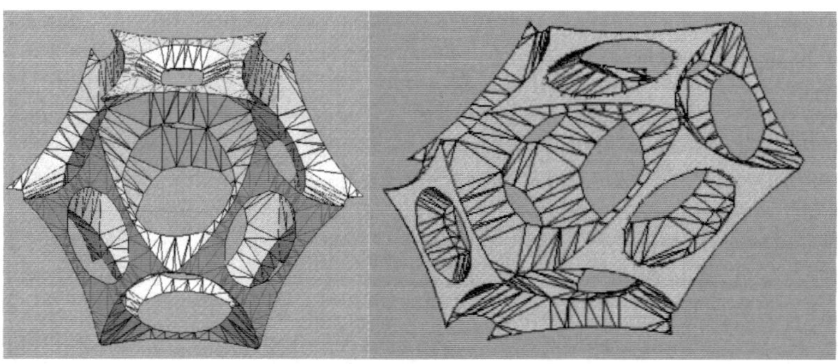

Figure 1. 0.90 porosity Kelvin's foam: isotropic (left); anisotropic with $a^* = 1.30$ (right).

Finally, $d_c/2$ long fictitious regions are assumed both upflow of the entrance section and downflow of the exit section of the foam domain, to help numerical convergence.

Mass, momentum and energy equations for the air flowing through the foam, under the assumptions of incompressible steady-state laminar flow, are:

$$\nabla \cdot \mathbf{u} = 0 \tag{4}$$

$$\rho\, \mathbf{u} \cdot \nabla \mathbf{u} = -\nabla p + \mu\, \nabla^2 \mathbf{u} \tag{5}$$

$$\rho\, C_p\, \mathbf{u} \cdot \nabla T = k\, \nabla^2 T \tag{6}$$

2.2. Boundary conditions and numerical modeling

Boundary conditions for the problem described by equations (1-3), similar to those in [7, 8, 13], are resumed in the following. Air at a 20°C uniform temperature is assumed to enter the domain at different velocities and zero-gauge pressure while outflow boundary conditions are assumed at its exit section. Symmetry conditions for both fluid flow and heat transfer are assumed at the side walls. No-slip condition and a q_{in} = 1000 W/m^2 uniform heat flux from the solid to the fluid phases are assumed at the their interface. The value of the heat flux is such as to make negligible radiation and temperature dependence of the thermo-physical properties.

Governing equations with the appropriate boundary conditions are solved with the finite element method. The boundary layer tetrahedral mesh with about 300,000 elements chosen for computations had been checked with a higher number of elements; the comparison showed no significant differences in the predicted convective heat transfer coefficient and pressure drop. A fully coupled direct PARDISO solver is employed for the governing equations with a 10^{-3} RMS convergence criterion, while crosswind and streamline stabilization are employed together with Lagrangian first-order elements in order to reach the code solution.

Predictions of the model are presented in terms of pressure drop, Δp, computed as the difference between the surface averaged pressure at the inlet section and the pressure at the outlet section, and of interfacial convective heat transfer coefficient, h_c, defined by the expression:

$$h_c = \frac{q_{in}}{T_s - T_f} \tag{7}$$

where T_s is the solid/fluid interface averaged temperature and T_f is the fluid phase volumetric averaged temperature.

2.3. Multi-objective optimization

After setting up the mathematical model, a multi-objective optimization procedure is required to derive optimum values based on two assigned objective functions: i) the convective heat transfer coefficient at the interface to be maximized; ii) the pressure drop to be minimized, which are typically in contrast in convective heat transfer.

The design variables of the optimization problem are resumed in table 1; it points out that not equi-spaced values of the a^* variable are modeled because of the very large computational effort required.

Table 1. Design variables for the optimization problem.

Design variable	Number of design variables						
	1	2	3	4	5	6	7
u_{in} (m/s)	0.55	1.0	1.5				
d_c (mm)	1.0	1.5	2.0	2.5	3.0	3.5	4.0
ε (–)	0.85	0.90	0.95				
direction	x	y	z				
a^* (-)	1.00	1.25	1.30				

Air is assumed to flow in three principal directions, say x, y, z, with x the elongation direction. In anisotropic foams elongated in the x-direction ($a^* > 1.00$), changing the flow direction affects the performances of the foam, as it was pointed out by Iasiello et al. [13] and recalled in the previous section.

Table 1 shows that the required simulations are 567. Since each simulation takes about 5 minutes, a couple of days is necessary to run all simulations. Therefore, the multi-objective optimization problem is solved by performing a brute-force (or exhaustive) search, without invoking any numerical optimization techniques as well as the genetic algorithm employed in [23-25].

After the Pareto front is obtained, an optimum value based on the well-established Utopia optimum criterion is derived, as in [23-25].

3. Results and discussion

Results of the multi-objective optimization procedure are presented in the following. Brute-force search (BFS) solutions for pressure drop as a function of the interfacial heat transfer coefficient, with Pareto front, utopia point and utopia optimum, are reported in figure 2. The heat transfer coefficient objective function varies in the 110 W/m^2 K and 460 W/m^2 K range and the pressure drop objective function varies in the 0.77 Pa - 51 Pa range. This means that maximizing the heat transfer coefficient implies high pressure drops and, on the other hand, minimizing the pressure implies very low heat transfer coefficients. Based on the Pareto front limits, the utopia point is derived; then, the utopia optimum is found as the Pareto front point at the minimum dimensionless distance from the utopia point. It is worth remarking that the utopia optimum, with h_c = 434 W/m^2 K and Δp = 8.6 Pa, is not that far from both the maximum heat transfer coefficient and the minimum pressure drop. Besides, it is also worth underlining the BFS optimization importance, since the Pareto front is far from the majority of the BFS solutions.

Design input variables that allowed to achieve the above reported Pareto front are presented in figures 3-5. Pareto solutions for the air inlet velocity and the cell size are shown in figure 3. As to the inlet velocity, finding fewer solutions for the u_{in} = 1.0 m/s intermediate value means that optimum points are achieved mainly for limiting conditions. On the other hand, all the simulated points point out that the cell size distribution is quite spread. However, one can remark that the majority of solutions are those for the 1.0 mm smallest cell sizes. As it was already reported in [27, 28], such values allow to maximize both the heat transfer coefficient and the pressure drop.

Pareto solutions for the porosity and the flow direction are presented in figure 4. The majority of optimum solutions for the porosity are achieved for its highest simulated value, likely because of the poor effect of porosity on heat transfer coefficients especially if compared to cell sizes [10, 27]; on the other hand, the same strut shape being the same in different Kelvin's foams, pressure drop decreases at increasing porosity [28]. As far as the flow directions are concerned, though optimum values are exhibited in each direction through the Pareto front, there is a dominance of points referred to the fluid

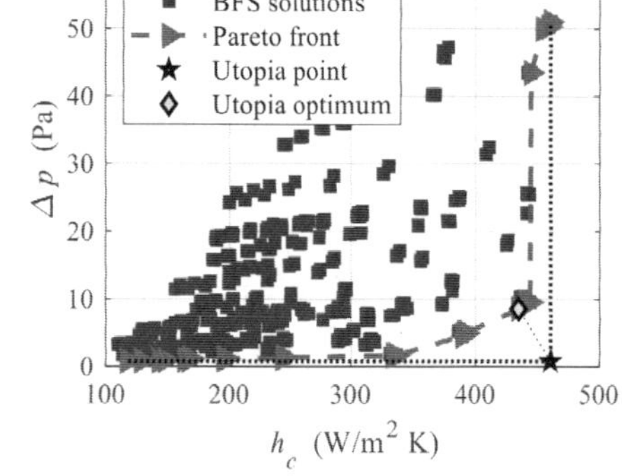

Figure 2. Pressure drop solutions *vs.* interfacial heat transfer coefficient, with Pareto front and Utopia point.

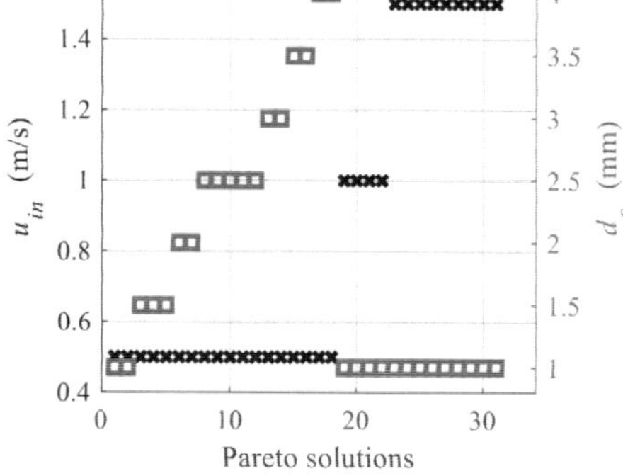

Figure 3. Pareto solutions for the air inlet velocity (left) and the cell size (right).

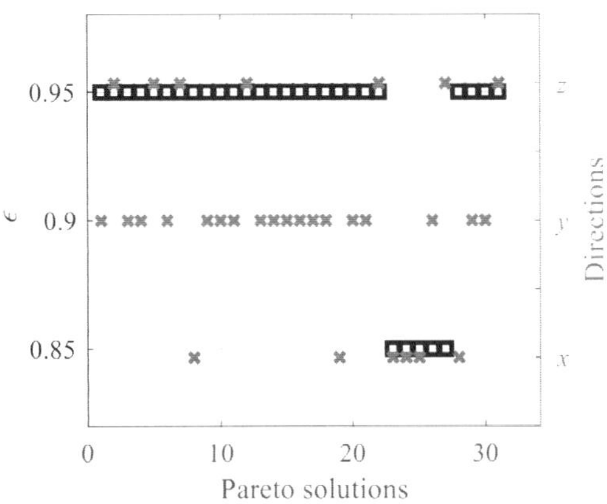

Figure 4. Pareto solutions for the porosity (left) and the flow direction (right).

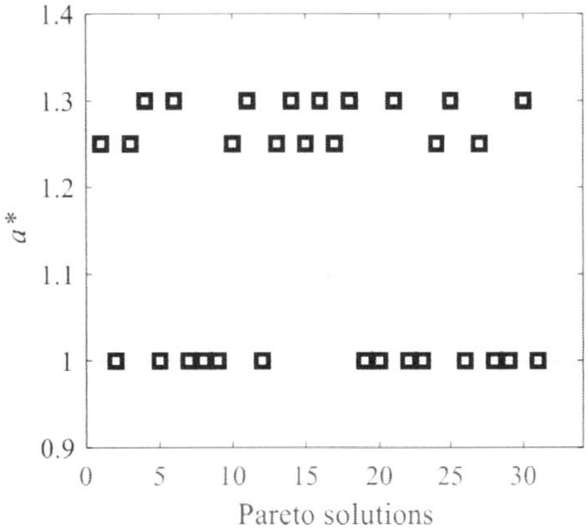

Figure 5. Pareto solutions for the cell anisotropy factor.

flow in the y-direction, which is furtherly discussed in the comments to the next figure.

Pareto solutions for the cell anisotropy factor are reported in figure 5. Anisotropy decreases both heat transfer coefficient and pressure drop per unit length when the foam is stretched in the x direction [13]. The figure also shows that optimum values are achieved for both non-elongated and elongated cells.

Pressure drop and interfacial heat transfer coefficient in the investigated Kelvin's foams are given by the following expressions:

$$\frac{\Delta p}{d_c} = \frac{(1039 - 1002\ \varepsilon)}{d_c^2} \mu\ u_{in} + \frac{0.5138\ \varepsilon^{-5.739}}{d_c} \rho\ u_{in}^2 \tag{8}$$

$$h_c = \frac{\left\{ 1.594 \left[\frac{\rho\ u_{in} (d_c / \varepsilon)}{\mu} \right]^{0.454} \varepsilon^{0.1644} \right\} k}{d_c / \varepsilon} \tag{9}$$

with u_{in} = 0.5 m/s, 1.0 m/s, 1.5 m/s; ε = 0.85, 0.90, 0.95; d_c = 1.0 mm, 1.5 mm, 2.0 mm, 2.5 mm, 3.0 mm, 3.5 mm, 4.0 mm.

Finally, in order to underline the importance of running optimization, the Pareto front from present work and solutions from [28] and [29] are presented in figure 6. The figure shows that most solutions on the Pareto 2D front are non-dominated with respect to those presented in [28] and [29]. This means that optimization plays an important role in finding the best mix of input variables to accomplish a certain objective function.

4. Conclusions

A brute-force search multi-objective optimization of Kelvin's foams pore-scale convective heat transfer and pressure drop is reported. The geometry of the foam was built up by assuming porosity, cell size and anisotropy stretching factor as design variables. Governing equations with appropriate boundary conditions as well as uniform solid to fluid heat flux have been solved with a numerical approach. The brute-force search multi-objective optimization has been carried out with the interfacial heat transfer coefficient and pressure drop as the objective functions and the porosity, cell size, anisotropy factor, flow main direction and air inlet velocity as the design variables.

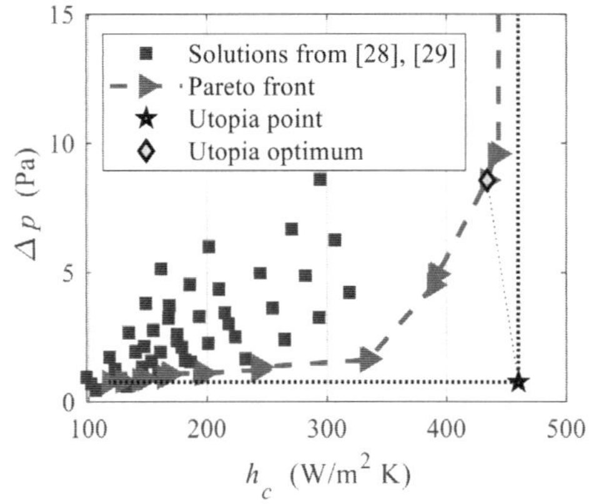

Figure 6. Pareto front from present work and solutions from [28] and [29].

Results showed that the Pareto front varies in the h_c = 110 W/m^2 K - 460 W/m^2 K range and in the Δp = 0.77 Pa - 51 Pa range; where the values of the utopia optimum are h_c = 434 W/m^2 K and Δp = 8.6 Pa. The above values are useful to foams designers depending on whether the heat transfer coefficient or the pressure drop are prioritized. Finally, the comparison of correlations for pressure drop and heat transfer with correlations taken from the literature underlines that running an optimization analysis is very useful in finding the best performances of the system under investigation.

5. References

[1] Bianco N, Contento G, Cunsolo S, Iasiello M, Naso V and Oliviero, M 2017 Heat transfer enhancement in open-cell foams *Annual Review Heat Transfer* **20**

[2] Kaviany M 2012 Principles of heat transfer in porous media (New York: Springer Science & Business Media)

[3] Haussener S, Corey P, Lipiński W, Wyss P and Steinfeld A 2010 Tomography-based heat and mass transfer characterization of reticulate porous ceramics for high-temperature processing *J. Heat Transfer* **132(2)** 023305

[4] Plateau J A F 1873 Statique expérimentale et théorique des liquides soumis aux seules forces moléculaires (Paris: Gauthier-Villars)

[5] Thomson, W. 1887 On the division of space with minimum partitional area *Acta mathematica* **11(1)** 121-134

[6] Weaire D, and Phelan R 1994. A counter-example to Kelvin's conjecture on minimal surfaces. *Philosophical Magazine Letters* **69(2)** 107-110

[7] Iasiello M, Cunsolo S, Oliviero M, Harris W M, Bianco N, Chiu W K, and Naso V 2014 Numerical analysis of heat transfer and pressure drop in metal foams for different morphological models *J. Heat Transfer* **136(11)** 112601

[8] Cunsolo S, Iasiello M, Oliviero M, Bianco N, Chiu W K, and Naso V 2016 Lord Kelvin and Weaire–Phelan foam models: heat transfer and pressure drop *J. Heat Transfer* **138(2)** 022601

[9] Bhattacharya A, Calmidi V V and Mahajan R L 2002 Thermophysical properties of high porosity metal foams *Int. J. Heat Mass Transfer* **45(5)** 1017-1031

[10] Ambrosio G, Bianco N, Chiu W K, Iasiello M, Naso V, and Oliviero M 2016 The effect of open-cell metal foams strut shape on convection heat transfer and pressure drop *Applied Thermal Engineering* **103** 333-343

[11] Pusterla S, Barbato M, Ortona A, and D'Angelo C 2012 Numerical study of cell morphology effects on convective heat transfer in reticulated ceramics *Int. J. Heat Mass Transfer* **55(25-26)** 7902-7910

[12] Kumar P, and Topin F 2014 Micro-structural impact of different strut shapes and porosity on hydraulic properties of Kelvin-like metal foams *Transport in Porous Media* **105(1)** 57-81

[13] Iasiello M, Bianco N, Chiu W K S and Naso V 2017 Anisotropy effects on convective heat transfer and pressure drop in Kelvin's open-cell foams *J. of Physics: Conference Series* **923(1)** 012035
[14] Moon C, Kim H D, and Kim K C 2018 Kelvin-cell-based metal foam heat exchanger with elliptical struts for low energy consumption *Applied Thermal Engineering* **144** 540-550.
[15] Calati M, De Monte E, and Mancin S 2021 Numerical analysis of the effects of the structure shafpe and orientation of Kelvin cell porous structures during air forced convection *Applied Sciences* **11(13)** 6189.
[16] Yilmaz M, Comakli O, Yapici S, and Sara O N 2005 Performance evaluation criteria for heat exchangers based on first law analysis *J. Enhanced Heat Transfer* **12(2)**
[17] Safikhani H, and Eiamsa-ard S 2016 Pareto based multi-objective optimization of turbulent heat transfer flow in helically corrugated tubes *Applied Thermal Engineering* **95** 275-280
[18] Chamoli S, Yu P, and Yu S 2017 Multi-objective shape optimization of a heat exchanger tube fitted with compound inserts *Applied Thermal Engineering* **117** 708-724
[19] Liu C, Bu W, and Xu D 2017 Multi-objective shape optimization of a plate-fin heat exchanger using CFD and multi-objective genetic algorithm *Int. J. Heat Mass Transfer* **111** 65-82
[20] Du S, He Y L, Yang W W, and Liu Z B 2018 Optimization method for the porous volumetric solar receiver coupling genetic algorithm and heat transfer analysis *Int. J. Heat Mass Transfer* **122**, 383-390
[21] Du S, Ren, Q, and He Y L 2017 Optical and radiative properties analysis and optimization study of the gradually-varied volumetric solar receiver *Applied Energy* **207** 27-35
[22] Siavashi M, Bahrami H R T, and Aminian E 2018 Optimization of heat transfer enhancement and pumping power of a heat exchanger tube using nanofluid with gradient and multi-layered porous foams *Applied Thermal Engineering* **138** 465-474
[23] Bianco N, Iasiello M, Mauro G M, and Pagano L 2021 Multi-objective optimization of finned metal foam heat sinks: tradeoff between heat transfer and pressure drop *Applied Thermal Engineering* **182** 116058
[24] Bianco N, Busiello S, Iasiello M, and Mauro G M 2021 Finned heat sinks with phase change materials and metal foams: Pareto optimization to address cost and operation time *Applied Thermal Engineering* **197** 117436
[25] Mauro G M, Iasiello M, Bianco N, Chiu W K, and Naso V 2022 Mono-and multi-objective CFD optimization of graded foam-filled channels *Materials* **15(3)** 968
[26] Brakk, K A 1992 The surface evolver *Experimental mathematics* **1(2)** 141-165
[27] Wu Z, Caliot, C, Flamant G, and Wang Z. 2011 Numerical simulation of convective heat transfer between air flow and ceramic foams to optimise volumetric solar air receiver performances *Int. J. Heat Mass Transfer* **54(7-8)** 1527-1537
[28] Wu Z, Caliot C, Bai F, Flamant G, Wang Z, Zhang J, and Tian C. 2010 Experimental and numerical studies of the pressure drop in ceramic foams for volumetric solar receiver applications *Applied Energy* **87(2)** 504-513
[29] Iasiello M, Cunsolo S, Bianco N, Chiu W K S, and Naso V 2017 Developing thermal flow in open-cell foams *Int. J. Thermal Sciences* **111** 129-137

Acknowledgments
This work was carried out within a Cooperation Agreement between the University of Connecticut (USA) and the Università degli Studi Federico II (Italy) and was supported by Italian Government MIUR Grant No. PRIN-2017F7KZWS.

Stability analysis of dual viscous flows saturating a vertical porous pipe

M Celli[1], A Barletta[1], P V Brandão[1], S da C Hirata[2], M N Ouarzazi[2]
[1]Department of Industrial Engineering, Alma Mater Sturiorum Università di Bologna, Viale Risorgimento 2, 40136, Bologna, Italy
[2]Unité de Mécanique de Lille, URL 7512, Université de Lille, Bd. Paul Langevin, CEDEX, 59655 Villeneuve d'Ascq, France

E-mail: michele.celli3@unibo.it

Abstract. The linear stability analysis of a mixed convection viscous flow in a vertical porous pipe is here investigated. The contribution of viscous heating is assumed to be non negligible. A fully developed flow regime is assumed for the basic state. The local balance equations for this state display dual stationary solutions. The dual branches of stationary solutions are determined numerically. Since the pipe is characterised by an isothermal lateral surface, the viscous heating is the sole cause of the buoyancy force. In order to investigate the stability of the basic dual solutions, small amplitude disturbances with the form of normal modes are superposed to the basic state. The solution of the eigenvalue problem obtained allows one to determine the growth rate associated to both the basic solution branches. The sign of the growth rate determines whether the particular basic solution is stable or unstable.

1. Introduction

In the last decades, several papers have been published reporting stability analyses of buoyant flows in porous media [1]. Most of these studies deal with the Rayleigh–Bénard type of instabilities occurring in porous layers saturated by a fluid and heated from below. The pioneering papers on this topic were authored by Horton, Rogers [2] and Lapwood [3]. Prats [4] expanded these earlier analyses by including a stationary horizontal throughflow across the layer. Barletta et al. [5] further extended this problem by considering the contribution of the viscous heating. The viscous heating becomes relevant when the throughflow is sufficiently intense and/or the fluid is characterised by high Prandtl numbers, namely it is both highly viscous and poorly conductive. When viscous heating is considered, the basic stationary solution of the governing equations may lose its uniqueness. More effort has been devoted to the investigation of dual solutions for clear fluids [6, 7, 8] compared to the convection of viscous fluids saturating porous layers [9]. Very recently, a few papers containing the stability analysis of dual stationary solutions have been published both for clear fluids [10, 11] and fluid saturated porous media [12]. The last paper deals with a two–dimensional flow in a plane vertical channel where the viscous dissipation effect is taken into account.

The paper is focussed on the three–dimensional linear stability analysis of a mixed convection flow in a vertical porous channel with circular cross–section when the contribution of viscous heating is non–negligible. The basic state whose stability is investigated feature dual stationary

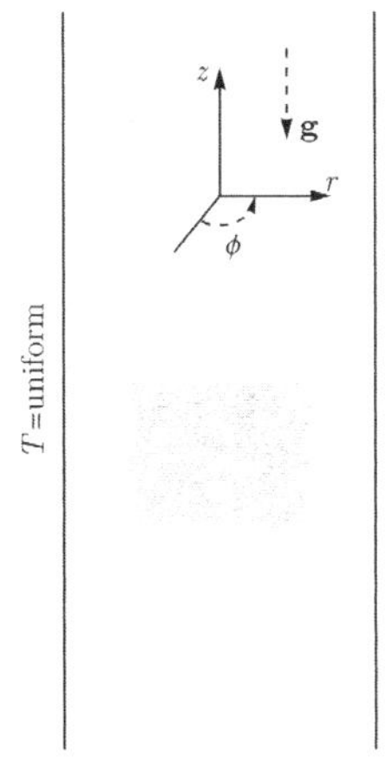

Figure 1. Sketch of the vertical porous layer.

solutions. The analysis aims to obtain the threshold values of the governing parameters for the onset of modal instability.

2. Mathematical model

A mixed convection in a vertical porous pipe characterised by a non–vanishing mass flow rate is assumed. The fluid saturating the porous pipe is in a fully developed flow regime and the mass flow rate depends on the imposed pressure gradient. The pipe has radius r_0 characterised by a uniform and constant temperature. Figure 1 displays a sketch of the porous duct.

The mass, momentum (Darcy's law) and the energy local balance equations employed to investigate the system just described are the following:

$$\begin{aligned}
&\frac{1}{r}\frac{\partial(r\,u)}{\partial r} + \frac{1}{r}\frac{\partial v}{\partial \phi} + \frac{\partial w}{\partial z} = 0,\\
&\frac{\mu}{K}u = -\frac{\partial p}{\partial r},\\
&\frac{\mu}{K}v = -\frac{1}{r}\frac{\partial p}{\partial \phi},\\
&\frac{\mu}{K}w = -\frac{\partial p}{\partial z} + \rho\, g\, \beta\,(T - T_0),\\
&\sigma\frac{\partial T}{\partial t} + u\frac{\partial T}{\partial r} + \frac{v}{r}\frac{\partial T}{\partial \phi} + w\frac{\partial T}{\partial z}\\
&\qquad = \alpha\left(\frac{\partial^2 T}{\partial r^2} + \frac{1}{r}\frac{\partial T}{\partial r} + \frac{1}{r^2}\frac{\partial^2 T}{\partial \phi^2} + \frac{\partial^2 T}{\partial z^2}\right) + \frac{\mu}{\rho\, c\, K}(u^2 + v^2 + w^2),\\
&r = r_0: \quad u = 0,\\
&r = r_0: \quad \frac{\partial T}{\partial z} = 0, \quad \frac{\partial T}{\partial \phi} = 0,
\end{aligned} \tag{1}$$

where (r, ϕ, z) are the cylindrical coordinates, (u, v, w) are the cylindrical components of the velocity vector, μ is the dynamic viscosity of the saturating fluid, K is the permeability of the porous medium, ρ is the density of the fluid evaluated at the reference temperature T_0, g is the modulus of the gravity acceleration vector $\mathbf{g}$, β is the thermal expansion coefficient of the fluid, c the specific heat of the fluid, σ is the ratio between the average volumetric heat capacity of the porous medium and the volumetric heat capacity of the fluid and α is the thermal diffusivity relative to the fluid saturated porous medium. The reference temperature T_0 is the average temperature on the pipe cross–section,

$$T_0 = \frac{1}{\pi r_0^2} \int_0^{r_0} \int_0^{2\pi} T\, r\, \mathrm{d}r \mathrm{d}\phi. \tag{2}$$

The last term of the local energy balance in Eq. (1) identifies the contribution of the viscous dissipation when Darcy's law is considered [1]. By employing the scaling

$$\frac{(r,z)}{r_0} \to (r,z), \quad \frac{\alpha}{\sigma r_0^2} t \to t, \quad \frac{Ge\, r_0}{\alpha}(u,v,w) \to (u,v,w), \quad \frac{T - T_0}{\Delta T} \to T, \quad \frac{Ge\, K}{\mu\, \alpha} p \to p, \tag{3}$$

the dimensionless form of Eq. (1) is obtained,

$$\begin{aligned}
&\frac{1}{r}\frac{\partial (r\,u)}{\partial r} + \frac{1}{r}\frac{\partial v}{\partial \phi} + \frac{\partial w}{\partial z} = 0,\\
&u = -\frac{\partial p}{\partial r},\\
&v = -\frac{1}{r}\frac{\partial p}{\partial \phi},\\
&w = -\frac{\partial p}{\partial z} + T,\\
&\frac{\partial T}{\partial t} + \frac{1}{Ge}\left(u\frac{\partial T}{\partial r} + \frac{v}{r}\frac{\partial T}{\partial \phi} + w\frac{\partial T}{\partial z}\right) = \frac{1}{r}\frac{\partial}{\partial r}\left(r\frac{\partial T}{\partial r}\right) + \frac{1}{r^2}\frac{\partial^2 T}{\partial \phi^2} + \frac{\partial^2 T}{\partial z^2} + u^2 + v^2 + w^2,\\
&r = 1: \quad u = 0,\\
&r = 1: \quad \frac{\partial T}{\partial z} = 0, \quad \frac{\partial T}{\partial \phi} = 0,
\end{aligned} \tag{4}$$

where the Gebhart number Ge and ΔT are defined as

$$Ge = \frac{g\,\beta\, r_0}{c}, \quad \Delta T = \frac{\nu\, \alpha}{Ge^2\, K\, c}. \tag{5}$$

As a consequence of Eq. (2), the dimensionless temperature satisfies the constraint

$$\int_0^1 \int_0^{2\pi} T\, r\, \mathrm{d}r \mathrm{d}\phi = 0. \tag{6}$$

3. Stationary dual solutions

The basic stationary solutions of Eq. (4) are assumed to be fully developed flows. The basic solutions are characterised by zero radial velocity $u = 0$. Among all the possible fully developed flows, stationary axisymmetric solutions with zero angular velocity $v = 0$ are chosen.

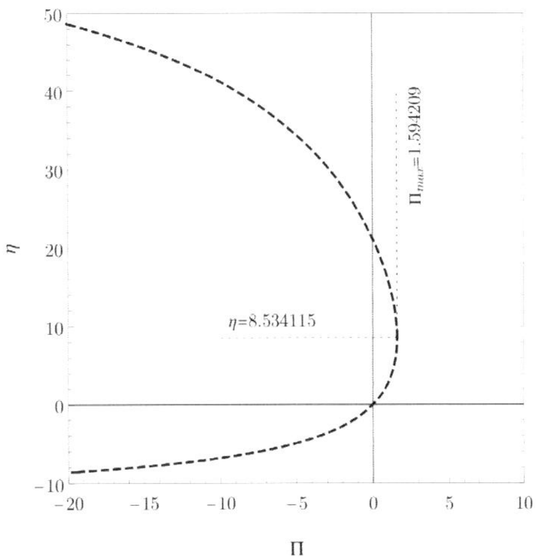

Figure 2. Dual basic solutions: for a given Π we obtains two possible velocity η on the pipe axis, $r \to 0$.

Equations (4) and (6) simplify to

$$\begin{aligned}
&u_b = 0, \\
&v_b = 0, \\
&w_b = T_b + \Pi, \\
&\frac{1}{r}\frac{\mathrm{d}}{\mathrm{d}r}\left(r\frac{\mathrm{d}w_b}{\mathrm{d}r}\right) + w_b^2 = 0, \\
&r = 0: \qquad \frac{\mathrm{d}w_b}{\mathrm{d}r} = 0, \\
&\int_0^1 w_b\, r \mathrm{d}r = \Pi,
\end{aligned} \tag{7}$$

where the subscript b denotes the basic state and the boundary condition $\mathrm{d}w_b/\mathrm{d}r = 0$ at $r = 0$ can be employed because the flow is axisymmetric. One can note that the assumption $u = v = 0$ yields, from Eq. (4), that p depends only on z. From Eqs. (6) and (7), it can be inferred that $\mathrm{d}^2 p/\mathrm{d}z^2$ is zero, so that one can define the dimensionless constant $\Pi = -\mathrm{d}p/\mathrm{d}z$.
The basic state is obtained by solving the initial value problem

$$\begin{aligned}
&\frac{1}{r}\frac{\mathrm{d}}{\mathrm{d}r}\left(r\frac{\mathrm{d}w_b}{\mathrm{d}r}\right) + w_b^2 = 0, \\
&r = \varepsilon: \qquad w_b = \eta, \qquad \frac{\mathrm{d}w_b}{\mathrm{d}r} = 0,
\end{aligned} \tag{8}$$

where η is the unknown velocity at $r = 0$ and ε is a small cutoff $\varepsilon \ll 1$, which is employed to overcome the numerical issue introduced by the singular behaviour of the ordinary differential equation (11) at $r = 0$. For any given value of the axial velocity η one may obtain a single value of Π by employing the last of equations (7), namely

$$\int_0^1 w_b\, r \mathrm{d}r = \Pi. \tag{9}$$

Table 1. Values of $\eta(\Pi_{max})$ and Π_{max} as functions of ε.

ε	$\eta(\Pi_{max})$	Π_{max}
10^{-1}	8.761257	1.672008
10^{-2}	8.529247	1.595070
10^{-3}	8.533982	1.594218
10^{-4}	8.534112	1.594209
10^{-6}	8.534115	1.594209
10^{-8}	8.534115	1.594209

The basic stationary solutions are displayed in Fig. 2: by fixing the basic imposed pressure gradient Π, two possible values of velocity at the pipe axis, $r = 0$, are obtained. One can conclude that a single value of the pressure gradient may produce more than one velocity profile. The curve in Fig. 2 displays a maximum value of Π,

$$\Pi_{max} = 1.594209, \qquad \eta(\Pi_{max}) = 8.534115. \tag{10}$$

The basic state solutions are sensitive to the values of ε employed. In Table 1 the values of $\eta(\Pi_{max})$ and Π_{max} as functions on ε are reported. The values are obtained by imposing 7 effective digits of accuracy. One can conclude that a reasonable value of ε to be employed for the computations is 10^{-4}.

4. Linear stability analysis

The stability of the dual basic solutions are here tested by superposing small amplitude disturbances to the basic state, namely

$$\mathbf{u} = \mathbf{u}_b(r) + \varepsilon \mathbf{U}(r,\phi,z,t), \quad p = -\Pi\, z + \varepsilon P(r,\phi,z,t), \quad T = T_b(r) + \varepsilon\Theta(r,\phi,z,t). \tag{11}$$

One may substitute Eq. (11) in Eq. (4) and then retain only terms $O(\varepsilon)$ to linearise the equations obtained, namely

$$\begin{aligned}
&\frac{1}{r}\frac{\partial(r\,U)}{\partial r} + \frac{1}{r}\frac{\partial V}{\partial \phi} + \frac{\partial W}{\partial z} = 0,\\
&U = -\frac{\partial P}{\partial r},\\
&V = -\frac{1}{r}\frac{\partial P}{\partial \phi},\\
&W = -\frac{\partial P}{\partial z} + \Theta,\\
&\frac{\partial \Theta}{\partial t} + \frac{1}{Ge}\left(U\frac{\mathrm{d}w_b}{\mathrm{d}r} + w_b\frac{\partial \Theta}{\partial z}\right) = \frac{1}{r}\frac{\partial}{\partial r}\left(r\frac{\partial \Theta}{\partial r}\right) + \frac{1}{r^2}\frac{\partial^2\Theta}{\partial \phi^2} + \frac{\partial^2 \Theta}{\partial z^2} + 2w_b\,W,\\
&r = 1: \quad U = 0,\\
&r = 1: \quad \frac{\partial \Theta}{\partial z} = 0, \quad \frac{\partial \Theta}{\partial \phi} = 0.
\end{aligned} \tag{12}$$

By employing the scaling

$$z\,Ge \to z, \qquad \frac{1}{Ge^2}P \to P, \tag{13}$$

Eq. (12) can be rewritten as follows:

$$\begin{aligned}
&\frac{1}{r}\frac{\partial(r\,U)}{\partial r}+\frac{1}{r}\frac{\partial V}{\partial \phi}+Ge\frac{\partial W}{\partial z}=0,\\
&U=-Ge^2\frac{\partial P}{\partial r},\\
&V=-\frac{Ge^2}{r}\frac{\partial P}{\partial \phi},\\
&W=-Ge^3\frac{\partial P}{\partial z}+\Theta,\\
&\frac{\partial \Theta}{\partial t}+\frac{1}{Ge}\left(U\frac{\mathrm{d}w_b}{\mathrm{d}r}+Ge\,w_b\frac{\partial \Theta}{\partial z}\right)=\frac{1}{r}\frac{\partial}{\partial r}\left(r\frac{\partial \Theta}{\partial r}\right)+\frac{1}{r^2}\frac{\partial^2 \Theta}{\partial \phi^2}+Ge^2\frac{\partial^2\Theta}{\partial z^2}+2w_b\,W,\\
&r=1:\quad U=0,\\
&r=1:\quad \frac{\partial \Theta}{\partial z}=0,\quad \frac{\partial \Theta}{\partial \phi}=0.
\end{aligned}\tag{14}$$

The value of the Gebhart number is typically much less than 1 (for instance water at 20°C flowing inside a 10 cm radius pipe yields $Ge \approx 10^{-7}$). For the limiting case $Ge \to 0$, the governing equations (14) can be drastically simplified to

$$\begin{aligned}
&U=V=0,\\
&W=\Theta,\\
&\frac{\partial \Theta}{\partial t}+w_b\frac{\partial \Theta}{\partial z}=\frac{1}{r}\frac{\partial}{\partial r}\left(r\frac{\partial \Theta}{\partial r}\right)+\frac{1}{r^2}\frac{\partial^2 \Theta}{\partial \phi^2}+2w_b\,W,\\
&r=1:\quad \frac{\partial \Theta}{\partial z}=0,\quad \frac{\partial \Theta}{\partial \phi}=0,
\end{aligned}\tag{15}$$

where the mass balance equation is not reported since it is identically satisfied. Equation (15) can be simplified by eliminating the presence of the temperature to obtain

$$\begin{aligned}
&\frac{\partial W}{\partial t}+w_b\frac{\partial W}{\partial z}=\frac{1}{r}\frac{\partial}{\partial r}\left(r\frac{\partial W}{\partial r}\right)+\frac{1}{r^2}\frac{\partial^2 W}{\partial \phi^2}+2w_b\,W,\\
&r=1:\quad \frac{\partial W}{\partial z}=0,\quad \frac{\partial W}{\partial \phi}=0.
\end{aligned}\tag{16}$$

By assuming the disturbances to have the form of normal modes,

$$W(r,t)=h(r)\cos(m\,\phi)e^{\lambda\,t}e^{i\,k\,z},\tag{17}$$

one obtains

$$\begin{aligned}
&\frac{1}{r}\frac{\mathrm{d}}{\mathrm{d}r}\left(r\frac{\mathrm{d}h}{\mathrm{d}r}\right)+\left(2\,w_b-\lambda-\frac{m^2}{r^2}-i\,k\,w_b\right)h=0,\\
&r=1:\quad h=0.
\end{aligned}\tag{18}$$

In Eq. (17) h is a complex eigenfunction, m is a non negative integer while $\lambda=\lambda_r-i\omega$, λ_r is the growth/damping rate, ω is the angular frequency and k is the wavenumber.

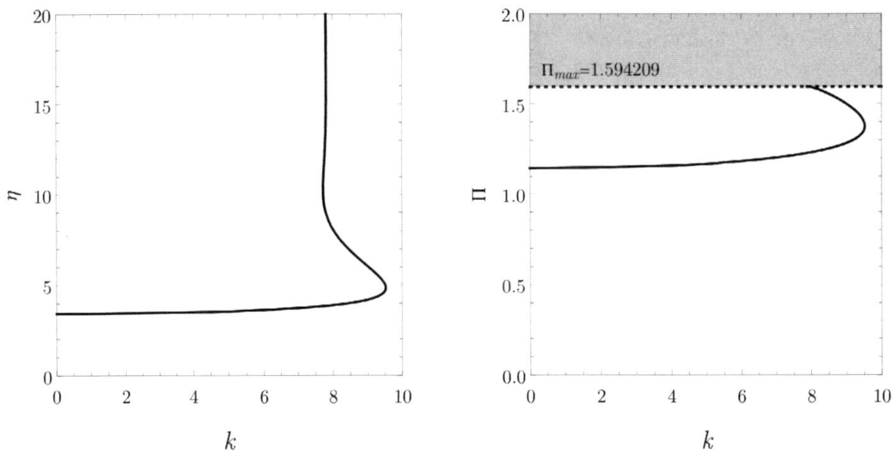

Figure 3. Neutral stability curve for $m = 0$: $\eta(k)$ on the left hand frame and $\Pi(k)$ on the right hand frame.

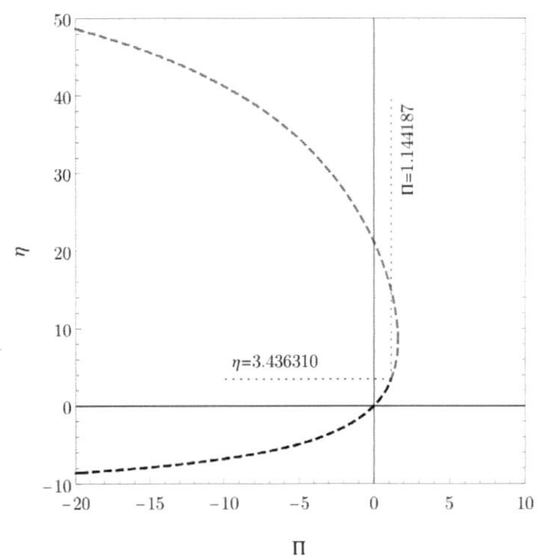

Figure 4. Possibly unstable basic solutions, red dashed curve, against the stable basic solutions, black dashed curve.

5. Discussion of the Results

The solutions of Eq. (18) are reported in Fig. 3. This figure displays the curves $\eta(k)$ and $\Pi(k)$ obtained by fixing the growth rates $\lambda_r = 0$ and for the modes defined by $m = 0$. For the case $m = 1$ and $m = 2$, no solutions charaterised by $\lambda_r = 0$ are obtained and the growth rates λ_r are always negative. One can conclude that, except for the axisymmetric modes characterised by $m = 0$, the other modes are stable.

The left and the right frame of Fig. 3 report the same neutral stability; they are reported by employing either η or Π. For the interpretation of Fig. 3 is necessary to look at Fig. 4 where the basic solutions that can possibly become unstable are identified by a red dashed line.

By comparing Fig. 3 and Fig. 4 one can conclude that the basic solutions characterised by $\eta < 3.436310$ cannot become unstable. These solutions are, in fact, endowed with negative growth rates. It is worth noting that right–hand frame of Fig. 3 displays a gray area for $\Pi > \Pi_{max}$. One has to disregard this area since the parameter Π shows a maximum, $\Pi_{max} = 1.594209$, in its range of variation.

6. Conclusions

The stability analysis presented in this paper aims to investigate the role of viscous dissipation for mixed convection flows in a fluid saturated vertical porous pipe. The problem presents dual stationary solutions whose linear stability is studied. The main conclusions one can draw from this analysis are the following:

- The basic stationary solutions are dual: for a given value of the imposed pressure gradient, two different possible velocity profiles exist;
- There exists a maximum value of basic pressure gradient beyond which no basic stationary solution is obtained;
- Part of the basic stationary solutions are always linearly stable, which belong to the lower branch of the dual flows;
- Only the axisymmetric disturbances can trigger the instability.

Acknowledgments

The authors acknowledge financial support from Grant No. PRIN 2017F7KZWS provided by the Italian Ministero dell'Istruzione, dell'Università e della Ricerca.

References

[1] Nield D A and Bejan A 2006 *Convection in porous media* 5th ed. Springer
[2] Horton C and Rogers F 1945 Convection currents in a porous medium *Journal of Applied Physics* **16**: 367–70
[3] Lapwood E 1948 Convection of a fluid in a porous medium *Mathematical Proceedings of the Cambridge Philosophical Society* **44**: 508–21
[4] Prats M 1966 The effect of horizontal fluid flow on thermally induced convection currents in porous mediums *Journal of Geophysical Research* **71**: 4835–38
[5] Barletta A, Celli M and Rees D A S 2009 The onset of convection in a porous layer induced by viscous dissipation: A linear stability analysis *International Journal of Heat and Mass Transfer* **52**: 337–44
[6] Barletta A, Magyari E and Keller B 2005 Dual mixed convection flows in a vertical channel *International Journal of Heat and Mass Transfer* **48**: 4835–45
[7] Barletta A, Lazzari S and Magyari E 2008 Buoyant Poiseuille–Couette flow with viscous dissipation in a vertical channel *Zeitschrift für angewandte Mathematik und Physik* **59**: 1039–1056
[8] Miklavčič M and Wang C Y 2011 Completely passive natural convection *Zeitschrift für angewandte Mathematik und Physik* **91**: 601–06
[9] Barletta A, Magyari E, Pop I and Storesletten L 2007 Mixed convection with viscous dissipation in a vertical channel filled with a porous medium *Acta Mechanica* **194**: 123–40
[10] Miklavčič M 2015 Stability analysis of some fully developed mixed convection flows in a vertical channel *Zeitschrift für angewandte Mathematik und Physik* **95**: 982–86
[11] Barletta A and Miklavčič M 2016 On fully developed mixed convection with viscous dissipation in a vertical channel and its stability *Zeitschrift für angewandte Mathematik und Physik* **96**: 1457–66
[12] Barletta A and Miklavčič M 2017 Instability of fully developed mixed convection with viscous dissipation in a vertical porous channel *Transport in Porous Media* **117**: 337–47

2509 (2023) 012016 doi:10.1088/1742-6596/2509/1/012016

Boundary element method for contactless estimation of spatially varying internal heat transfer coefficient in circular pipes

Alessandra Aimi[1], Fabio Bozzoli[2], Luca Cattani[2], Chiara Guardasoni[1]

[1]Department of Mathematical, Physical and Computer Sciences, University of Parma, Parco Area delle Scienze 53/A I-43124 Parma, Italy
[2]Department of Engineering and Architecture, University of Parma, Parco Area delle Scienze 181/A I-43124 Parma, Italy

luca.cattani1@unipr.it

Abstract. In the present study it is proposed a solution approach of the Inverse Heat Conduction Problem (IHCP) intended at assessing the local heat transfer coefficient on the inner wall of a tube, under a forced convection problem. The estimation method is established on the Boundary Element Method (BEM) combined with the Truncated Singular Value Decomposition (TSVD) methodology, employed to manage the ill-conditioned nature of the problem. The numerical results of the direct problem, built by the BEM, are firstly validated, and consequently adopted as synthetic data inputs to resolve the IHCP and corroborate the whole assessment procedure. This approach is also tested with experimental data about forced convection problem in coiled pipes. In these geometries, the convective heat transfer coefficient changes considerably along the wall periphery and for this reason, it constitutes a perfect example to test the ability of the presented method to infer a spatially varying internal heat transfer coefficient distribution.

1. Introduction

A very tricky issue in several engineering applications originates from the demand of determining what occurs inside a heat transfer device by measuring a physical quantity from outside. This could be extremely valuable when it is not feasible to implement invasive experimental techniques to examine the inside of the device, for mechanical or security purposes, as for example in nuclear or in any internal flow applications where the inner walls can't be directly investigated. A powerful answer for this remarkable task can be found in the establishment of the Inverse Heat Conduction Problem (IHCP) in the wall of the apparatus, by measuring, conceivably adopting contactless experimental techniques, the outer wall temperature. Between the various potential applications of this method, the estimation of the heat flux or the convective heat transfer coefficient distribution (h_{int}) along the inner surface of a tube, using the external surface temperature distribution as input data, has demonstrated to be noteworthy within the experimental research on heat transfer. Concerning IHCPs, many solution techniques have been proposed under both the parameter estimation and function estimation approach [1] using both numerical and analytical models. Between the numerical methods, a decidedly appealing technique for solving problems where the boundary is of principal relevance, as for instance in wave scattering by bounded obstacles in unbounded domains [2], where high precision is demanded, is the Boundary Element Method (BEM) [3,4]. BEM shows some benefits over the more classical domain approaches, such as the dimensionality reduction since there is no need for discretizing the domain under consideration, the necessity of only the boundary data as inputs, the limited meshing effort, the smaller dimension of linear system matrices, the consequent lower computational cost, and the higher achievable accuracy. Though, this method does not totally overcome all the complications embedded in the IHCP; particularly it does not resolve the difficulties due to the ill-conditioned nature of the problem. To overcome this critical task, unavoidably embedded in the inverse problem, several methods established on the processing of the experimental measurements have been proposed and validated in literature [1]. Between the most employed procedure, the Truncated Singular Value Decomposition (TSVD) has shown robust regularization performance to solve IHCP in steady [5] and transient regimes [6]. The stabilizing effect of this technique is based on the removal of the highest modes of the input signal [7]. In the present work the solution methodology, developed for a 2D model, is established on the BEM combined with the TSVD method. The numerical results of the direct problem, built by the BEM model, are firstly validated, and then adopted as synthetic data input to resolve the IHCP and corroborate the whole assessment procedure. This approach is also tested with experimental data about forced convection problem in coiled pipes. In these geometries, the h_{int} changes considerably along the wall periphery, due to the flow pattern that develops because of curvature and the demand of estimating its distribution frequently then occurs in industrial applications [8]. For this reason, it constitutes an ideal case to test the ability of the presented procedure to infer a spatially varying internal heat transfer coefficient distribution.

2. Problem's definition

The aim of present research is to produce and validate a technique to assess the local convective heat transfer coefficient h_{int} on the inner wall of a pipe's cross section, in a forced convection problem. The problem under test is outlined in Fig. 1 where the cross section of the tube is displayed: the fluid flows inside, while the outer wall of the tube is exposed to a uniform environmental temperature T_{env} and a uniform heat power q_g is generated in the solid wall. The test section is modeled as a 2D solid domain because the temperature gradient along the tube's axis is assumed to be negligible with respect to the one along the angular direction. The pipe is characterized by an internal radius r_{int}, an external radius r_{ext} and by a thermal conductivity k.

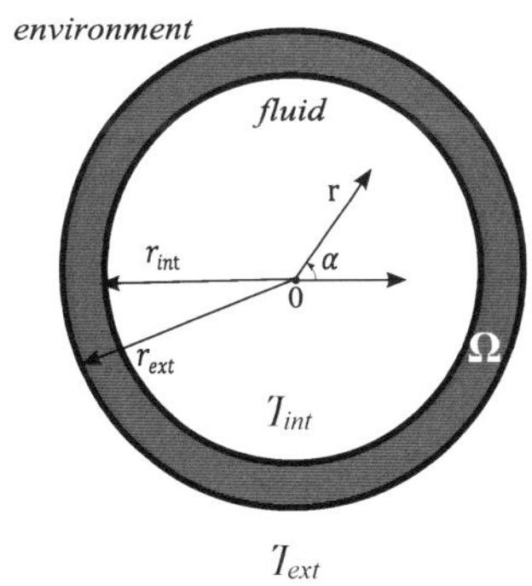

Fig. 1: Geometrical domain with coordinate system

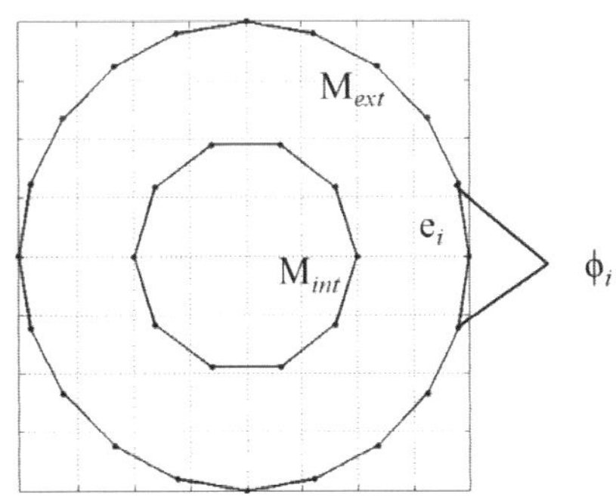

Fig.2: Mesh configuration.

2.1. Direct model

The local steady state energy balance equation in the wall domain is formulated in cartesian coordinates as [2,3]:

$$\nabla_\zeta^2 T(\zeta) = -\frac{q_g}{k} \qquad \zeta := (\zeta_x, \zeta_y) \in \Omega \tag{1}$$

The subsequent two boundary conditions completed the energy balance equation:

$$-k\frac{\partial T}{\partial \mathbf{n}}\bigg|_{\Gamma_{int}}(\zeta) = h_{int}(\zeta)(T(\zeta) - T_b) \quad \zeta \in \Gamma_{int} \tag{2}$$

$$-k\frac{\partial T}{\partial n}\bigg|_{\Gamma_{ext}}(\zeta) = h_{ext}(\zeta)(T(\zeta) - T_{env}) \quad \zeta \in \Gamma_{ext} \tag{3}$$

In the direct problem formulation of Eqs. (1–3) the geometrical parameters, wall thermal conductivity k, heat flux on the inner wall tube q, heat transfer coefficient with the environment h_{ext}, bulk T_b and ambient temperature T_{env} are considered known, while the temperature distribution on the outer surface of the pipe is determined. Considering:

$$\boldsymbol{\xi} := (\xi_x, \xi_y),\ \mathbf{r} := \sqrt{(\zeta_x - \xi_x)^2 + (\zeta_y - \xi_y)^2} \tag{4}$$

and

$$\mathrm{G}(\boldsymbol{\zeta}; \xi) = -\frac{1}{2\pi}\log(\mathbf{r}) \tag{5}$$

the fundamental solution, that satisfies the equation:

$$\Delta_\xi \mathrm{G}(\boldsymbol{\zeta}; \xi) = -\delta(\boldsymbol{\zeta} - \xi) \tag{6}$$

and the boundary condition:

$$\mathrm{G}(0; \xi) = 0 \quad \text{in } \partial\mathrm{B} = \{\xi \in \mathbb{R} \text{ such that } \|\boldsymbol{\xi}\|^2 = 1\} \tag{7}$$

Then the integral representation formula for the solution T of Eqs. (1–3) is:

$$T(\boldsymbol{\zeta}) = \int_{\Gamma_{int}\cup\Gamma_{ext}} G(\boldsymbol{\zeta};\boldsymbol{\xi})\nabla_\xi T(\boldsymbol{\xi})\cdot\mathbf{n}\,d\xi - \int_{\Gamma_{int}\cup\Gamma_{ext}} T(\boldsymbol{\xi})\nabla_\xi G(\boldsymbol{\zeta};\boldsymbol{\xi})\cdot\mathbf{n}\,d\xi + \int_\Omega \frac{q_g}{k}G(\boldsymbol{\zeta};\boldsymbol{\xi})\,d\xi \quad \forall\boldsymbol{\zeta}\in\Omega \tag{8}$$

In fact, multiplying Eq. (6) by T, integrating over Ω, applying twice the Green Theorem, and applying boundary conditions (Eqs. 2-3), we obtain:

$$\begin{aligned} T(\boldsymbol{\zeta}) = &-\int_{\Gamma_{int}} \frac{h_{int}(\xi)}{k}(T(\xi) - T_b)\,G(\boldsymbol{\zeta};\boldsymbol{\xi})\,d(\xi) - \int_{\Gamma_{ext}} \frac{h_{ext}(\xi)}{k}(T(\xi) - T_{env})\,G(\boldsymbol{\zeta};\boldsymbol{\xi})\,d(\xi) \\ &- \int_{\Gamma_{int}\cup\Gamma_{ext}} T(\boldsymbol{\xi})\nabla_\xi G(\boldsymbol{\zeta};\boldsymbol{\xi})\cdot\mathbf{n}\,d\xi + \int_\Omega \frac{q_g}{k}G(\boldsymbol{\zeta};\boldsymbol{\xi})\,d\xi \end{aligned} \tag{9}$$

The solution T in the entire domain as expressed in Eq. (9) can be obtained after having determined the temperature at the boundary. Hence, the boundary integral equations to be solved are:

$$\begin{aligned} &\frac{1}{2}T(\boldsymbol{\zeta}) + \int_{\Gamma_{int}} \frac{h_{int}(\xi)}{k}T_{int}(\xi)G(\boldsymbol{\zeta};\xi)\,d\xi + \int_{\Gamma_{ext}} \frac{h_{ext}(\xi)}{k}T_{ext}(\xi)G(\boldsymbol{\zeta};\xi)\,d\xi + \int_{\Gamma_{int}\cup\Gamma_{ext}} T(\xi)\nabla_\xi G(\boldsymbol{\zeta};\xi)\cdot\boldsymbol{n}\,d\xi \\ &= \int_{\Gamma_{int}} \frac{h_{int}(\xi)}{k}T_b\,G(\boldsymbol{\zeta};\xi)\,d\xi + \int_{\Gamma_{ext}} \frac{h_{ext}(\xi)}{k}T_{env}\,G(\boldsymbol{\zeta};\xi)\,d\xi + \int_\Omega \frac{q_g}{k}G(\boldsymbol{\zeta};\xi)\,d\xi \\ &\forall\zeta \in \Gamma_{\mathrm{int}} \bigcup \Gamma_{ext} \end{aligned} \tag{10}$$

Really, we are not interested to the domain temperature but only to the temperature T_{ext} trace of the domain temperature T on the exterior boundary Γ_{ext}. Assuming a discretization of the boundary in $M = M_{int} + M_{ext}$ straight elements $e_1, \ldots, e_M$ (as in Fig. 2) and a class of finite dimensional functional subspaces V_M such that $\dim(V_M) = M$ [2,3]:

$$V_M = \left\{\varphi \in H^1\left(\Gamma_{int}\bigcup\Gamma_{ext}\right) | \varphi|_{e_i} \in \mathbb{P}^1(e_i)\ i = 1, \ldots, M\right\} \tag{11}$$

a basis for this functional space is constituted by the classical piecewise linear hat functions ϕ_i for $i = 1, \ldots, M$ (see Fig. 2) that can be used either as test functions and for the approximation of T on the boundary. Given:

$$T\zeta \approx \sum_{i=1}^{M} T_i\,\phi_i(\boldsymbol{\zeta}) \text{ and } T_{int}(\boldsymbol{\zeta}) \approx \sum_{i=1}^{M_{int}} T_i\,\phi_i(\boldsymbol{\zeta}) \text{ and } T_{ext}(\boldsymbol{\zeta}) \approx \sum_{i=M_{int}+1}^{M} T_i\,\phi_i(\boldsymbol{\zeta}) \tag{12}$$

it is then possible to define a linear system of equations that can be formulated in compact form:

$$\begin{pmatrix} a_{1\,1} & \cdots & a_{1\,M_{int}} & a_{1\,M_{int+1}} & \cdots & a_{1\,M} \\ \vdots & \ddots & \vdots & \vdots & \ddots & \vdots \\ a_{M_{int}\,1} & \cdots & a_{M_{int}\,M_{int}} & a_{M_{int}\,M_{int+1}} & \cdots & a_{M_{int}\,M} \\ a_{M_{int+1}\,1} & \cdots & a_{M_{int+1}\,M_{int}} & a_{M_{int+1}\,M_{int+1}} & \cdots & a_{M_{int+1}\,M} \\ \vdots & \ddots & \vdots & \vdots & \ddots & \vdots \\ a_{M\,1} & \cdots & a_{M\,M_{int}} & a_{M\,M_{int+1}} & \cdots & a_{M\,M} \end{pmatrix} \begin{pmatrix} T_1 \\ \vdots \\ T_{M_{int}} \\ T_{M_{int}+1} \\ \vdots \\ T_M \end{pmatrix} = \begin{pmatrix} b_1 \\ \vdots \\ b_{M_{int}} \\ b_{M_{int}+1} \\ \vdots \\ b_M \end{pmatrix} \tag{13}$$

with:

$$\begin{aligned}
a_{ji} &= \int_{\Gamma_{int}} \frac{1}{2}\phi_i(\zeta)\phi_j(\zeta)d\zeta + \int_{\Gamma_{int}} \phi_j(\zeta)\int_{\Gamma_{int}} \frac{h_{int}(\xi)}{k}\phi_i(\xi)G(\zeta;\xi)\,d\xi d\zeta + \int_{\Gamma_{int}} \phi_j(\zeta)\int_{\Gamma_{int}} \phi_i(\xi)\nabla_\xi G(\zeta;\xi) \\
&\quad \cdot \boldsymbol{n}d\xi d\zeta \quad j,i = 1,\dots,M_{int} \\
a_{ji} &= \int_{\Gamma_{int}} \phi_j(\zeta)\int_{\Gamma_{ext}} \frac{h_{ext}(\xi)}{k}\phi_i(\xi)G(\zeta;\xi)\,d\xi d\zeta + \int_{\Gamma_{int}} \phi_j(\zeta)\int_{\Gamma_{int}} \phi_i(\xi)\nabla_\xi G(\zeta;\xi)\cdot \boldsymbol{n}d\xi d\zeta \quad j = 1,\dots,M_{int} \\
&\qquad i = M_{int}+1,\dots,M \\
a_{ji} &= \int_{\Gamma_{int}} \phi_j(\zeta)\int_{\Gamma_{int}} \frac{h_{int}(\xi)}{k}\phi_i(\xi)G(\zeta;\xi)\,d\xi d\zeta + \int_{\Gamma_{ext}} \phi_j(\zeta)\int_{\Gamma_{int}} \phi_i(\xi)\nabla_\xi G(\zeta;\xi)\cdot \boldsymbol{n}d\xi d\zeta \\
&\qquad j = M_{int}+1,\dots,M \quad i = 1,\dots,M_{int} \\
a_{ji} &= \int_{\Gamma_{ext}} \frac{1}{2}\phi_i(\zeta)\phi_j(\zeta)d\zeta + \int_{\Gamma_{ext}} \phi_j(\zeta)\int_{\Gamma_{ext}} \frac{h_{ext}(\xi)}{k}\phi_i(\xi)G(\zeta;\xi)\,d\xi d\zeta + \int_{\Gamma_{ext}} \phi_j(\zeta)\int_{\Gamma_{int}} \phi_i(\xi)\nabla_\xi G(\zeta;\xi) \\
&\quad \cdot \boldsymbol{n}d\xi d\zeta \quad j,i = M_{int}+1,\dots,M \\
b_j &= \int_{\Gamma_{int}} \phi_j(\zeta)\left(\int_{\Gamma_{int}} \frac{h_{int}(\xi)}{k}T_b\, G(\zeta;\xi)\,d\xi d\zeta + \int_{\Gamma_{ext}} \frac{h_{ext}(\xi)}{k}T_{env}\, G(\zeta;\xi)\,d\xi d\zeta + \int_{\Omega} \frac{q_g}{k}\, G(\zeta;\xi)\,d\xi d\zeta\right) \\
&\qquad j = 1,\dots,M_{int} \\
b_j &= \int_{\Gamma_{ext}} \phi_j(\zeta)\left(\int_{\Gamma_{int}} \frac{h_{int}(\xi)}{k}T_b\, G(\zeta;\xi)\,d\xi d\zeta + \int_{\Gamma_{ext}} \frac{h_{ext}(\xi)}{k}T_{env}\, G(\zeta;\xi)\,d\xi d\zeta + \int_{\Omega} \frac{q_g}{k}\, G(\zeta;\xi)\,d\xi d\zeta\right) \\
&\qquad j = M_{int}+1,\dots,M
\end{aligned} \tag{14}$$

2.2. Validation of the model

To corroborate the presented technique established on the BEM, the direct solution of Eq. (10) has been contrasted with the results of a numerical simulation under the finite element method (FEM) performed within Comsol Multiphysics environment where Eqs (1-3) were numerically solved by adopting a non-structured triangular mesh of about 3000 elements.

As input data in the direct formulation the analytical distribution of the local heat transfer coefficient distribution reported below has been adopted:

$$h_{int} = 1000 \cdot (sin(\alpha/\pi)^2 + 1) \tag{15}$$

This distribution constitutes a challenging test for the proposed method since it is characterized by a significant variation along the periphery and by a discontinuity in the first derivative at the end points (see Fig.3). The excellent correspondence between the outer surface temperature obtained by the here proposed model and the FEM simulation (see Fig. 4) validates both modeling approaches. Moreover, the BEM presents some advantages respect to FEM since it allows to reduce the dimensions of the domain and therefore to limit the computational cost. To assess the efficacy of the presented approach for various number of mesh elements and different number of Gauss quadrature nodes a comparative study on the direct problem was performed by computing the restored temperature distribution obtained with different combinations of the two parameters. This analysis is reported in Fig. 5 and it shows that 8 Gauss quadrature nodes and 64 mesh boundary elements allow to obtain satisfactory results.

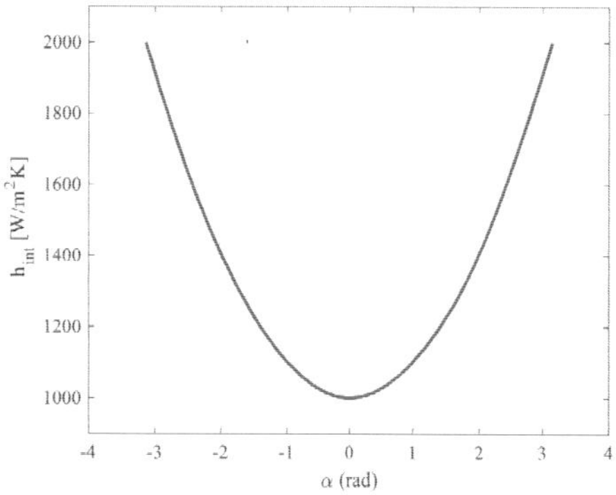

Fig. 3: h_{int} distribution adopted for the numerical validation of the estimation procedure.

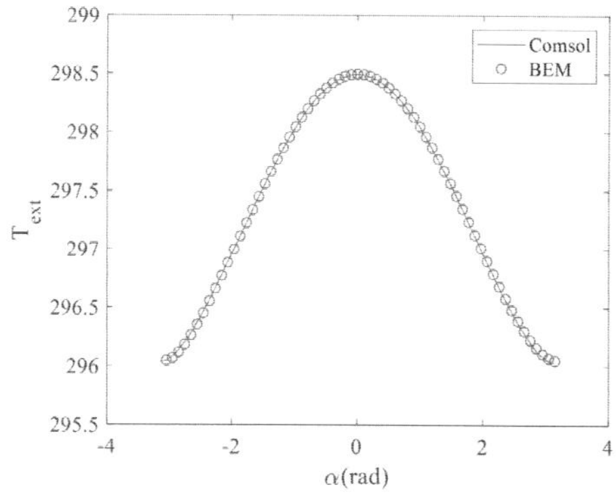

Fig. 4: Comparison between the external wall temperature derived by the BEM and Comsol models

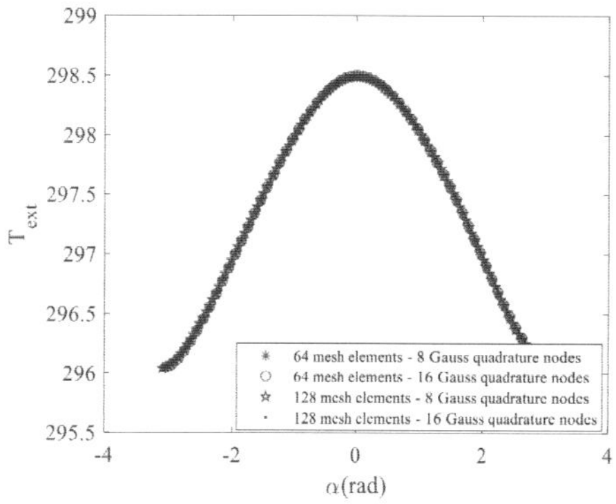

Fig.5: Temperature distribution for different Gauss quadrature nodes and boundary mesh elements.

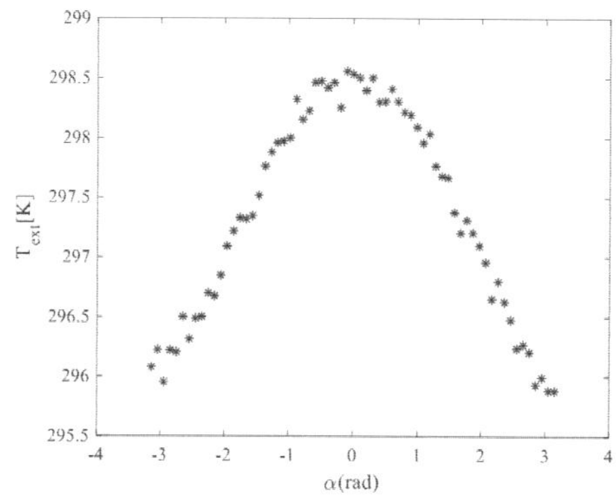

Fig.6: Synthetic noisy temperature on the outer wall: case of noise level $\sigma = 0.1$ K.

3. Inverse problem

Under the inverse approach, the convective heat flux distribution on the inner surface of the tube is considered the unknown quantity in Eqs. (1-3) while the outer wall temperature distribution represents the starting data, and all the other parameters are known.

Starting from the same integral representation formula (Eq. (8)) it is possible to write [2,3]:

$$\begin{aligned}\frac{1}{2}T(\zeta) + \int_{\Gamma_{int}} \frac{h_{int}(\xi)}{k}(T_{int}(\xi) - T_b)G(\zeta;\xi)\, d\xi + \int_{\Gamma_{int}} T_{int}(\xi)\nabla_\xi G(\zeta;\xi)\cdot \mathbf{n}\, d\xi \\ = \int_{\Gamma_{ext}} \frac{h_{ext}(\xi)}{k}(T_{env} - T_{ext}(\xi))\, G(\zeta;\xi)\, d\xi - \int_{\Gamma_{ext}} T_{ext}(\xi)\nabla_\xi G(\zeta;\xi)\cdot \mathbf{n}\, d\xi \\ + \int_\Omega \frac{q_g}{k} G(\zeta;\xi)\, d\xi \qquad \forall\zeta \in \Gamma_{int} \bigcup \Gamma_{ext}\end{aligned} \tag{16}$$

that is the same boundary integral equation written in Eq. (10) but with different unknown functions and a different known term denoted by:

$$f(x) := \int_{\Gamma_{ext}} \frac{h_{ext}(\xi)}{k}(T_{env} - T_{ext}(\xi))\, G(\zeta;\xi)\, d\xi - \int_{\Gamma_{ext}} T_{ext}(\xi)\nabla_\xi G(\zeta;\xi)\cdot \mathbf{n}\, d\xi + \int_\Omega \frac{q_g}{k} G(\zeta;\xi)\, d\xi \tag{17}$$

In this case, using basis functions already introduced before, unknowns can be written as:

$$T_{int}(\zeta) \approx \sum_{i=1}^{M_{int}} T_i\,\phi_i(\zeta) \text{ and } \frac{\partial T_{int}}{\partial \mathbf{n}}(\zeta) = -\frac{h_{int}(\xi)}{K}(T_{int} - T_b)(\zeta) \approx -\sum_{i=1}^{M_{int}} q_i\,\phi_i(\zeta) \tag{18}$$

It is possible to write a linear system of equations that can be formulated in compact form:

$$\begin{pmatrix} a_{1\,1} & \cdots & a_{1\,M_{int}} & a_{1\,M_{int+1}} & \cdots & a_{1\,2M_{int}} \\ \vdots & \ddots & \vdots & \vdots & \ddots & \vdots \\ a_{M_{int}\,1} & \cdots & a_{M_{int}\,M_{int}} & a_{M_{int}\,M_{int+1}} & \cdots & a_{M_{int}\,2M_{int}} \\ a_{M_{int+1}\,1} & \cdots & a_{M_{int+1}\,M_{int}} & a_{M_{int+1}\,M_{int+1}} & \cdots & a_{M_{int+1}\,2M_{int}} \\ \vdots & \ddots & \vdots & \vdots & \ddots & \vdots \\ a_{2M_{int}\,1} & \cdots & a_{2M_{int}\,M_{int}} & a_{2M_{int}\,M_{int+1}} & \cdots & a_{2M_{int}\,2M_{int}} \end{pmatrix} \begin{pmatrix} T_1 \\ \vdots \\ T_{M_{int}} \\ q_1 \\ \vdots \\ q_{M_{int}} \end{pmatrix} = \begin{pmatrix} b_1 \\ \vdots \\ b_{M_{int}} \\ b_{M_{int+1}} \\ \vdots \\ b_{2M_{int}} \end{pmatrix} \tag{19}$$

$$\begin{aligned} a_{ji} &= \int_{\Gamma_{int}} \frac{1}{2}\phi_i(\zeta)\phi_j(\zeta)d\zeta + \int_{\Gamma_{int}} \phi_j(\zeta) \int_{\Gamma_{int}} \phi_i(\xi)\nabla_\xi G(\zeta;\xi)\cdot \mathbf{n}\,d\xi d\zeta \quad j,i = 1,\dots,M_{int} \\ a_{ji} &= -\int_{\Gamma_{int}} \phi_j(\zeta)\int_{\Gamma_{int}} \phi_i(\xi)G(\zeta;\xi)\,d\xi d\zeta \quad j = 1,\dots,M_{int} \;\; i = M_{int}+1,\dots,2M_{int} \\ a_{ji} &= \int_{\Gamma_{ext}} \phi_j(\zeta)\int_{\Gamma_{int}} \phi_i(\xi)\nabla_\xi G(\zeta;\xi)\cdot \mathbf{n}\,d\xi d\zeta \quad j = M_{int}+1,\dots,2M_{int} \;\; i = 1,\dots,M_{int} \\ a_{ji} &= -\int_{\Gamma_{int}} \phi_j(\zeta)\int_{\Gamma_{int}} \phi_i(\xi)G(\zeta;\xi)\,d\xi d\zeta \quad j,i = M_{int}+1,\dots,2M_{int} \\ b_j &= \int_{\Gamma_{int}} \phi_j(\zeta)f(\zeta)d\zeta \quad j = 1,\dots,M_{int} \\ b_j &= \int_{\Gamma_{ext}} \phi_j(\zeta)\left(f(\zeta) - \frac{1}{2}T_{ext}(\zeta)\right)d\zeta \quad j = M_{int}+1,\dots,M \end{aligned} \tag{20}$$

Nevertheless, this method does not totally overcome all the complications intrinsic in the IHCP. Particularly it does not bypass the difficulties related to the ill-conditioned nature of the inverse problem. To overcome this critical task, intrinsic in the inverse problem approach, TSVD has been adopted. To select the proper truncation parameter, the discrepancy principle was applied [9]. Once the heat flux distribution is determined on the inner wall of the pipe it is possible to compute the h_{int} by dividing the flux by the difference between the internal wall temperature and the fluid bulk temperature.

4. Results

4.1. Numerical validation

One of the aims of the present work was to validate the proposed h_{int} estimation procedure. By adopting a known distribution of h_{int} and by numerically resolving the direct problem stated by Eqs. (1–3) with the BEM model, a synthetic temperature distribution on the outer surface was achieved with 8 Gauss quadrature nodes and 64 mesh boundary elements. The physical and geometrical parameters adopted in the present analysis relate to a stainless-steel pipe AISI 304 characterised by an inner and outer diameter of 14 and 16 mm, respectively. Regarding the h_{int}, the analytical distribution defined in Eq. (15) and shown in Fig. 5 was used. Then the synthetic temperature distribution on the outer wall, intentionally spoiled by random noise (see for instance Fig. 6), was adopted as the starting data of the inverse problem implemented within Matlab environment. A white noise characterized by a standard deviation σ in the range 0.01 - 1 K was studied. From the solution of the IHCP the distribution of the local heat flux on the inner wall and then the corresponding h_{int} were determined. The estimated distribution of the latter on the internal wall surface was then compared to the data employed to produce the starting data (Eq. (14) and Fig.5). The comparison is reported in Fig. 7 for the representative case of $\sigma = 0.1$ K while in Fig. 8 the residuals between exact and estimated temperature on the outer surface of the pipe are shown.

It is possible to see that the estimation procedure almost perfectly restored the h_{int} distribution apart from the borders where the presence of a discontinuity point in the first derivative prevents the exact reconstruction of the local heat transfer coefficient. Moreover, the random distribution of the residuals shown in Fig. 8 confirms the appropriateness of the proposed strategy. To quantify the efficacy of the proposed method at various noise levels a residual study was carried out by plotting the estimation error on h_{int}, defined as follows:

$$E_h = \frac{\|(h_{int})_{exact} - (h_{int})_{restored}\|_2}{\|(h_{int})_{exact}\|_2} \tag{21}$$

versus the standard deviation present in the input data.

The very limited estimation errors shown in Fig.9 demonstrate the efficient performance of the proposed procedure. Moreover, it must be considered that in the case of $\sigma = 1$ K the E_h is below 10 % even if the noise to signal ratio is very high $(\sigma/(T_{max}-T_{min}) \approx 0.4)$.

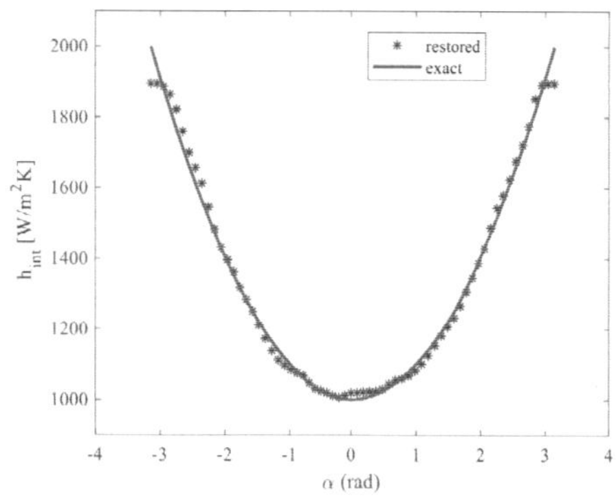

Fig. 7: Exact and restored h_{int} distribution for noise level $\sigma = 0.1$ K.

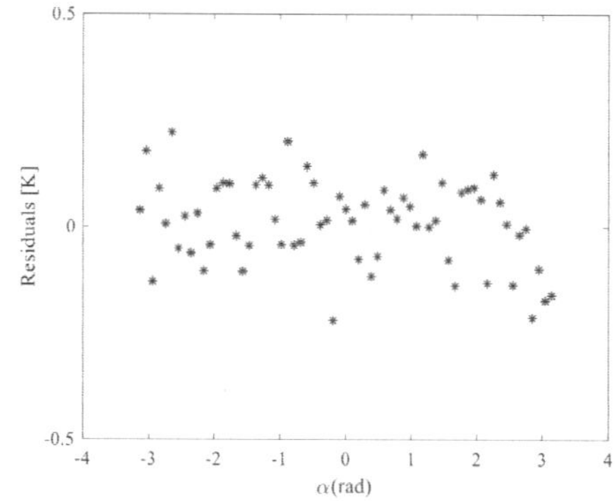

Fig.8: Residuals between the noisy and the restored temperature for $\sigma = 0.1$ K.

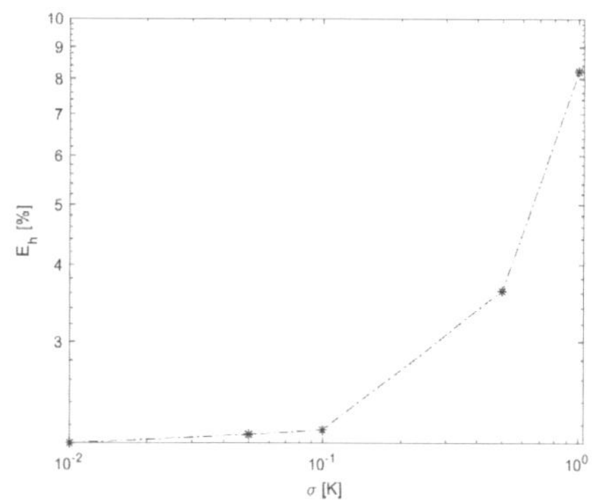

Fig. 9: Estimation error on h_{int} at various noise levels.

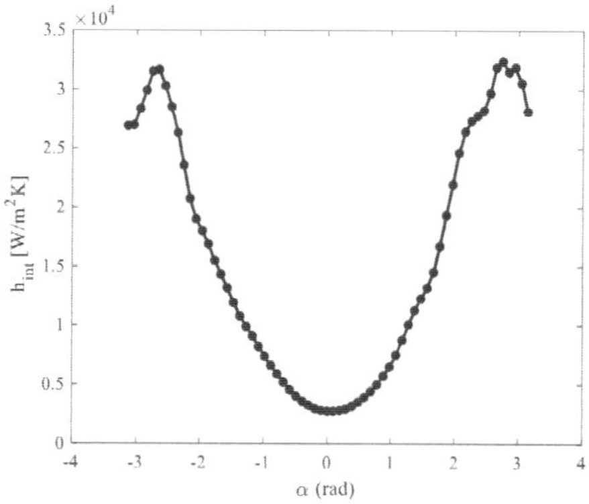

Fig.10: Experimental h_{int} distribution, $Re = 12218$

4.2. Experimental study case

The here proposed estimation methodology was then applied to an experimental case, previously studied by Cattani et al. [10] and concerning forced convection problem in helically coiled tubes. This study case is remarkably appropriate for assessing the proposed parameter estimation procedure, since it presents a heat transfer coefficient that changes considerably along the cross-section perimeter because of the complicated flow pattern produced by the centrifugal force. The studied tube, sketched in Fig.11, presented a helical profile comprised of eight coils: the diameter and the pitch of the helix were 310 mm and 200 mm, respectively while the inner and the outer diameters were 14 and 16 mm, respectively. The outer wall temperature distribution was experimentally measured adopting a FLIR SC7000 infrared camera, with a 640 x 512 pixels detector array and with a sensitivity and an absolute accuracy of 20 mK at 303 K and ±1 K, respectively. Turning the infrared camera around the pipe, several pictures of the test section were obtained: successively referring to a position reference placed on the pipe wall, the several infrared images were suitably cut, processed by perspective algorithms [11] and merged in Matlab. With this data processing method, a continuous temperature distribution on the tube wall versus the angular coordinate was obtained. The case reported in [10] is about the study of heat transfer in turbulent regime by adopting water as a working fluid. For a representative Reynolds number value (Re = 12218), the distribution of the convective heat transfer coefficient restored by the BEM method is reported in Fig.10. The obtained distribution is in line with the results provided in literature [10,12].

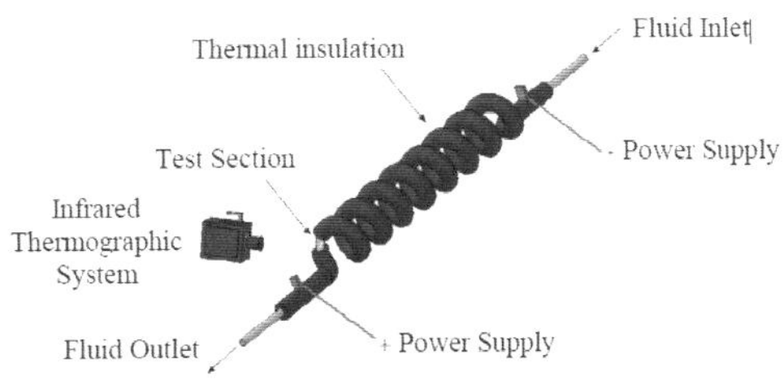

Fig.11: Sketch of the experimental set up adopted by Cattani et al. [10].

5. Conclusions

The main goal of the present analysis was to propose and validate a procedure to estimate the local convective heat transfer coefficient on the inner wall of a tube, under a forced convection problem. The methodology is established on the solution of IHCP in the solid wall domain by using the outer surface temperature distribution as starting data. The proposed approach is based on the BEM approach combined with the TSVD method, adopted to manage the ill-conditioned character of the problem. The model is formulated and then validated by means of numerical simulations within Comsol environment. The numerical data are then employed as starting synthetic data to resolve the IHCP under the BEM-TSVD approach. The validation of the methodology is completed by its application to experimental input data concerning a forced convection problem in a helically coiled pipe since it represents an appropriate and challenging study case for assessing the performance of the presented estimation methodology. In fact, it is characterized by a heat transfer coefficient that changes considerably along the cross-section perimeter because of the complicated flow pattern produced by the centrifugal force.

References

[1] J.V. Beck, K. Arnold, Parameter Estimation in Engineering and Science, Wiley Interscience, New York, 1977.

[2] Aimi A., Diligenti M., Guardasoni C. (2021). Energetic boundary element method for accurate solution of damped waves hard scattering problems. J. Eng. Math., 127(1), 1-30.

[3] Chen G., Zhou J. (1992). Boundary Element Methods, Academic Press.

[4] Y. Gu, W. Chen, Z.J. Fu, Singular boundary method for inverse heat conduction problems in general anisotropic media, Inverse Prob. Sci. Eng. 22 (6) (2014) 889–909.

[5] Rouizi Y., Al Haddad W., Maillet D., Jannot Y. (2015). Experimental assessment of the bulk fluid temperature profile in a minichannel through inversion of external surface temperature measurements, Int. J. Heat Mass Transf., 83 522–535.

[6] M.García J. A., Cabezas J. M. G., Rodríguez A. C. (2009). Two-dimensional non-linear inverse heat conduction problem based on the singular value decomposition, Int. J. Therm. Sci., 48(6) 1081–1093.

[7] Shenefelt J.R., Luck R., Taylor R.P., Berry J.T. (2002). Solution to inverse heat conduction problems employing singular value decomposition and model-reduction, International Journal of Heat and Mass Transfer, 45(1) 67-74.

[8] Rainieri S., Bozzoli F., Cattani L., Pagliarini G. (2013). Compound convective heat transfer enhancement in helically coiled wall corrugated tubes, Int. J. Heat Mass Transf., 59 353-362.

[9] Morozov, V. A. (2012). Methods for solving incorrectly posed problems. Springer Science & Business Media.

[10] Cattani, L., Bozzoli, F., Rainieri, S. (2017). Experimental study of the transitional flow regime in coiled tubes by the estimation of local convective heat transfer coefficient. Int. J. Heat Mass Transf., 112, 825-836.

[11] B. Cyganek, J.P. Siebert, An introduction to 3D Computer Vision Techniques and Algorithms, Wiley, New York, 2011.

[12] J.S. Jayakumar, S.M. Mahajani, J.C. Mandal, N.I. Kannan, P.K. Vijayan, CFD analysis of single-phase flows inside helically coiled tubes, Comput. Chem. Eng. 34 (2010) 430–446.

Fouling effect in a shell-and-tube heat exchanger with twisted tape inserts applied to a small-scale biomass gasification power plant

Giulio Allesina[(a,b)], Simone Pedrazzi[(a,b)], Filippo Ottani[(a)], Paolo Tartarini[(a,b)]

[(a)] BEELab - University of Modena and Reggio Emilia – Department of Engineering "Enzo Ferrari" Modena

[(b)] INTERMECH - Inter-departmental Center, University of Modena and Reggio Emilia, Via Vivarelli 2 – 41125 Modena, Italy

[*] Corresponding author e-mail: giulio.allesina@unimore.it

Abstract. In this work the over-time behaviour of a shell-and-tube type heat exchanger applied to a commercial and small-scale wood biomass gasification system was investigated. The heat exchanger, equipped with twisted tape turbulators, was used to cool down the syngas produced by the power plant. An experimental campaign was conducted to evaluate the performance trend in the first 15 hours of operation. The results showed an increase in heat transfer which led to a progressive reduction of the gas outlet temperature from 105.4 °C to 90.0 °C. The data collected confirm the literature studies on the positive effect that the deposits of particulate matter and tars have in reducing the clearance between the heat exchanger pipes and the turbulators, providing information for the temporal optimization of the cleaning strategy of the heat exchanger itself.

1. Introduction

A gasification power plant consists in a system that, through a series of thermochemical reactions, decomposes the molecules of the biomass into simpler ones generating a fuel gas, commonly known as *producer gas* or *syngas*. This is a multiphase fluid, mainly formed by a gaseous phase of H_2, CO, CO_2, H_2O, CH_4, N_2 [1] that also carries a series of pollutants in vapour phase, i.e. tars and in a solid phase represented by soot, ashes and particulate matter in general [2]. In fixed bed reactors, the higher heating value of the clean producer gas ranges from 4 to 7 MJ/Nm^3 depending on the fuel used and the gasification conditions [3]. This gas mixture is mostly used to fuel internal combustion engines to produce combined electrical and thermal power [4].

Several architectures and technologies can be employed in the design of a biomass power plant but the *fil rouge* that links all of them is the presence of crucial components such as a gasification reactor (gasifier) where the thermochemical reactions occur, a gas conditioning system that includes cooling and filtering stages and, at the end, a power unit that converts the fuel gas into electrical and thermal power.

Since the producer gas exits from the reactor at temperatures close to 600 °C, the cooling stage is needed in order to cool down the gas and to avoid damages to the filtration stage and the internal combustion engine. This work focuses on the assessment of the performances of a shell-and-tube heat exchanger applied to a commercial small scale gasification power plant: the Power Pallet 30 manufactured by the US company All Power Labs [5]. In particular, a series of tests were conducted in order to investigate the effectiveness of the heat exchanger after several hours of running. The heat exchanger, specifically prototyped for this study, is made of 7 straight tubes 38 mm diameter and 1.5 m long where the producer gas passes in. Twisted tape inserts, 35 mm in width, and 200 mm of pitch (based on 180°), were inserted into the tubes to enhance the heat exchange. The fluid flowing in the shell is the same water running in the engine cooling system, therefore the engine electronic thermostat and radiator are used to maintain the cooling water temperature in the range between 85 °C and 90 °C.

Temperatures of the producer gas were logged at the heat exchanger inlet and outlet during tests. The experimental campaign is carried out through a series of runs at constant engine power output, and so, at constant producer gas mass flow rate. During the tests, an increase in the gas temperature difference between inlet and outlet was monitored. This difference led to an increased effectiveness of the heat exchanger during the first hours of running. During the heat exchanger inspections and maintenance after the tests, a reduction in the clearance between the twisted tapes and the tubes was recorded due to the deposition of soot and tars contained in the gas stream. According to the literature findings [6, 7], the fouling seems thus to generate a tight-fit configuration that tends to increase the heat transfer coefficient (HTC).

2. Materials and methods

This section describes the experimental setup used to evaluate the performance of the heat exchanger.

2.1. The gasification facility

The gasification power plant used as a test bench for the heat exchanger is the Power Pallet 30 by All Power Labs [5].

The Power Pallet unit can be divided into 3 main subsystems that are:

- the biomass feeding system and the gasification reactor, which converts the solid wood into syngas. The Power Pallet 30 can be fuelled with both wood pellet and wood chips [8,9] and, in this work, standard A2 pellets were used for the test to guarantee a proper repeatability of the experiments.
- the gas conditioning stage, which is composed by a gas/water heat exchanger (henceforth called HX) object of the study, and a filter unit that is used to reduce the particulate matter content and the tar amount in the syngas, avoiding its later condensation in the pipelines downstrams [10]. The original cooling system was replaced with the shell-and-tube HX [11,12] described in the following section.
- the power generation unit consists of an internal combustion engine connected to a 30 kVA generator capable of generating both electrical and thermal power, recovering the heat from the engine block and exhaust gases.

The Power Pallet 30 original system is shown in Figure 1.

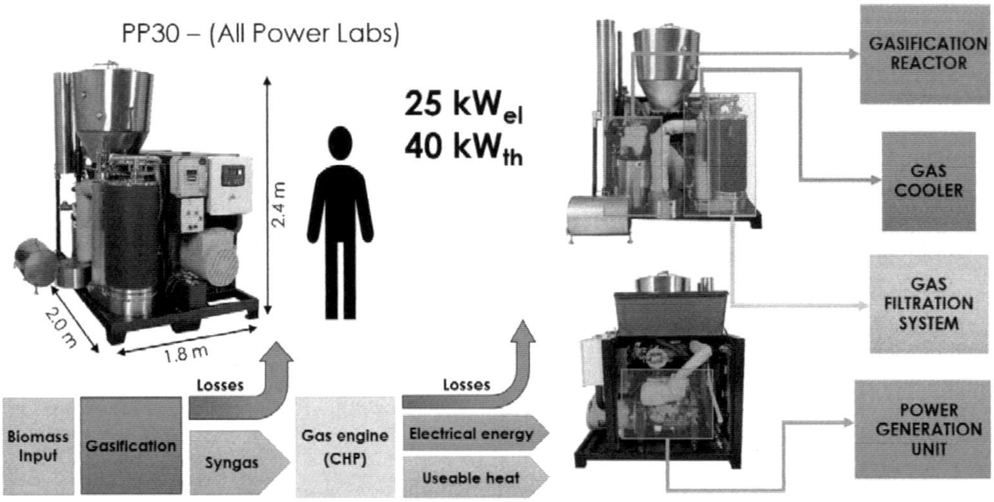

Figure 1. Block diagram of the PP30 and representation of the several subsystems.

2.2. *The gas cooling system of the Power Pallet 30*

2.2.1. *Syngas cooling strategy*

The gasification process of wood biomass takes place at temperatures that ranges from 700 to 1000 °C [1, 3] and the syngas outlet temperature depends on the reactor geometry and type. The Power Pallet 30 is powered by a single-throat downdraft reactor where the biomass and syngas run through the reactor in the same direction. In this type of reactor, the intake air, used for combustion, is generally preheated using the sensible heat of the syngas, which leaves the reactor at a temperature close to 600°C and drop at 350-400°C after the cyclonic separator used to collect the coarse fraction of the particulate matter [13].

The stages following gasification (i.e. syngas filtration and power generation) have necessarily lower working temperatures since, on the one hand, the filter often consists of polyester filter bags with a maximum operating temperature of 105 °C [14]; on the other hand, the internal combustion engine needs to operate with the lowest possible mixture temperature in order to increase its volumetric efficiency and to prevent detonation [15,16].

However, during the cooling down of the syngas, it is necessary to consider that the fractions of tars and water in the syngas in the form of vapour may condensate. For this reason, in order to avoid the accumulation of tarry water, it is necessary to keep the whole system above the syngas dew point. Previous studies estimated this value to range between 50 and 60 °C [17,9].

For this reason, in the Power Pallet 30 the syngas is cooled using the same coolant loop of the internal combustion engine, the fluid has a temperature between 80 and 90°C, and it prevents the system from going below the syngas dew point.

2.2.2. *The tubes-in-shell heat exchanger*

The original PP30 HX has been replaced with a custom HX in which, as previously described, the syngas flows inside the pipes, while the cooling water flows between the baffles inserted in the shell (Figure 2). This configuration allows testing of different solutions for the heat transfer enhancement and, in this case, twisted tape inserts were inserted in each of the 7 tubes of the HX.

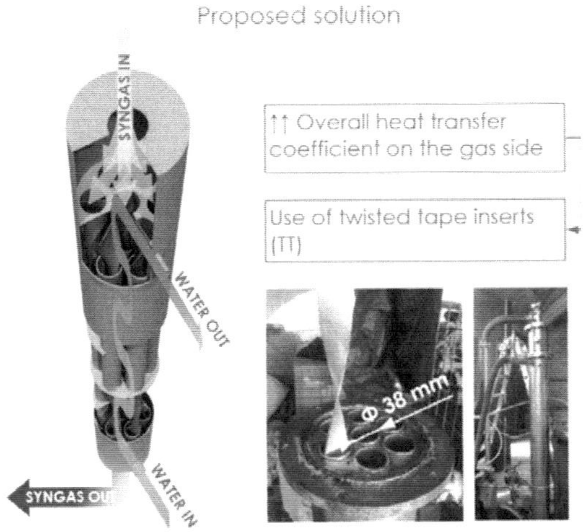

Figure 2. custom HX used for this test.

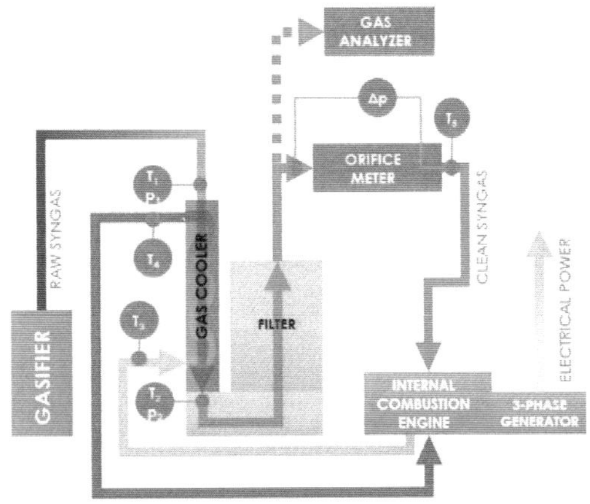

Figure 3. Block diagram of the test rig.

The custom HX is the same used in the previous work by Morselli et al., 2019 [11] and its dimensions are reported in the following scheme (Figure 4 and Figure 5).

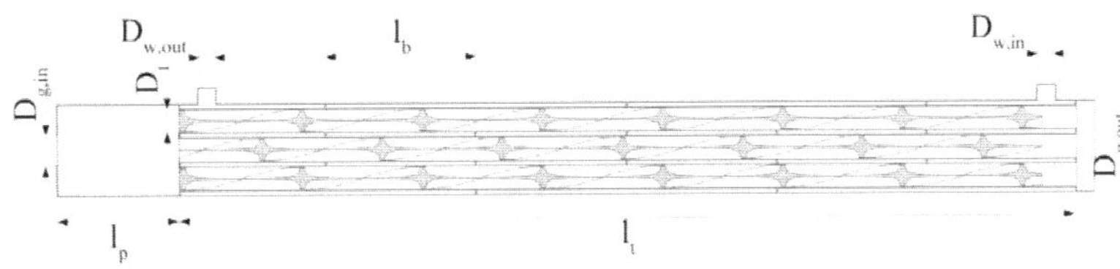

Figure 4. Longitudinal section of the HX.

Figure 5. Dimensions of the twisted tape inserts used.

l_b = 300 mm
$D_{g,in}$ = 60.3 mm
$D_{g,out}$ = 168.3 mm
l_t = 1500 mm
l_p = 250 mm
D_t = 38 mm
$D_{w,in}$ = $D_{w,out}$ = 33.7 mm
w = 35 mm
p = 200 mm

It is important to underline that a clearance of 1.5 mm has been maintained between the width of the twisted tape and the internal diameter of the pipe to guarantee a proper sliding of the twisted tapes for tube cleaning.

Due to tar and particulate matter deposition, this clearance is progressively reduced during the power plant uptime, the aim of this work is to evaluate the performance over time of the HX according to the test setup described in the next section.

2.3. *Test rig description*

To evaluate the performance of the HX, it is necessary to measure the thermal power exchanged between syngas and cooling water (Figure 3). For this purpose, K-type thermocouples were installed at the inlet (T_1) and at the outlet (T_2) of the syngas side of the HX. T-type thermocouples were used to measure the mean cooling water temperature (T_3 and T_4). In fact, in previous experiments [10] it was noted how difficult it is to properly measure the power absorbed by the cooling water, as the ΔT measured is in the order of 1 °C and comparable to the accuracy of the T-type thermocouples used

[18]. The data logging system used consists in a Picotech TC-08 [19] which has an integrated cold-junction compensation system that, via software, corrects the temperatures read. The measurement uncertainty for T-type thermocouples is ± 1°C while for K-type thermocouples it is ± 2°C.

The mass flow rate of syngas was measured through a calibrated orifice located downstream of the filtering section to reduce fouling and therefore reduce the reading error. The density of the gas was calculated by measuring the chemical composition of the syngas through a micro-gas-chromatograph [20]. To evaluate the performance of the HX over time, 3 tests were performed over three consecutive days. Each test lasted for 5 hours. The tests included the running of the internal combustion engine at a constant power of 11 kW for the entire duration of the test. Since the power required by the engine is maintained constant, assuming that the efficiency of the engine as well as the gasification conditions within the reactor are invariant during the tests, it is possible to consider the mass flow rate of syngas as a constant. For each test the flow rate of syngas was monitored continuously and gas samplings were performed in order to verify the stability of the gas composition. Whenever a variation of the composition was recorded, the density and the syngas' higher heating value were corrected in the calculation of the gas flow rate as well as in the air-fuel energy content drawn by the engine. Working at constant mass flow, it was possible to directly compare the ΔT generated by the HX in the three different tests and, in the next section, the results obtained in the various tests are reported.

3. Results

In Figure 6 the temperature trend during the three tests is reported, showing the inlet and outlet temperatures of the gas from the HX, the average coolant temperature and the temperature drop of the syngas (Gas ΔT).

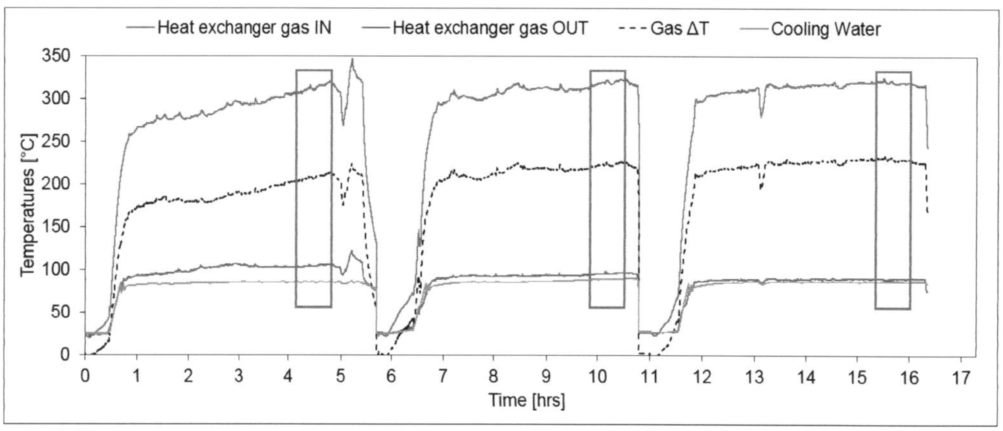

Figure 6. Trend of the temperatures over time for the 3 tests at constant syngas mass flow rate.

It can be noted how the inlet temperature to the HX tends to increase during the test. This phenomenon is, on the one hand, due to the stabilisation of the thermochemical reactions inside the gasification reactor, and on the other hand due to the accumulation of sensible heat in the metal structures of the different stages of the gasifier. Nevertheless, comparisons between the different tests were performed in the test portions under steady state conditions, highlighted by the red rectangles in Figure 6. The average values of these test portions are shown in Figure 7.

The average values of the syngas inlet temperatures vary from 316 °C for the first test to 320 °C for the and third tests; a variability that remains within the uncertainty range of the K-type thermocouples used (± 2°C [4]). The same result is obtained for the cooling water temperature which does not vary significantly in the inlet side (85-86°C). The only data that varies beyond the measurement uncertainty is the syngas outlet temperature from the exchanger and therefore the ΔT of the syngas associated with each test. The syngas outlet temperature from the HX were respectively for the three tests: 105.4 °C, 96.0 °C and 90.0 °C, leading to a ΔT equal to 210 °C in the first, 224 °C in the second and 230 °C in the third test respectively, confirming a significant increase in the effectiveness of the HX during the three tests. Figure 8 shows the syngas composition values during the 3 tests. At the same time, the HHV and syngas density values are reported. No variations in the syngas properties were found such as to justify a possible reduction in the mass flow rate of syngas processed by the HX.

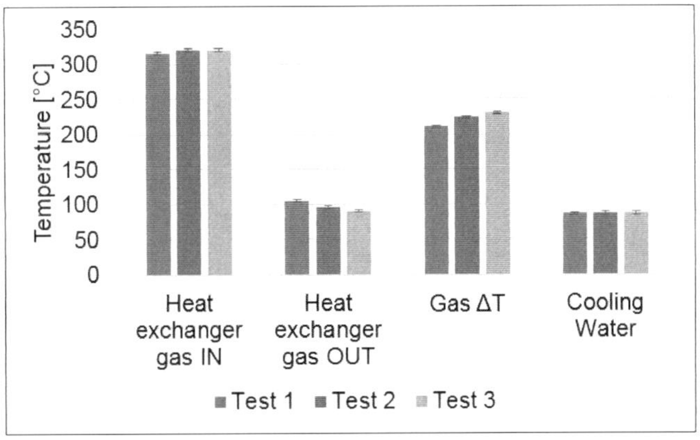

Figure 7. Average temperatures of the syngas and water side of the HX.

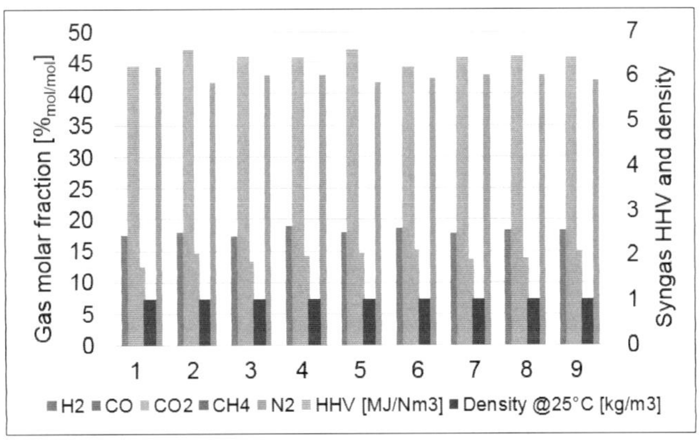

Figure 8. Syngas quality and properties.
(Test #1 samples from 1 to 3; Test #2 from 4 to 6; Test #3 from 7 to 9)

At the end of the third test, the HX was disassembled and inspected (Figure 9a,b), observing a considerable deposit of tar and particulate matter on the surfaces of the twisted tapes and pipes. The deposit had completely eliminated the clearance between the twisted tape and the pipe, reducing leakage and forcing the syngas into the swirl motion induced by the twisted tape.

According to the literature findings [6,7], the fouling seems to generate a tight-fit configuration that tends to increase the HTC, more than counterbalance its reduction due to soot layer formation as reported by SERI [21] (Figure 9b). Future work will be focused on the evaluation of the HX effectiveness (ε). This parameter is usually determined as the ratio between q and q_{max}.
The value q_{max} is the maximum heat that could be transferred between the fluids and it depends only on the fluid properties and the inlet temperatures. Therefore q_{max} can be considered as a constant in all the cases analysed. The value q is the measured heat exchange by the HX. This work proved that the temperature difference of the gas flowing through the HX progressively increases test after test, leading to a higher value for the heat transferred. Therefore it is possible to state that also the effectiveness ε increases during the tests.

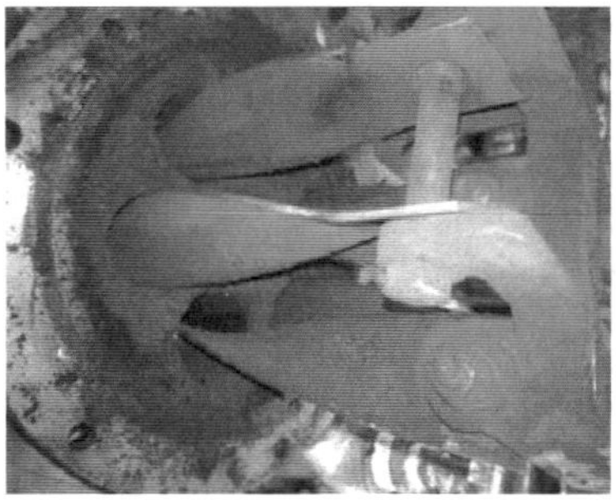

Figure 9a. Soot deposition on the top of the HX: rack of twisted tapes after the tests.

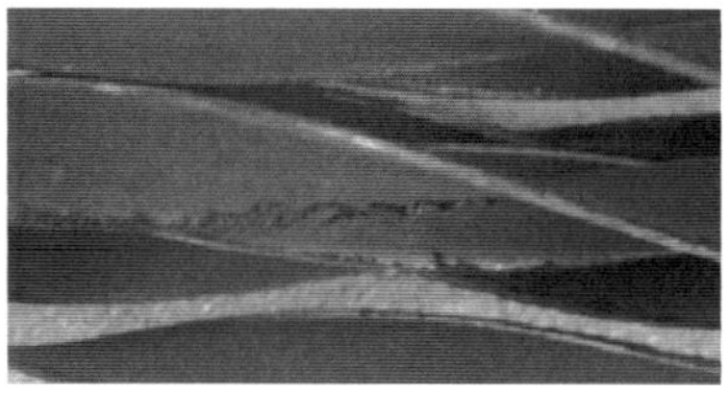

Figure 9b. Soot and tars deposited over the twisted tape inserts.

4. Conclusions

In this work, the performance of a shell-and-tube type HX for cooling the syngas produced by a small commercial gasification plant was tested over time. The HX was instrumented and the inlet and outlet temperatures of the syngas were monitored, conducting various tests at constant syngas mass flow rates. The results showed that, under the same working conditions, the effectiveness of the HX increases during the first 15 hours of operation, as a consequence of the reduction of the outlet temperature of the syngas. The phenomenon was ascribed to the reduction in the clearance between the twisted tape inserts and the tubes of the HX, which reduces the leakages and increases the swirl motion of the gas thus reducing the thickness of the thermal boundary layer.

Further studies should be conducted to address the duration of this phenomenon over the lifetime of the HX, investigating the possibility to optimise the cleaning strategy of the HX before the thickening of the soot layer insulation completely cancels out the gained advantages.

References

[1] Basu P. Chapter 7 - Gasification theory, 3rd ed. Academic Press; 2018. p. 211–62. doi:10.1016/B978-0-12-812992-0.00007-8.

[2] Reed TB, Das A. Handbook of biomass downdraft gasifier engine systems; 1988.

[3] FAO Forestry Paper. Wood gas as engine fuel; 1986.

[4] Rinaldini C.A., Allesina G., Pedrazzi S., Mattarelli E., Tartarini P. Modeling and optimization of industrial internal combustion engines running on Diesel/syngas blends (2019) Energy Conversion and Management, 182, pp. 89 - 94 DOI: 10.1016/j.enconman.2018.12.070

[5] ALL Power Labs, website: www.allpowerlabs.com

[6] S. AI-Fahed and W. Chakroun, Effect of tube-tape clearance on heat transfer for fully developed turbulent flow in a horizontal isothermal tube. Int. J. Heat and Fluid Flow 17: 173-178, 1996.

[7] Z. H. Ayub and S. F. AI-Fahed, The effect of gap width between horizontal tube and twisted tape on the pressure drop in turbulent water flow. Int. J. Heat and Fluid Flow, Vol. 14, No. 1, March 1993.

[8] Allesina G., Pedrazzi S., Rinaldini C. A., Morselli N., Savioli T., Mattarelli E., Tartarini P., "Experimental-analytical evaluation of sustainable syngas-biodiesel CHP systems based on oleaginous crops rotation." Proceedings of the International Conference on Power Engineering-15 (ICOPE-15), code 118770. November 30- December 4, 2015, Yokohama, Japan DOI: 10.1299/jsmeicope.2015.12._ICOPE-15-_1

[9] Puglia, M., Rizzo, A., Morselli, N., Tartarini, P. Efficiency and economical assessment of a solar powered dryer combined with a biomass gasification system (2019) International Journal of Heat and Technology, 37 (3), pp. 705-709. DOI: 10.18280/ijht.370306

[10] Morselli, N., Barbolini, F., Ottani, F., Parenti, M., Pedrazzi, S., Allesina, G., Tartarini, P. Modelling of tar and water vapors condensation in a commercial micro-scale gasification power plant (2021) European Biomass Conference and Exhibition Proceedings, (29th e-EUBCE) April 26-19, 2021, Marseille (France) attended online due to COVID19 restrictions, pp. 480-483. DOI: 10.5071/29thEUBCE2021-2CO.9.3

[11] Morselli, N., Puglia, M., Mason, J., Parenti, M., Ottani, F., Tartarini, P. Experimental and modelling evaluation of possible solutions for compact design of producer-gas heat exchangers (2020) Journal of Physics: Conference Series, 1599 (1), art. no. 012034, . DOI: 10.1088/1742-6596/1599/1/012034

[12] Morselli, N., Ottani, F., Parenti, M., Tartarini, P. Enhanced heat transfer in tubes-in-shell heat exchanger for syngas cooling: A comparison between conventional and perforated twisted tape inserts (2020) European Biomass Conference and Exhibition Proceedings, (28th e-EUBCE) July 6-9, 2020, Marseille (France) attended online due to COVID19 restrictions, pp. 288-294. DOI: 10.5071/28thEUBCE2020-2BO.2.3

[13] Pedrazzi, S., Allesina, G., Sebastianelli, L., Puglia, M., Morselli, N., Tartarini, P. Chemically enhanced char for syngas filtering purposes (2018) European Biomass Conference and Exhibition Proceedings, 2018 (26thEUBCE), 14-15 May 2018, Copenhagen (Denmark), pp. 694-698. DOI: 10.5071/26thEUBCE2018-2CV.2.7

[14] Morselli, N., Parenti, M., Puglia, M., Tartarini, P. Use of fabric filters for syngas dry filtration in small-scale gasification power systems (2019) AIP Conference Proceedings, 2191, art. no. 020117. DOI: 10.1063/1.5138850

[15] Heywood J 1988 Internal Combustion Engine Fundamentals, McGraw-Hill

[16] Pedrazzi, S., Allesina, G., Morselli, N., Puglia, M., Rinaldini, C.A., Savioli, T., Mattarelli, E., Giorgini, L., Tartarini, P. Modified diesel engine fueled by syngas: Modeling and experimental

validation (2016) European Biomass Conference and Exhibition Proceedings, 2016 (24thEUBCE) – 6-9 June 2016, Amsterdam (Netherlands), pp. 880-883. DOI: 10.5071/24thEUBCE2016-2CV.3.25

[17] Allesina G, Pedrazzi S, Puglia M, Tartarini P. A psychrometric approach to fixed bed biomass gasifier design. 23th European Biomass Conference and Exhibition, Wien, Austria, 2015.

[18] TC-direct thermocouples datasheet. Website https://www.tcdirect.it/ accessed 9 March 2022.

[19] PicoTech. Tc-08 datasheet; 2021. URL: https://www.picotech.com/data-logger/tc-08/usb-tc-08-manuals, accessed on 2021.10.27.

[20] Pollution S.r.l. Pollution website; 2020. URL: http://www.pollution.it, accessed on 2021.01.08.

[21] SERI Generator Gas: Te Swedish Experience from 1939-1945, Di Ingenjörsvetenskapsakademien (Sweden). Solar Energy Research Institute, United States. Dept. of Energy, Golden, Colorado, 1979.

Analysis of Variance of the Heat and Mass Transfer Coefficients in an Evaporative Condenser

Michael Giovannini, Marco Lorenzini

Università di Bologna, DIN – Dipartimento di Ingegneria Industriale, Via Fontanelle 40, I-47121, Forlì (FC) Italy

E-mail: marco.lorenzini@unibo.it

Abstract. Modelling of evaporative condensers relies heavily on the correct choice of the correlations for transport coefficients to give an accurate estimate of the dissipated thermal load and evaporated water mass flowrate, yet the literature does not cover all flow arrangements, nor operating conditions. In order to avoid expensive, and often impractical, experimental campaigns devoted to the determination of these quantities for a specific piece of equipment, a statistical approach based on the analysis of variance can be adopted: by analysing the sensitivity of the dependent variables on the choice of the transport coefficients, the best set of correlations can be determined, depending on whether a more accurate estimate is desired for the heat flux or evaporated mass flowrate, or if both are equally important.

1. Introduction

Evaporative condensers are employed in industrial refrigeration, as they allow, among others, a reduction in the condensation pressure, with a decrease in electric power consumption. In industrial plants, the reference value for the evaporation pressure can vary little to keep the goods at the desired storage temperature, the condensation pressure can be modified depending on ambient conditions in order to minimize the overall energy consumption of the plant, without compromising the process. It is difficult, when not impossible, to investigate the influence of the choice of the set point on the total energy demand of a real-life, operating refrigeration loop, lest the goods be damaged during the tests, but this is not the case for numerical simulations, which can be run for several ambient and load conditions. In this perspective, the condenser is a crucial component of any control-oriented model. After defining the system's geometry, the governing equations for heat and mass transfer must be derived and the proper correlations for transport coefficients chosen. Transport coefficients are strongly dependent on the tube arrangement and flow field around them, therefore applicability of the many correlations available is normally restricted only to the conditions under which they were obtained, as e.g. a different arrangement of the fluid flow (co-flow instead of counter-flow) might yield incorrect results, [1]. On the other hand, though, obtaining the values of heat and mass transfer coefficients through experimental campaigns is usually cost-intensive and often impractical, [2].

The discussion above highlights the importance of evaluating the applicability of extant correlations under conditions different from the ones for which they were obtained, in order to use them confidently. In this paper, a sensitivity analysis of a model of an evaporative condenser to the correlations for transport coefficients is carried out through a statistical tool based on the analysis of variance (ANOVA) method. In this way, the influence on those process variables which are amenable to on-site measurement even in an industrial scenario, such as the thermal power removed and the flow rate of evaporated water, of several correlations available was evaluated. The results were compared with the nominal operating data provided by the manufacturer and some criteria for choosing the most suitable set of correlations suggested.

2. Modelling of evaporative condensers and correlations investigated

There is rather large body of studies on evaporative cooling originating with the work by Merkel [3], which laid the foundations for all later analyses and modelling. Poppe and Rögener [4] further developed Merkel's model, dropping the assumptions of unit Lewis factor, *Le*, and negligible evaporated water flowrate.

The doctoral work by Dreyer [5] used both approaches to develop a bi-dimensional, distributed-parameter model for evaporative heat transfer in a tube bundle. The work considers co-flow, counter flow and cross-flow of water and air relative to one another and to the heat and mass transfer surface. The results were compared with those obtained from a lumped-parameter model, relying on the assumption that water temperature over the tube bundle can be considered as constant, [6].

Zalewsky [7, 8] developed a one-dimensional, distributed parameter model exploiting the periodicity of evaporative heat transfer in the tube bundle. This approach allows the investigation of co-flow and counterflow of water and air, but not their cross-flow.

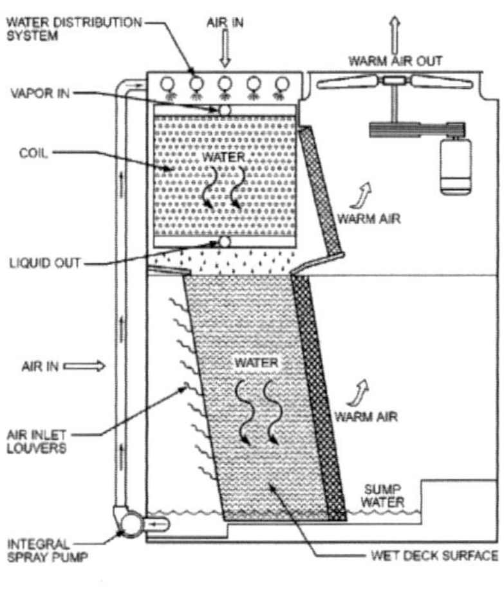

Figure 1: Evaporative condenser with fill-pack. [9]

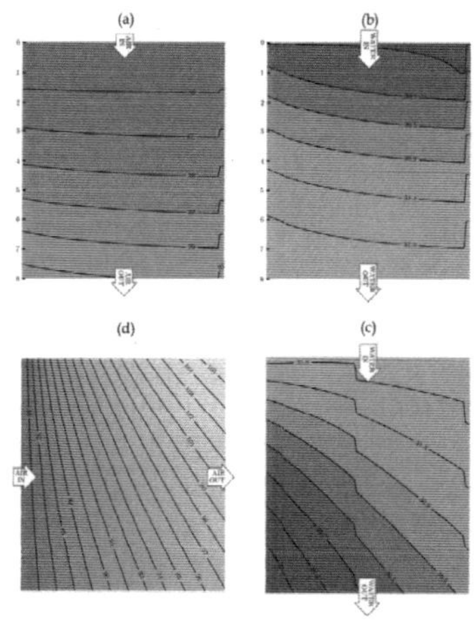

Figure 2: Moist air enthalpy and water temperature in the tube bundle (a-b) and in the fill-pack (d-c), of the condenser [10].

An extensive survey of evaporative condensers and cooling towers is offered by Kröger [11], who applied both Merkel's and Poppe's approaches to several cases, favouring the integral solution whenever possible.

Heat transfer in evaporative condensers involves three fluids at the same time: the refrigerant, water and moist air. Condensation of the refrigerant takes place inside the tube bundle, while water is sprinkled over the outer pipe wall through nozzles located above the bundle and the spent fluid is collected in a sump below, from which it is pumped back to the nozzles, Fig. 1. Contrary to the usual arrangement, air and water over the tube bundle are in co-flow.

More recent types of evaporative condensers have one more section, called fill-pack, which promotes heat transfer between the warmer water from the tube bundle and moist air, as happens in cooling towers, to further sink the water temperature, before it is collected in the sump. The fill-pack is usually made of polymeric materials and consists of a number of surfaces packed together, and shaped so as to maximise heat and mass transfer between moist air and water.

A schematic of an evaporative condenser similar to the one considered in this work is shown in Fig. 1: the fill-pack is below the tube bundle, so as to catch the water flow dripping from the tubes before it is collected in the sump below; air is sucked into the condenser by axial fans and is split into two independent streams, one flowing in co-current with water through the tube bundle, the other in cross-flow with water through the fill-pack.

A two-dimensional, distributed parameter model following Poppe's approach was chosen in this study because of two main reasons:

- the mass flowrate of evaporated water can be computed and used for validation of the model;
- the distributed-parameter model can be applied also in the case of fluids in cross-flow, as happens in the fill pack.

Evaporative heat and mass transfer over the tube bundle is computed solving the system of five differential equations below, which derive from mass and energy balances [10] :

$$dh_a = \frac{\beta_t}{\dot{m}_a}[Le(h_{asw} - h_a) - (Le - 1)(w_{asw} - w_a)h_v]dA_0 \tag{1}$$

$$dw_a = \frac{\beta_t}{\dot{m}_a}(w_{asw} - w_a)dA_o \tag{2}$$

$$d\dot{m}_w = -\dot{m}_a dw_a \tag{3}$$

$$dT_w = -\frac{1}{\dot{m}_w \cdot c_{p,w}}(\dot{m}_a dh_a + c_{pw}T_w d\dot{m}_w + \dot{m}_r dh_r) \tag{4}$$

$$dh_r = \frac{1}{\dot{m}_r}\left(\frac{1}{\alpha_r}\left(\frac{d_o}{d_i}\right) + \frac{d_o}{2\lambda}\ln\left(\frac{d_o}{d_i}\right) + \frac{1}{\alpha_w}\right)^{-1}(T_r - T_w)dA_o \tag{5}$$

A full description of the symbols is omitted due to space constraints, but is available in [10]. In the fill-pack, Eq. (5) is dropped, and in Eq. (4) the enthalpy variation of the refrigerant dh_r is zero. Finally, the heat and mass transfer area of the fill-pack is evaluated multiplying the front cross sectional area for the airflow by a surface area density coefficient, a.

The two systems of differential equations are integrated numerically with a fixed time-step Euler's method using an in-house code developed in Python; for the tube bundle, an iterative approach must be used to determine the actual refrigerant flowrate condensed, [10].

Shah's correlation, [12], was used to compute the convective heat transfer coefficient for condensation, α_r, within the tubes. To compute the evaporative heat transfer over the tube bundle, both the convective heat transfer for the water film over the tubes, α_w, and the mass transfer coefficient between water and moist air, β_t, are needed. Similarly, the mass transfer coefficient β_f must be provided to calculate the evaporative heat transfer in the fill-pack.

The correlations found in the literature and used in this work to compute α_t and β_t are reported in Table 1: none of them was devised for co-flow of water and air over the tube bundle, and only Eqs. (12) and (13) are for a cross-flow arrangement, whilst those remaining are for counterflow. This is the reason why a sensitivity analysis to the correlations employed of the model's response was deemed necessary in order to estimate their applicability outside the parametric range for which they were obtained.

For what concerns the mass transfer coefficient β_f, the surface morphology and the transport characteristics of the fill-pack are difficult to measure on site and are mostly proprietary. Since no detailed information on the issue was available, the same set of correlations was used as for the tube bundle. In the present case, the correlations are intentionally used beyond the scope of applicability provided by the authors, so the uncertainty of the calculated transfer coefficients can be evaluated only roughly with the data provided in the respective references; indeed, the aim of this work is to explore an extended field of applicability of these correlations with a statistical approach.

Table 1: Correlations considered.

Authors	Heat Transfer		Mass Transfer	
Parker and Treybald [13]	$\alpha_w = 704(1.3936 + 0.02214T_{wm})\left(\frac{\Gamma_m}{d_o}\right)^{\frac{1}{3}}$	(6)	$\beta_t = 0.04935\left(\frac{\dot{m}_a}{A_c}\right)^{0.905}$	(7)
Mizushina [14]	$\alpha_w = 2102.9\left(\frac{\Gamma_m}{d_o}\right)^{\frac{1}{3}}$	(8)	$\beta_t = 5.5439 \times 10^{-8} Re_{am}^{0.9} Re_{wm}^{0.15} d_o^{-1.6}$	(9)
Nitsu [15]	$\alpha_w = 990\left(\frac{\Gamma_m}{d_o}\right)^{0.46}$	(10)	$\beta_t = 0.076\left(\frac{\dot{m}_{am}}{A_c}\right)^{0.8}$	(11)
Erens and Dryer [11]	$\alpha_w = 2843\left(\frac{\Gamma_m}{d_o}\right)^{0.384}$	(12)	$\beta_t = 5.5749 \times 10^{-5}\, \mathrm{Re}_{am}^{0.64} Re_{wm}^{0.2}$	(13)
Zheng [16]	$\alpha_w = 350.3(1 + 0.0169T_w)\left(\frac{\dot{m}_a}{A_c}\right)^{0.59}\left(\frac{\Gamma}{d_o}\right)^{\frac{1}{3}}$	(14)	$\beta_t = 0.034\left(\frac{\dot{m}_a}{A_c}\right)^{0.977}$	(15)
Leidenfrost [17]	$\alpha_w = 2064\left(\frac{\Gamma_m}{d_o}\right)^{0,252}$	(16)	-	
Hasan [18]	-		$\beta_t = 0.065\left(\frac{\dot{m}_a}{A_c}\right)^{0.773}$	(17)

The model sensitivity to the various parameters was investigated through the analysis of variance (ANOVA) [19], a method which assesses the influence of one or more independent variables on a dependent variable. For the latter, the thermal power removed from the refrigerant and the amount of water evaporated under steady-state conditions were chosen, whose value were supplied by the manufacturer at two operating points, which had the same condensation temperature $T_{cond} = 35\,°C$, but different wet and dry bulb temperatures, namely $T_{wb,1} = 27\,°C$, $T_{wb,2} = 26\,°C$, $T_{db,1} = 37\,°C$, $T_{db,2} = 36\,°C$. The values of the dependent variables were consequently different: $\dot{Q}_{f,1} = 1610\,kW$, $\dot{Q}_{f,2} = 1783\,kW$, $\Delta\dot{m}_{w,1} = 0.655\,kg \cdot s^{-1}$ and $\Delta\dot{m}_{w,2} = 0.706\,kg \cdot s^{-1}$; no uncertainties were provided for the data, but this is a minor drawback, given the nature of the analysis carried out. The first operating point was used to evaluate the sensitivity of the two quantities, the second as verification.

3. Results and discussion

The first stage in the statistical analysis of the data computed was carried out using so-called box-plots, which relate the effects of the correlations employed for transport coefficients on the two dependent variables. Box-plots are a means of representing a data distribution: the central, horizontal line indicates the median, the upper and lower side of the box correspond to the 25th and the 75th quantile, with the box thus containing 50% of all data; the whiskers cover the extent of the dispersion in the remaining data which are not classified as outliers. The latter, where present, are plotted as lozenges.

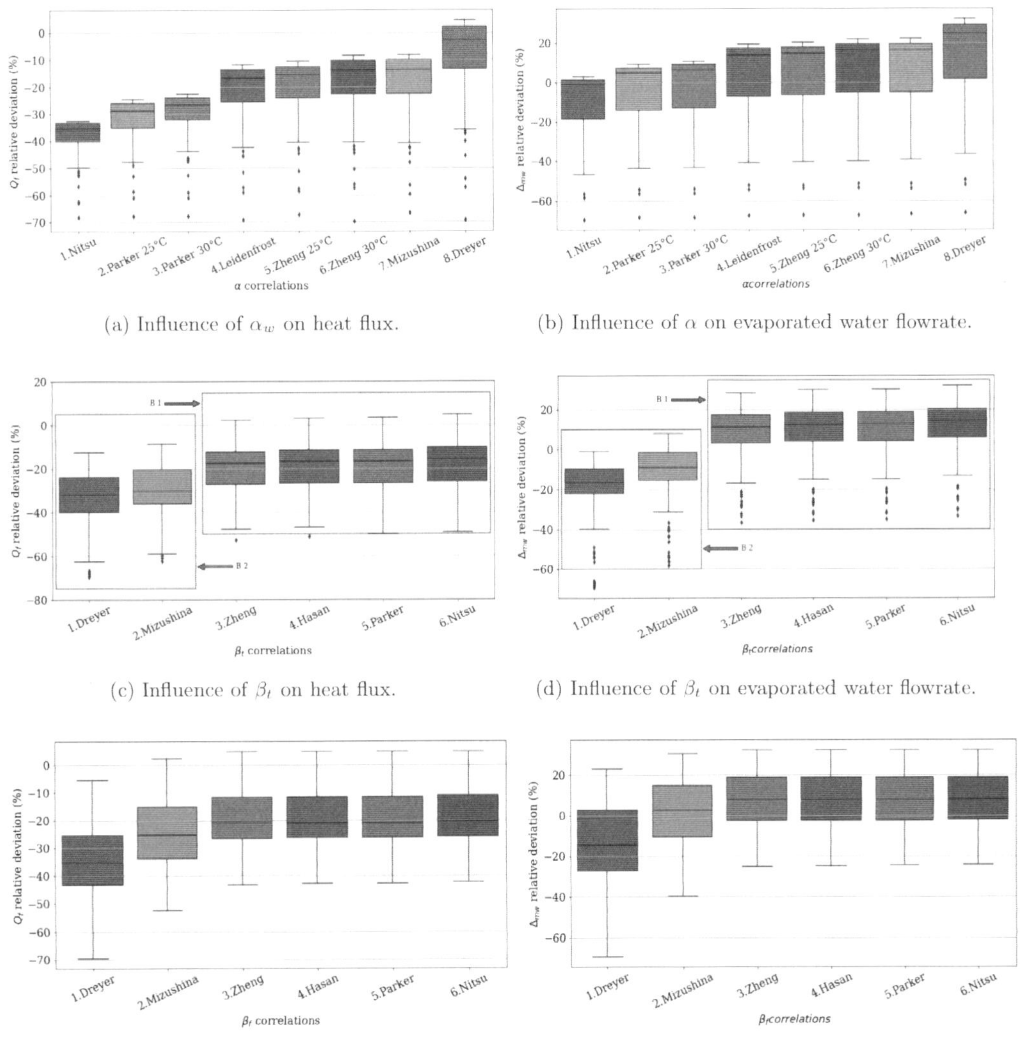

(a) Influence of α_w on heat flux.

(b) Influence of α on evaporated water flowrate.

(c) Influence of β_t on heat flux.

(d) Influence of β_t on evaporated water flowrate.

(e) Influence of β_f on heat flux.

(f) Influence of β_f on evaporated water flowrate.

Figure 3: Influence of the correlations chosen on cooling power and evaporated water flowrate.

Figure 3(a) shows the influence of the convective heat transfer coefficient in the film on the dissipated heat flux: it turns out that all correlations but Dreyer's underestimate it. The effect of this correlation on heat transfer from the tube bundle is markedly different from that of the remaining ones, which can be divided into three groups of similar effects.

This behaviour is confirmed by two further tests not shown here, namely the single-factor

ANOVA, consisting in the analysis of variance for a data set related to one dependent variable (dissipated heat flux or flowrate of evaporated water) under the influence of a single independent variable, and the Tukey's HSD (honestly significant difference) test, which allow multiple comparisons in order to pinpoint any significant difference due to the factor applied on the dependent variable.

The correlations used for α_w does not appear to influence the mass flowrate of evaporated water (see Fig. 3(b)); the values obtained for all correlations fall at 50% within a ±20% relative error on the expected value; Dreyer's correlation overestimates the flowrate the most.

The choice of the correlation used to compute β_t is shown in Fig. 3(c-d); these can be grouped into two sets, $B1$ and $B2$. The name value of the dissipated thermal power falls within the upper quartile of group $B1$. For the evaporated water flowrate, the data sets within the box-plots fall short of the name value of little less than ±20% or more. Correlations of group $B1$ on the other hand, tend to overestimate the actual evaporated mass flowrate.

Figures 3(e-f) show the effect of the correlations used for the mass transfer coefficient in the fill-pack. Dreyer's correlation stands out among the others in this case too, for both dependent variables. All correlations generally underestimate the actual dissipated heat flux, and the name value is attained in the upper quartile at most. For the evaporated water flowrate, 50% of the data lie within ±20% of the name value; Dreyer correlation tends to underestimate it, whilst the other overestimate it.

Figure 4 shows three plots of the relative error in the data for the estimate of the dissipated heat flux and evaporated mass flowrate. Colours refer to the correlation used. It is seen that no combination of correlations was able to achieve a relative error below ±10% for both dependent variables. Yet data cluster can be collected, which minimise the relative error for either of the dependent variables.

Box 1 encircles the tests which yield a low error (±5% from the name value) on the dissipated heat flux, whilst the evaporated mass flowrate is captured with a relative error between 25% and 35%. Box 2, on the other hand, represents the data which minimise the error on the evaporated mass flowrate (±7% of name value), with the corresponding dissipated heat flux falling within a relative error between −18% and -8%.

Also, all data in both boxes share Dreyer's correlation to compute α_w: it can therefore be concluded that this is the one suggested to obtain data accurate to within a reasonable error. From Fig. 4(b), it can be noticed that the correlations used to compute the mass transfer coefficient in the tube bundle are clearly distinguishable. Indeed, box 1 contains correlations from the $B1$ group only, which ANOVA detected, whilst box 2 only has data obtained using the correlations by Mizushina and Dryer, which were referred to as group $B2$. The choice of Dreyer's correlation for α_w in combination with one correlation from group $B1$ for β_t grants the minimum deviations on the dissipated heat flux, at the expenses of a somewhat overestimated evaporated mass flowrate. Conversely, using a correlation from group $B2$ for β_t underestimates the dissipated thermal power, whilst yielding the best results for the evaporated mass flowrate.

Lastly, concerning the choice of the correlation for β_f in the fill-pack, Fig. 4(c) shows that all correlations but Dreyer's are contained within the two boxes: as demonstrated in Fig. 5 (c)-(d), the latter is the only correlation which underestimates both dependent variables in the fill-pack. To wrap it all up, the set of correlations recommended for the distributed-parameter of the type of evaporative condenser considered should use Dreyer's correlation to compute α_t in Eq. (12), but the correlation by same author should not be employed to determine β_f, Eq. (13). For β_t over the tube bundle, two options are available:

- if an accurate estimate of the dissipated heat flux is sought, the correlation should be chosen among those of group $B1$, thus obtaining one of the points in box 1;
- if the evaporated mass flowrate is more important, correlations from group $B2$, which keep the deviation in $\dot{Q}_f$ to ±10%, and one of the data in box 2 is obtained.

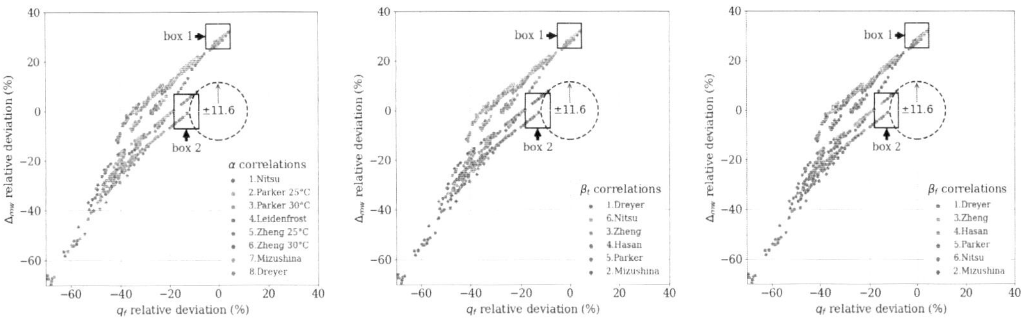

(a) Influence of α_w (b) Influence of β_t (c) Influence of β_f

Figure 4: Influence of the correlations chosen on cooling power and evaporated water flowrate, operating point 1.

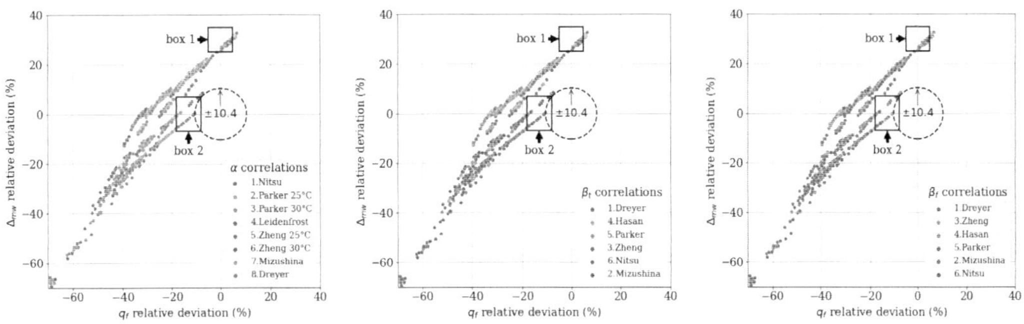

(a) Influence of α_w (b) Influence of β_t (c) Influence of β_f

Figure 5: Influence of the correlations chosen on cooling power and evaporated water flowrate, operating point 2.

The results for operating point 2, Fig. 5, show a similar behaviour.

Three optimal sets of correlations for each of the two operating points, and the results are reported in Table 2 and Table 3 respectively. Set 1 minimises the deviation of $\dot{Q}_f$ and belongs to box 1, set 2, which relates to box 2, minimizes the deviation of both $\Delta\dot{m}_w$ and $\dot{Q}_f$. Set 3, finally, minimizes the mean deviation of both dependent variables and is computed as

$$\delta_{tot} = \sqrt{\delta_{mw}^2 + \delta_{qf}^2} \tag{18}$$

If the results for both operating points according to the latter criterion are averaged, the optimal deviation obtained is $\delta_m \simeq 11.4\%$ for the choice of Dreyer, Mizushina and Nitsu for α, β_t and β_{fp} respectively.

Table 2: Optimal correlation sets for operating point 1.

Set of correlations Deviation considered	Set 1 $\delta_{qf} \simeq 0.02\%$	Set 2 $\delta_{mw} \simeq 0.8\%$, $\delta_{qf} \simeq 12.6\%$	Set 3 $\delta_{tot} \simeq 11.6\%$
α	Dreyer	Dreyer	Dreyer
β_t	Zheng	Dreyer	Mizushina
β_f	Hasan	Nitsu	Nitsu

Table 3: Optimal correlation sets for operating point 2.

Set of correlations Deviation considered	Set 1 $\delta_{qf} \simeq 0.35\%$	Set 2 $\delta_{mw} \simeq 0.09\%$, $\delta_{qf} \simeq 10.6\%$	Set 3 $\delta_{tot} \simeq 10.4\%$
α	Dreyer	Dreyer	Dreyer
β_t	Hasan	Dreyer	Dreyer
β_f	Hasan	Nitsu	Zheng

4. Conclusions

This paper suggested some criteria to choose the most suitable set of correlations for transport coefficients to be used in the distributed-parameter model of an evaporative condenser without the need of a complex and expensive experimental campaign to obtain them on site. The model, which uses the approach by Poppe and Rögener was tested with correlations available from the literature, none of which had been obtained for the same fluid flow arrangement. The suggested statistical investigation, based on the analysis of variance, evidenced the model sensitivity to the correlations used with respect to two dependent variables, namely the dissipated heat flux and the evaporated water mass flowrate. Based on the results, different criteria were suggested to choose the triplet of correlations which minimise the absolute deviation from the reference value for either $\dot{Q}_f$ or $\Delta\dot{m}_w$ or from both. With these choices, the maximum deviations between the model's computations and the reference values are slightly larger than $\pm 11\%$. While the numerical results pertain to a specific realisation, the method proposed can be extended to any other evaporative coolers, provided a suitable model for their description is available.

5. Referecens

[1] Dreyer A A and Erens P J 1996 *International Journal of Heat and Mass Transfer* **39** 109–123
[2] Lorenzini M, Fabbri G and Salvigni S 2007 *Applied Thermal Engineering* **27** 969–975
[3] Merkel F 1925 *VDI-Zeitchrift* **70** 123–128
[4] Poppe M and Rögener H 1991 *VDIWärmeatlas* Mi 1–Mi 15
[5] Dreyer A A 1988 *Analysis of Evaporative Coolers and Condensers* Ph.D. thesis Stellenbosch University
[6] Webb R L 1984 *Ashrae Transactions* **90** 398–415
[7] Zalewski W 1993-01-01 *International Journal of Refrigeration* **16** 23–30
[8] Zalewski W and Gryglaszewski P A 1997-07-01 *Chem. Eng. and Proc.: Process Intensification* **36** 271–280
[9] ASHRAE 2016 *Handbook—HVAC Systems and Equipment* (ASHRAE, Atlanta, GA)
[10] Giovannini M and Lorenzini M 2022-04-01 *Journal of Physics: Conference Series* **2177** 012009
[11] Kröger D G 2004 *Air-Cooled Heat Exchangers and Cooling Towers* vol 1 (PennWell Books)
[12] Shah M M 1979 *International Journal of Heat and Mass Transfer* **22** 547–556
[13] Parker R and Treybal R 1961 *Chemical Engineering Progress Symposium Series* **57** 138–149
[14] Mizushina T, Ito R and Miyashita H 1967 *International Chemical Engineering* **7** 727–732
[15] Niitsu Y, Naito K and Anzai T 1969 *Journal of SHASE, Japan* **43** 581–590
[16] Zheng W Y, Zhu D S, Song J, Zeng L D and Zhou H j 2012-03-01 *Applied Thermal Engineering* **35** 233–239
[17] Leidenfrost W and Korenic B 1979-03 *Wärme- und Stoffübertragung* **12** 5–23
[18] Hasan A and Sirén K 2002-06-01 *Energy and Buildings* **34** 477–486
[19] Montgomery D C 2013 *Design and Analysis of Experiments* eighth edition ed (Hoboken, NJ: Wiley)

Transient analysis of SIRIO using RELAP5/MOD3.3 system code

Martina Molinari[a*], Vincenzo Narcisi[a], Marco Caramello[b], Mariano Tarantino[c], Fabio Giannetti[a]

[a]DIAEE – Nuclear Section, "Sapienza" University of Rome, Corso Vittorio Emanuele II, 244, 00186, Rome, Italy
[b]Ansaldo Nucleare S.p.A., Corso F.M. Perrone, 25, 16152, Genova, Italy
[c]ENEA, Department of Fusion and Technology for Nuclear Safety and Security, 40032, Camugnano, Bologna, Italy

*corresponding author: martina.molinari@uniroma1.it

Abstract. The main outcome of the present paper is the feasibility analysis of SIRIO (Sistema di rimozione della Potenza di decadimento per Reattori InnOvativi) facility with conditions based on those of its reference facility. The aim of SIRIO project is to study an innovative Decay Heat Removal System (DHRS) for liquid metal reactor and advanced Light Water Reactor (LWR). Such system must ensure passive control of the power removed from the primary system in abnormal condition, and must ensure reactor cooling in both short and long term. This study present numerical simulations developed with RELAP5/MOD3.3, of two operational procedures: the first one is a steady-state and the second one is a transient phase with decay heat generation. The thermal-hydraulic model, developed with RELAP5/MOD3.3, simulates the whole facility including lines, valves, water and gas tanks, and the Molten Salts (MS) gap. Since there is not experimental data, the present paper is a pre-test study based on SIRO facility design.

Key words: *natural circulation, in-pool heat exchange, isolation condenser, passive safety systems, passive self-regulation, heat exchanger, lead cooled fast reactor*

1. Introduction

The SIRIO project is partially funded by the Italian Ministry of Economic Development and is developed by a consortium of research institution (ENEA) and industrial companies (Ansaldo Nucleare, SIET, SRS). The aim of the project is to study an innovative Decay Heat Removal System (DHRS) for liquid metal reactors and advanced LWRs, able to ensure passive control of the power removed from the primary system, both in short and long term.

When a rector is shutdown, the core still produces the decay heat. This decay heat must be removed by the cooling system to prevent the overheating of the core and the reactor pressure and temperature increasing. From the Fukushima Daiichi Accident, the interest in safety system at different levels of passivity has grown rapidly [1]. This happens because during abnormal conditions, a reactor provided with Passive Safety Systems (PSSs) can reach safety conditions without the need of external energy supply or human intervention, according to the degree of passivity, for a pre evaluated and known period.

As shown in Ref. [2], the PSSs can be accumulators, evaporative or condensation heat exchangers, and gravity driven safety injection systems. Thanks to the presence of PSSs the maintenance and the operation of active components can be eliminated, increasing the safety systems reliability.

There are different types of PSSs, for both the core and the reactor containment. Different technologies have been studied, some of these are the effectiveness of the Core Makeup Tank in Ref. [3] through which borated water flows to the core. The Passive Residual Heat Removal System is used to remove the decay heat through a natural circulation loop and a single phase liquid, in Ref. [4] and Ref. [5] has been studied for the Molten Salts Reactors. The same type of DHRS has been studied for liquid metal reactor in Ref. [6] and for light water reactor in Ref [7]. In the Passive Isolation Condenser studied in Ref. [8] and Ref. [9], the steam goes towards the isolation condenser and the condensate return to the core by gravity. Another type of Passive Isolation Condenser, shown in Ref.[10], foresees the passive regulation of the power removed. Another example of PSS is the Sump Natural Circulation shown in Ref. 10[11], here the reactor cavity is filled with water which can be the reservoir in case of damage of the primary system. The reactor cavity can collect the coolant in case of loss of water from the primary system.

About the reactor containment PSSs, the Pressure Suppression Pool studied in Ref. [12], is a system able to condense the steam vented from the containment, avoiding the over pressurization of the system.

The DHRS actuates to remove decay heat produced in the core. The safety system must be able of self-regulating the power removed from the primary circuit, to avoid primary coolant freezing in case of liquid metal coolants, and the excessive thermal stresses of the primary system in case of LWR.

To ensure the passive self-regulation of the power removed, Ansaldo Nucleare patented a DHRS which allows the control of heat transfer rate through the injection of Non-Condensable (NC) gas into the circuit, passively regulated by the pressure difference of NC gas with steam [13].

The SIRIO facility is a non-nuclear test facility, scaled from the DHRS of the Advanced Lead Cooled Fast Reactor European Demonstrator (ALFRED) [14]. The aim of the experiment is to assess feasibility of passive self-regulation of heat removal with the proposed solution.

In Figure 1 it can be seen the ALFRED DHRS. The bayonet tube heat exchanger (i.e., Steam Generator) is the one always used during the normal operation. The Steam Generator (SG) has the Hot Leg (HL) and the Cold Leg (CL) intercepted by the DHRS in case of abnormal condition through the steam isolation valve and the condensate isolation valve, shown upstream and downstream of the Isolation Condenser (IC) in Figure 1. The IC is composed of vertical tubes connected with the NC gas line to NC gas tank.

When the DHRS is isolated from the circuit, the NC gas fill also the IC tube side, at a pressure of about 110 bar. When DHRS is activated, the SG HL and CL are intercepted, and when the steam flows into the IC tube side the NC gases are confined and then pressurized into the NC gas tank. The steam into the IC is condensed by the heat exchange with a water pool in environmental conditions. When the IC tube side pressure is reduced to the NC gas pressure, the NC gas flows into the IC tube side. The purpose of using NC gas is to degrade the Heat Transfer Coefficient (HTC) of the steam in condensation phase with the water pool. In this way passive self-regulation of the power removed by DHRS is done.

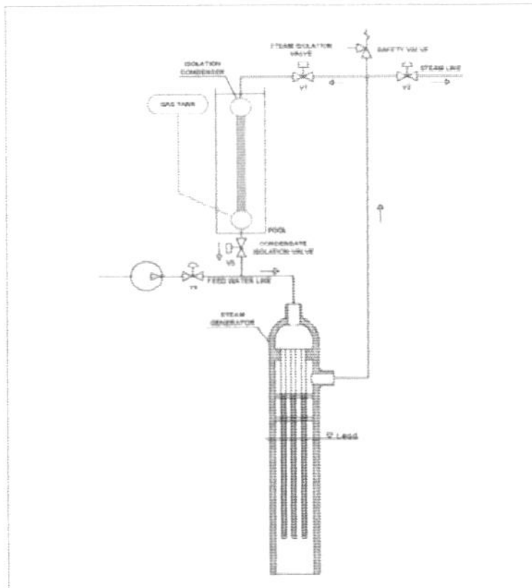

Figure 1: ALFRED DHRS schematic diagram [2]

The Department of Astronautical Electrical and Energy Engineering (DIAEE) of "Sapienza" University of Rome, has developed a thermal-hydraulic model of the SIRIO facility, which includes all the described components. To study the operational procedures a modified version of the system thermal-hydraulics (STH) code RELAP5/MOD3.3 [15] was used. The fundamental equation of RELAP5/MOD3.3 are based on a two-fluid, nonequilibrium, nonhomogeneous model for transient behavior with six-equations (the mass continuity, energy conservation and the momentum conservation for both the phases), as explained in Ref. [15].The modified version has different implementation, which consist in: lithium-lead (LiPb) and HITEC© working fluids with their thermophysical properties, new heat transfer correlations for liquid metals and MS, helical tubes heat transfer correlations and two-phase flow maps, module for MHD modelling capabilities and new heat transfer correlation for the evaluation of HTC in nucleate pool boiling and film condensation. The modified code has already been validated for different models such as helical tubes heat transfer correlations shown in Ref. [16], module for MHD modelling capabilities shown in Ref. [17] and correlations for the evaluation of HTC in nucleate pool boiling and film condensation, through submerged heat exchangers in water pools [18] and is able to simulate this type of system.

In this framework the aim of this study is to verify the feasibility of experiments based on the conditions of the reference facility ALFRED. Since there is no experimental data to be compared with numerical simulations, this study is a pre-test phase based on the design of the SIRIO facility.

2. Facility overview and scaling procedure

The SIRIO facility is a non-nuclear test facility, scaled form the ALFRED DHRS shown in Figure 1. The scaling approach is "power-to-volume": power density is kept constant, elevation and operational pressure are the same of the reference facility. The scaling factor has been defined with the NC gas tank as a reference component, and the volume scaling factor is 1:47 [14].

The primary circuit is composed by the SG, steam line, feedwater line, IC tube side and bypass HX tube side. The secondary circuit is composed by the bypass HX pool side and the IC pool side. The primary circuit is fully driven by natural circulation.

As it can be seen in Figure 2, the main difference between SIRIO configuration and the reference system shown in Figure 1, is the bypass Heat eXchanger (HX) line which provides for steady state conditions before transient testing.

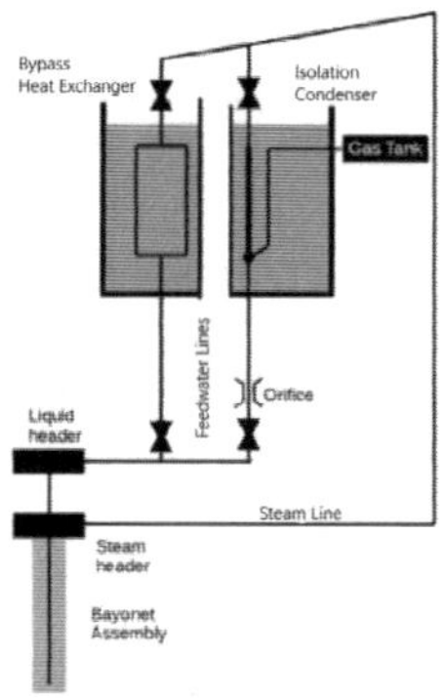

Figure 2: SIRIO schematic diagram [2]

The SG is composed of 11 tubes, each one composed of 6 components as shown in Figure 3. The component 1 in Figure 3 is the feedwater header which takes the fluid from the feedwater line; the component 2 in Figure 3 is the inner tube which role is the feedwater downcomer; the component 3 in Figure 3 is the steam trap which role is to limit regeneration phenomena for the initial downcomer region, in normal operation conditions; the component 4 in Figure 3 is the steam riser, is the most external tube through which the steam goes up towards the steam header; the component 5 in Figure 3is the steam header which collect the steam of the SG; the last component the is the Molten Salt (MS) gap (component 6 in Figure 3). The MS gap is a coaxial tube, used to simulate the power supplied by natural circulation in a same way of the reference facility ALFRED. Thermal power is supplied to the SG through electrically heated MS gap. The feedwater goes through the SG inner tube, and then goes up through the outer tube towards the steam header becoming steam. The steam, exiting the SG header, rises through the steam line towards the bypass HX or the IC.

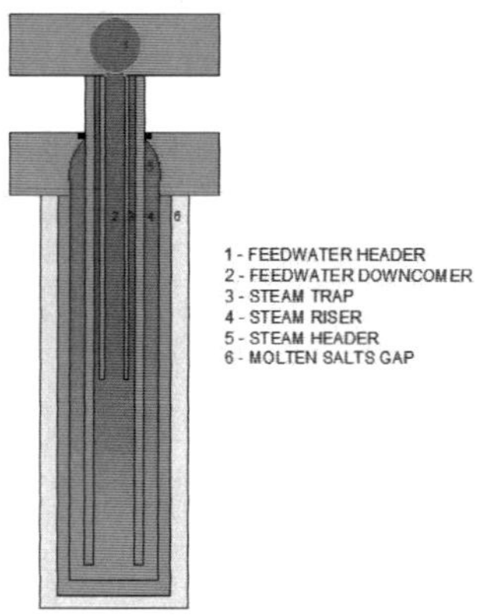

Figure 3: SIRIO - bayonet tube notional view

The SIRIO IC consists of a single tube immersed in the water pool. From the scaling procedure it was chosen to simulate the piping keeping the internal diameter (ID) and the outer diameter (OD) equal to those of ALFRED DHRS, with a pipe length of 0.68 m (compared to 2m of ALFRED DHRS [19]), thus having a deformation of the scaling factor of 0.39%.

This choice has an impact on natural circulation. However, considering the height of the facility (about 15 m), the distortion has been considered acceptable [14].

As it can be seen in Figure 1, for ALFRED DHRS configuration the IC has a top header and a bottom header. From scaling procedure, to simulate the volume occupied by these two headers, a pipe with ID of 2½'', downstream of IC is used.

The bypass line is not present in the reference system, but its presence is necessary because its role is to bring the facility to steady-state conditions, allowing the natural circulation of the fluid. In the reference configuration the fluid circulates thanks to the conditions resulting from the normal operations of the plant. As it can be seen in Figure 2, the bypass line is an independent line parallel to the IC and is completely isolated when the IC comes into operation. When bypass HX comes into operation, the IC is completely isolated from the rest of the circuit. The bypass HX or IC activation or isolation takes place through a one upstream valve and one downstream valve for each bypass HX and IC. The bypass HX has the function of being the heat sink during the steady state. This HX is made up of two parallel pipes immersed in a water pool of 2 m high. The steam from the steam line, goes downward inside these tubes and is condensed.

Components for both feedwater and steam line, were chosen respecting the scaling procedure and the economic/scientific characteristics (i.e., avoiding the flow pattern variations, keeping the pressure drops comparable to those of ALFRED, minimization of heat losses). Into the feedwater line coming from the IC, there is an orifice of 5 mm, which role is to prevent the NC from flowing outside the IC tube side.

The gas tank is connected downstream of the IC tube side, via the NC gas line.

3. Models

The SIRIO nodalization is divided in 3 hydrodynamic regions: the primary circuit, the IC pool and the bypass HX pool. The model is composed of hydrodynamic components, Heat Structures (HS), and control systems. The model is developed with the sliced approach for the whole facility, and the length ratio of the adjacent nodes is kept into the range of 1 ÷ 1.25.

Steam line and feedwater line are simulated with 5 and 8 pipe components, respectively. Actual bends, section variations, tees, and orifice are included in the modeling. The bypass HX has been simulated with two parallel pipe components for tube side and two parallel pipe components for pool side, to allow the simulation of fluid mixing. The isolation condenser was simulated with the single tube on the tube side and also in this case two pipe components for pool side to simulate the mixing of the fluid.

The NC gas line has been simulated considering all the curves present in the circuit.

The hydrodynamic characteristics of the SG are modeled collapsing all the 11 tubes into a single bayonet element. The SG flow area is the total flow area, while the hydraulic diameter is evaluated according to the geometry of the single subchannel.

Active and passive HS are included in the modeling. The active HS are those that provide for the heat exchange between the components, while the passive HS are those used to thermally insulate the components. Active HS involve heat exchange in the SG, bypass HX and IC. As shown in Figure 4, for the SG three different HS have been simulated, for each of these the heat exchange area used is the total heat exchange area. The first one (named as HS 13 in Figure 4) has the purpose of simulating the power supplied to the SG through a multilayer HS, thanks to which it is possible to simulate the heat exchange by conduction through the MS; the second HS (HS 14 in Figure 4) simulates the heat exchange between the steam riser and steam trap (component 3 in Figure 4) and the second region of the water downcomer; the third HS (HS 15 in Figure 4) simulates the heat exchange between the steam trap and the water downcomer. For the bypass HX there is a HS for each pipe component, which simulates the heat exchange between the tube side and the pool side. For the IC, there is a HS that simulates the heat exchange between the tube side and the pool side.

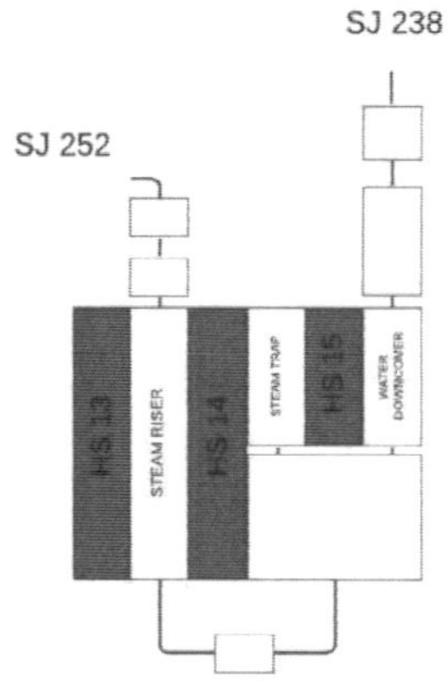

Figure 4: SIRIO SG thermal-hydraulic model

There are passive HS for all the insulated pipes. The pipe components are insulated with Rockwool, and are simulated the environmental temperature of 303 K, and environmental HTC of 15 W/m^{2}K.

4. Test: steady-state phase

The test phase is divided into two parts: a steady state phase and a transient phase with decay heat generation.

For the steady state phase, a constant power is provided to the SG, equal to 55 kW. To reach a steady state phase, an equilibrium between power removed by bypass HX and power supplied to SG is necessary, at rated pressure. To reach this balance condition shown in Figure 5, the bypass HX pool side has been initialized as empty as shown in Figure 6. Through a time dependent junction the level was quickly increased to 0.76 m (38% of the total height). After the initial phase, the flow rate has been adjusted to maintain a constant level of the bypass HX pool side, at around 0.63 m (31.5% of the total height).

Into the primary circuit there aren't control system. Bypass HX upstream and downstream valve are kept fully open. IC upstream and downstream valve are closed for the whole test, keeping this line isolated.

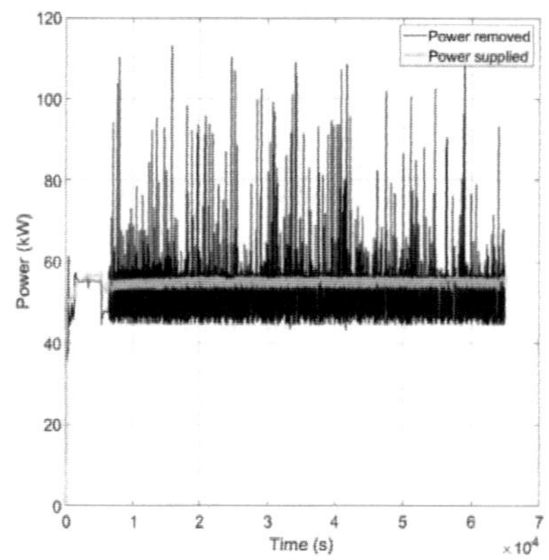

Figure 5: power removed and supplied during steady-state phase

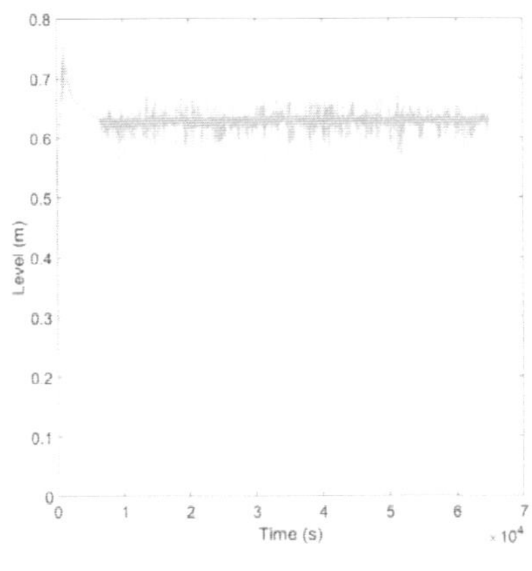

Figure 6: Bypass HX pool side - Water level

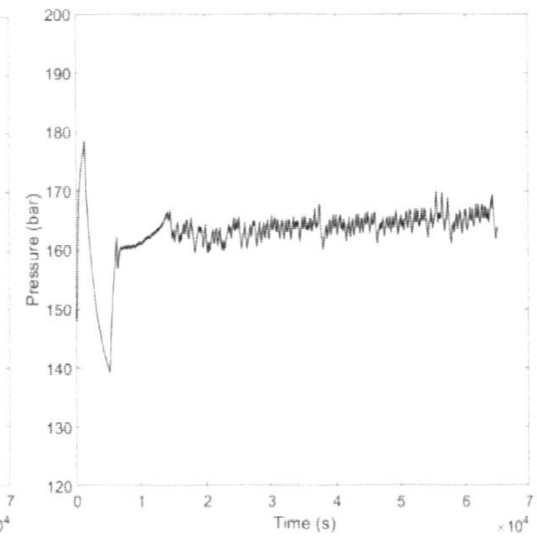

Figure 7: Primary side pressure trend

Looking at the power trend in Figure 5 and the bypass HX pool side water level in Figure 6, it is possible to notice that an equilibrium condition has been reached between power removed and supplied, and how the water pool level stability of the bypass HX has been achieved, albeit with oscillations. However, the pressure trend (Figure 7) in the primary circuit of the facility is characterized by strong oscillations, and its trend is less stable when compared to Figure 6 one.

The nature of these oscillations observed in Figure 6, Figure 7 and Figure 8 may be due to: the detail of the model, either of a computational nature or of a thermal-hydraulic nature. The first cause can be excluded, since the magnitude of level variations are all within the same control volume. If the oscillation had not remained confined within the same control volume as the IC component, this cause could not be excluded. The other two possibilities have to be investigated. The computational nature of the oscillation cannot be excluded. The last possibility is thermal-hydraulic instability characteristic of parallel pipes. The latter is possible due to the fact that both the IC pool side and the bypass HX have been modeled with parallel pipes connected through multiple junctions. Nevertheless, the results can be considered satisfactory for the purpose of the present activity.

5. Test: transient phase - decay heat generation

The second part of the test is a transient phase with decay heat generation. For the test: the steam line is isolated from the heat sink, so pressure in the primary circuit increases reaching the DHRS setpoint pressure of 190 bar. When DHRS is activated, the power supplied follows a decay heat generation trend[1], simulated as shown in Table 1, from 55 kW of the steady-state phase down to 4.84 kW in around 24 h. The time zero represents the bypass HX isolation instant, starting from 34800 s of the steady state test.

Table 1: Decay heat trend

Time (s)	Power (kW)
0	55.00
2000	55.00
2060	31.43
2600	20.64
5600	12.38
12800	8.89
38000	6.03
88400	4.84

[1] Full power is provided for 2000 seconds from begin of test to account metal structures heat capacity and reduce its impact on global power balance.

When the steam line is isolated, there is a sharp increase in pressure up to 190 bar, shown in Figure 8. This represents the opening set point for the upstream valve of the IC. With a delay of 60 s from the opening of the upstream valve, also the downstream valve of the IC opens, as shown in Table 2.

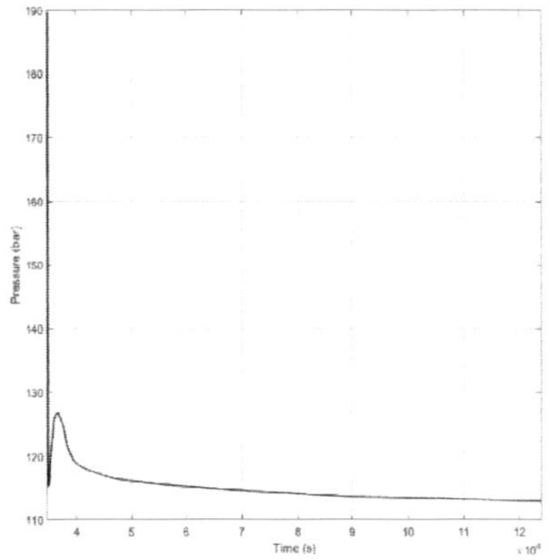

Figure 8: Pressure trend after IC activation

When IC upstream valve is opened, the NC gas are confined and pressurized int the NC line and NC gas tank. In this way steam condenses in absence of NC gas and as it can be seen in Figure 8, there is a sharp drop of pressure right after the DHRS activation.

After 115s from the IC downstream valve opening, the differential pressure of steam into IC tube side and NC gas vanishes, and the NC gas flows into the primary circuit. When NCs flows into the circuit, they also enter in the SG feedwater downcomer for about 13800 s, and in the SG steam riser for about 3300 s, during which there are maximum NC quality of 0.87 e 0.008514 respectively. After 14468 s from IC downstream valve opening, the NC gas are confined into the IC tube side. In fact, IC tube NC quality is around 0.9, as shown in Figure 9.

Table 2: Valves operation

Valve	Open/Close condition	Opening/Closing Time
Bypass HX – upstream/downstream	Time	0 s
IC – upstream	Pressure – 190 bar	152 s
IC - downstream	Time – 60 s after IC upstream	212 s

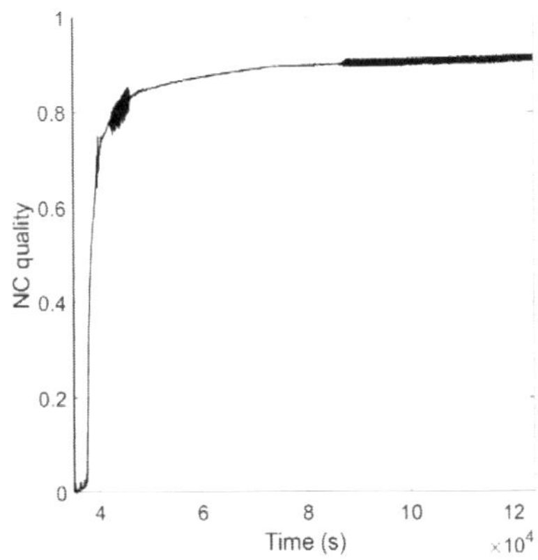

Figure 9: NC gas quality in IC tube side

As the objective is to verify the anti-freezing feature of the system in liquid metal-cooled reactors, and thermal stresses mitigation in case of liquid water reactors, Figure 10 shows temperature profiles, from MS side to SG steam riser wall temperature respectively: 1 h after DHRS activation (Figure 10a), 12 h after DHRS activation (Figure 10b) and 24 h after DHRS activation (Figure 10c).

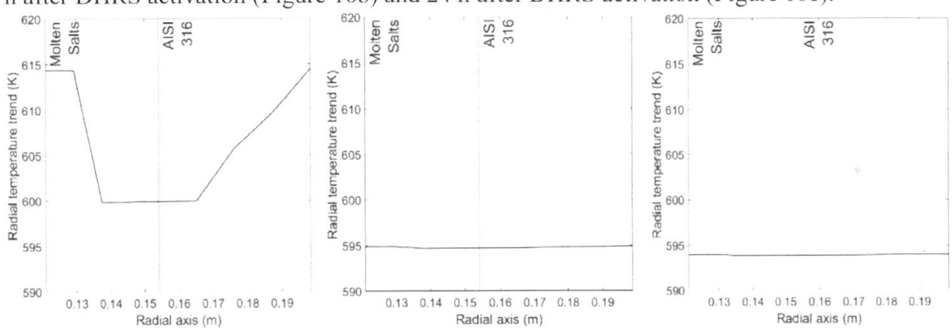

(a) (b) (c)

Figure 10: Temperature profiles from MS to SG after 1h (a), 12 h (b) and 24 h (c) of DHRS activation

Even during the transient phase – decay heat generation test, the results are satisfactory. The purpose of confining the NC gas thanks to the orifice is achieved, thus confining the NCs. The purpose of HTC degradation during heat removal is achieved. From the facility point of view, the temperature reached by the MS is not an issue because the MS used (Dynalene MS-2) has a melting point of 403 K.

6. Conclusions

Both test phases were carried out in respecting all the required boundary conditions, and as much as possible the constructive characteristics of the test facility. In the steady-state case, despite the oscillations present on the studied parameters in the primary circuit, a satisfactory result was obtained with the achievement of a quasi-steady-state.

In the study of the second test phase, the behavior of the passive self-regulation system of the power removed was seen. In this case the behavior of all active and passive components has been simulated.

The NC gas confinement with an orifice is achieved, and so the passive self-regulation of the power removed.

Due to these results, future sensitivity analyses (nodalization, boundary conditions and MS components simulation) are required and comparisons with experimental results, to better estimate differences of numerical simulations results with experimental data, are required.

References

[1] International Atomic Energy Agency *Safety related terms for advanced nuclear plants* (Vienna, Austria) IAEA-TECDOC-626

[2] International Atomic Energy Agency *Passive Safety Systems and Natural Circulation in Water Cooled Nuclear Power Plants* (Vienna, Austria) IAEA-TECDOC-1624

[3] Wang M, Tian W, Qiu S, Su G and Zhang Y 2013 An evaluation of designed passive Core Makeup Tank (CMT) for China pressurized reactor (CPR1000) *Ann. Nucl. Energy* **56** 81-86

[4] Sun L, Sun L, Yan C, Fa D and Wang N 2014 Conceptual design and analysis of a passive residual heat removal system for a 10 MW molten salt reactor experiment *Prog. Nucl. Energy* **70** 149-158

[5] Wang C, Guo Z, Zhang D, Qiu S, Tian W, Wu Y and Su G 2013 Transient behavior of sodium – potassium alloy heat pipe in passive residual heat removal system of molten salt reactor *Prog. Nucl. Energy* **68** 142-152

[6] Giannetti F, Vitale Di Maio D, Naviglio A and Caruso G 2016 Thermal-hydraulic analysis of an innovative decay heat removal system for lead-cooled fast reactors *Nucl. Eng. Des.* **305** 168-178

[7] Vitale Di Mario D, Naviglio A, Giannetti F and Caruso G 2012 An innovative pool with passive heat removal system *Energy* **45** 296-303

[8] Burgazzi L 2017 Passive System Reliability Analysis: A Study on the Isolation Condenser *Nucl. Technol.* **139** 3-9

[9] Alcaro F, Bersano A, Bertani C and Mascari F 2021 BEPU analysis of a passive decay heat removal system with RELAP5/MOD3.3 and RELAP5-3D *Prog. Nucl. Energy* **136** 103724

[10] Caramello M, Gregorini M, Bertani, De Salve M, Alemberti A and Panella B 2017 Thermal hydraulic analysis of a passively controlled DHR system *Prog. Nucl. Energy* **99** 127-139

[11] Modro S M, Fisher J E, Weaver K D, Reyes J N, Groome J T, Babka P and Carlson T M 2003 *Multi Application Small Light Water Reactor* (United States: Idaho National Laboratory) INEEL/EXT-04-01626

[12] Kataoka Y, Fukui T, Hatamiya S, Murase M, Naitoh M and Sumida I 1990 Thermal-Hydraulic Characteristics and Heat Removal Capability of Containment Cooling System with External Water Wall *J. Nucl. Sci. Technol.* **27** 802-814

[13] Narcisi V, Giannetti F, Caramello M and Caruso G 2020 Preliminary evaluation of ALFRED revised concept uder station black out *Nucl. Eng. Des.* **364** 110648

[14] Nitti F S, Marinari R and Valdiserri M 2018 *SIRIO Facility Conceptual Design* (TH-I-R-323)

[15] The RELAP5/MOD3.3 Code Development Team 2003 *RELAP5/MOD3.3 Code Manual Volume I: code structure, system models, and solution methods* (Washington DC: U.S. Nuclear Regulatory Commission) NUREG/CR-5535/Rev P3-Vol I

[16] Molinari M, Narcisi V, Ciurluini C and Giannetti F 2022 Transient analysis of OSU-MASLWR with RELAP5 J. *Phys.: Conf. Ser.* **2177** 012018

[17] Melchiorri L, Narcisi V, Giannetti F, Caruso G and Tassone A 2021 Development of a RELAP5/MOD3.3 Module for MHD pressure drop analysis il liquid metals loops: verification and validation *Energies* **14** 5538

[18] Narcisi V, Melchiorri L and Giannetti F 2021 Improvement of RELAP5/MOD3.3 heat transfer capabilities for simulation of in-pool passive power removal *Ann. Nucl. Energy* **160** 108436

[19] Alamberti A, Caramello M, Frignani M, Grasso G, Merli F, Morresi G and Tarantino M 2020 ALFRED reactor coolant system design *Nucl. Eng. Des.* **370** 110884

The effect of different sediment conditions on CO_2-CH_4 replacement in natural gas hydrates

Beatrice Castellani[1,*], Alberto Maria Gambelli[1], Giorgio Minelli[2], Federico Rossi[1]

[1]Department of Engineering, University of Perugia, Italy
Via G. Duranti 67, 06125, Perugia Italy
[2]Department of Physics and Geology, University of Perugia, via Alessandro Pascoli, Perugia, 06123, Italy

*Corresponding Author: beatrice.castellani@unipg.it

Abstract. Natural gas hydrates are the largest reservoir of natural gas on earth and represent an important solution for the transition from the actual energy scenario to a renewable one. Methane, contained in hydrate reservoirs, can be replaced with carbon dioxide, so that the obtained fuel is neutral in terms of climate-changing emissions and therefore equivalent to renewable energy sources. The experimental tests presented in this paper are devoted to study the CO_2-CH_4 replacement process in natural gas hydrates in presence of water salt. Test are carried out in a lab scale reactor of 1 l volume. The investigation aims at obtaining data on methane recovery rate, CO_2 sequestration, composition of the obtained outlet CO_2/CH_4 mixture, as well as spatial distribution of hydrates thanks to the analysis of temperature profiles. Temperature analysis is used to determine hydrates' spatial distribution within the sediment. The number of methane hydrate moles formed with salt water is lower than that with pure water. During CO_2 injection, not only CH_4-CO_2 replacement took place, but also new CO_2 hydrate formation. The exchange factor, in fact, ranges from 0.11 to 0.4 for tests with salt.

1. Introduction

Clathrate hydrates are crystalline structures where water molecule cages host gas molecules [1]. Natural deposits of methane hydrates are found both in continental sedimentary rocks in polar areas and in marine sediments. The main deposits have been identified at the margins of the continental platforms, where the formation is favored by the presence of organic material and the appropriate pressure and temperature conditions. Natural gas hydrates (NGH) deposits in underwater sediments, generally found along the continental shelves and slopes, with water depths exceeding 500 m [2-3].

NGH constitute the largest reservoir of natural gas on the planet: it is estimated that the organic carbon stored in NGH is twice the natural gas from all available fossil sources [4-5].

NGH deposits are classified in four categories. Class 1 hydrates, which are formed by a hydrate bearing interval overlying a free gas interval, are the most desirable deposit for exploitation because of the favorable permeability regime and the thermodynamic proximity to the hydrate equilibrium line, necessitating only small changes in p and T to induce dissociation. The wide range of p-T conditions and gas availability in these natural systems yield to a wide range of possible hydrate saturations, from 5% up to 80% of pore space [6].

For the production of natural gas from oceanic sediments, the following methods are proposed: thermal stimulation, depressurization, injection of chemical inhibitors for hydrate equilibrium curve shift [7-8].

If CO_2 is injected into NGH sediments, it causes CH_4 release and CO_2 hydrate formation. The replacement can potentially occur by total or partial CH_4 hydrate dissociation and new formation of CO_2 hydrate in liquid water. Given a constant pressure, there is a region between the curves where gaseous CH_4 is released while CO_2 still forms hydrates. Such thermodynamic window ranges up to 35°C, depending on the system pressure. Operating within this region, the release of methane from the cages and the simultaneous formation and preservation of CO_2 hydrate is obtained [9-11]. This process, known as CH_4-CO_2 replacement, creates a unique opportunity to recover an energy resource, CH_4, while entrapping a greenhouse gas, CO_2. Hydrate reservoirs can hence function as both CH_4 sources and CO_2 storage sites, enhancing the idea of a carbon neutral fuel source.

Since most of methane hydrates occur in oceanic sediments, pore water involved in formation has a salinity similar to seawater; about 3.5%. It is known that the stability of methane hydrate in saline water decreases as the salinity of water increases [12]. It is therefore important to investigate the CO_2-CH_4 replacement process in saline water to investigate if with the inhibitor effect of salt replacement, it is more efficient.

The experimental tests presented in this paper are devoted to study the CO_2-CH_4 replacement process in natural gas hydrates with a salinity of 37 g/L. Test are carried out in a lab-scale reactor of 1 l volume. The investigation aims at obtaining data on methane recovery rate, CO_2 sequestration, composition of the obtained outlet CO2/CH4 mixture, as well as spatial distribution of hydrates thanks to the analysis of temperature profiles. Temperature analysis is used to determine hydrates' spatial distribution within the sediment. It helps in distinguishing new CO_2 hydrate formation from CO_2 replacement in cages previously formed.

2. Materials and methods

2.1 Experimental apparatus

The lab – scale reactor used for experimental tests was realized with 316SS stainless steel and has an internal cylindrical volume of 949 cm^3; the diameter is equal to 7.3 cm while internal height is about 22.1 cm. The experimental apparatus is shown in Figure 1.

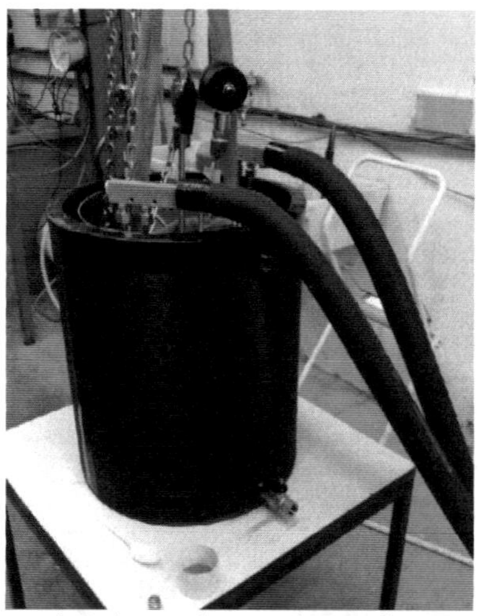

Figure 1. Lab – scale experimental apparatus

Both the upper and the lower surface of that volume are sealed with two 316SS stainless steel flanges (both of them composed by two 4.7 cm thick plates). The reactor is positioned inside a thermostatic bath to ensure temperature control. That vessel is filled with water and glycol, and is refrigerated by a chiller, model GC – LT (by Euro – chiller). The lower flange has one center 3/8" port, used for methane injection. The upper flange has several ports: i) one center port is used for CO_2 injection, ii) one port is used for positioning four thermocouples, iii) one port is used for check valve and pressure gauge, iv) one port is used for gas recovery. A detailed description of the reactor is available elsewhere in literature [9-10]. The CO_2 port is equipped with ¼" tube with a length of mm to inject the CO_2 inside the hydrate sediment. The four 1-mm K type thermocouples are inserted in the reactor at different depths through a high pressure multi-sensor fitting (supplied by Spectite), to precisely monitor temperature and well identify differences due to hydrate formation or dissociation. The uncertainty of the thermocouples is ±0.75%; since the experimental tests are carried out in the range from 0.2°C to 3.3°C, the uncertainties are lower than 0.03 °C and they are acceptable. A pressure release valve and a digital manometer (supplied by Kobold) are inserted in the third port to monitor the headspace pressure and provide depressurization in case of pressure increase above the nominal value. The accuracy of the manometer is ±0.5% and, for a pressure of 60 bar, the uncertainty is 0.3 bar. The fourth port is used to recovery methane after dissociation and CO_2-CH_4 replacement. All devices are connected with a data acquisition system manufactured by National Instruments for monitoring and recording experimental data, which is operated with Labview software. Figure 2 shows a schematic drawing of the experimental apparatus. Gascromatographic analyses are carried out with a gascromatographer supplied by VARIAN (model VARIAN CP 4900 Micro-GC). Measurements, periodically carried out with certified mixtures, show that uncertainties are always lower than 1%, with an excellent repeatability on sequences of samples.

The internal volume is filled with silica quartz sand and water. Only 0.01 m headspace is left over the sand deposit for accumulation of gas and pressurization. Water is added to the sand to reach the targeted initial water saturation. Silica quartz sand has an average of 500 µm diameter and a porosity equal to 34% (measured with a porosimeter, model Thermo Scientific Pascal 140).

Methane and carbon dioxide (99.97% purity and 99.99% purity respectively) are provided in gas bottles by Air Liquide Italia Service. Pure sodium chloride (NaCl) was added to water with a concentration of 37 g/l.

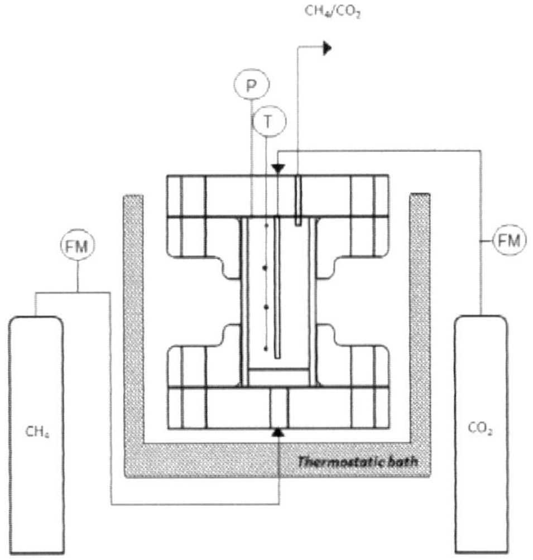

Figure 2. Schematic drawing of the experimental apparatus

2.2 Experimental procedures

The experimental procedure consisted of two steps: CH_4 hydrate formation followed by CO_2 injection. Methane injection starts when the internal temperature reaches the equilibrium value of 2.5°C at a pressure of 0.1 MPa. Methane pressurizes the vessel to the initial pressure P_i, and gas injection is stopped. Because of the exothermic nature of the hydrate formation process, temperature increases noticeably. After the inlet flow is stopped, the internal pressure gradually decreases, since gaseous methane is entrapped into the hydrate phase.

The formation test is considered completed when all the thermocouples measure 2.5°C. The number of hydrate moles formed in the system once it reaches the equilibrium given here are obtained following the Equation (1) given in [13]:

$$n_{hyd} = \frac{V_{pore}\left(P_i Z_f - P_f Z_i\right)}{Z_f\left(RT - \frac{P_f}{\rho_{hyd}}\right)} \tag{1}$$

Where P_i and P_f are the initial and final system pressure respectively; Z_i and Z_f are the compressibility factors, calculated using the Peng-Robinson equation. ρ_{hyd} is the ideal molar density of hydrate, calculated assuming 100 % cage occupancy. Pore space saturation can be obtained from Equation (2) given in [13]:

$$s_H = \frac{n_{hyd}\frac{1}{\rho_{hyd}}}{V_{pore}} \tag{2}$$

where n_{hyd} is the number of formed hydrate moles and V_{pore} is the total volume of sand pores. CO_2 is injected at the end of methane hydrate formation. The replacement is carried out through depressurization. At the end of methane hydrate formation, pressure is rapidly decreased to a value equal to 30 bar. At this value, CO_2 is injected, with outlet valve kept open to flush gaseous methane out of the reactor. Then the outlet valve is closed while the CO2 injection continues to maintain pressure constant.

Once the process is completed, the gas phase is quickly expelled, then the reactor was closed. The dissociation of hydrates and the substantial increase in pressure allows to quantify the total moles of CH_4 and CO_2 involved. Some values are then calculated: moles of CH_4 hydrates involved in the replacement process and new CO_2 hydrates were formed.

3. Results

Three tests on methane hydrate formation and CO_2 replacement were carried out in presence of water with a salinity equal to 37g l^{-1}. Test 4 was carried out with pure water for sake of comparison. Table 1 shows experimental parameters for each tests.

Table 1. Experimental tests on methane hydrate formation

Test	Initial Pressure P_i (bar)	Initial temperature T_i (°C)	Final Pressure P_f (bar)
1	63	2.8	59
2	61	3.3	56
3	61	1.6	57
4	57	2.5	52
Test	**Final temperature T_f (°C)**	**Injected CH_4 moles (mol)**	**CH_4 moles in hydrates (mol)**
1	0.2	0.799	0.056
2	0.6	0.762	0.065
3	0.7	0.775	0.051
4	2.2	0.704	0.082

Test 1 was carried out at the initial pressure and temperature respectively of 63.0 bar and 2.8 °C. To reach the initial pressure, 0.799 moles were injected. The finale pressure achieved at the end of the hydrate formation phase was 59 bar, with a temperature of 0.2 °C. The number of CH_4 moles in the hydrate phase was equal to 0.056.

Test 2 was carried out at the initial pressure and temperature respectively of 61.0 bar and 3.3 °C. To reach the initial pressure, 0.762 moles were injected. The finale pressure achieved at the end of the hydrate formation phase was 56 bar, with a temperature of 0.6 °C. The number of CH_4 moles in the hydrate phase was equal to 0.065.

Test 3 was carried out at the initial pressure and temperature respectively of 61.0 bar and 1.6 °C. To reach the initial pressure, 0.775 moles were injected. The finale pressure achieved at the end of the hydrate formation phase was 57 bar, with a temperature of 0.7 °C. The number of CH_4 moles in the hydrate phase was equal to 0.051.

Test 4 was carried out at the initial pressure and temperature respectively of 57.0 bar and 2.5 °C, without salt. To reach the initial pressure, 0.704 moles were injected. The finale pressure achieved at the end of the hydrate formation phase was 52 bar, with a temperature of 2.2 °C. The number of CH_4 moles in the hydrate phase was equal to 0.082.

This results confirm literature data about the inhibitor effect of salt on methane hydrate formation [14-16]. The number of CH_4 moles entrapped in hydrates is smaller by 25-30% in salt water with respect to pure water.

Figure 3 shows the pressure profile for Test 3. The hydrate formation causes a pressure decrease due to the methane adsorption inside the hydrates. Once pressure is stabilized, the reactor was depressurized to allow CO_2 replacement, during which pressure started to decrease again. Then the vessel was depressurized until atmospheric pressure. Dissociation started and pressure increased due to the hydrate dissociation and the releasing of gaseous mixture from solid to gaseous phase. Figure 4 shows the temperature vs. time diagram during methane hydrate formation, in which the temperature increase due to the exothermic nature of the formation is visible.

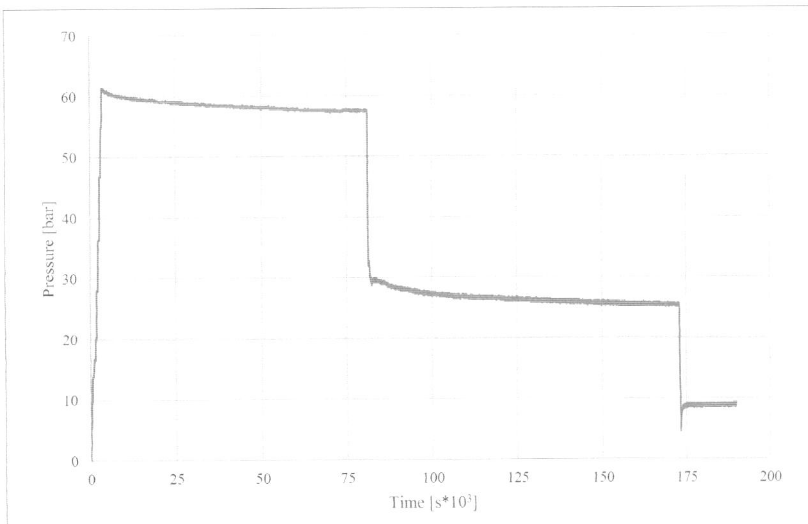

Figure 3. Pressure profile-Test 3.

Thermocouple T02 is positioned in the headspace, where there is only gas, temperature does not change over time, because the headspace is not involved in hydrate formation.

CH_4 hydrate formation starts in the upper part of the sediment (thermocouple T07), where the first peak due to the exothermic nature of methane hydrate formation is recorded. Then it moves downwards, to the T11 and T16 regions. The most significant temperature increase occurs at the center of the sediment (corresponding to thermocouples T11). Table 2 shows experimental parameters for the three tests regarding CO_2 replacement.

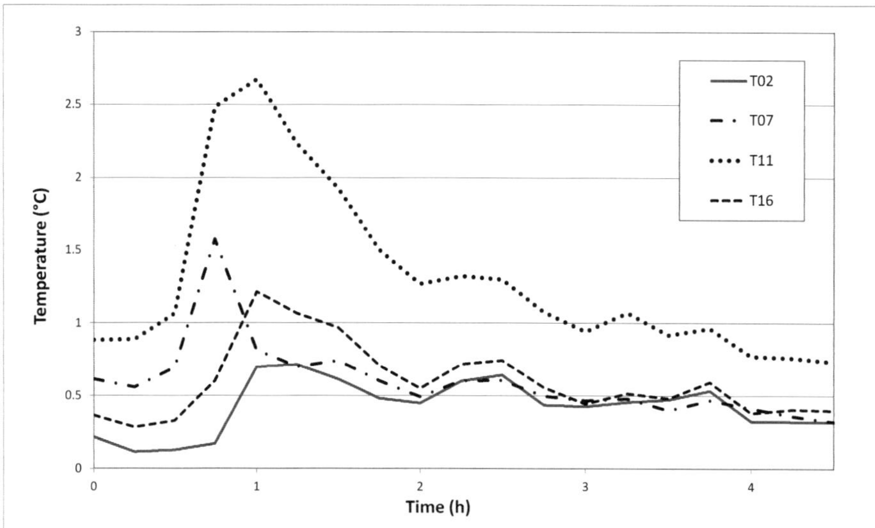

Figure 4. Temperature profiles during methane hydrate formation -Test 3.

Table 2. Experimental tests on CO_2 replacement.

Test	P_i (bar)	P_f (bar)	CH_4 moles in hydrate (mol)	CO_2 moles in hydrate (mol)	Total moles in hydr. (mol)
1	30.03	24.91	0.048 (43%)	0.065 (57%)	0.113
2	31.43	24.99	0.053 (45%)	0.064 (55%)	0.116
3	29.16	25.08	0.023 (23%)	0.073 (76%)	0.096
4	38.22	34.38	0.056 (20%)	0.107 (80%)	0.163

The CO_2 replacement was carried out by depressurization, at an internal pressure of 30.03 bar in Test 1, 31.43 bar in Test 2, 29.16 bar in Test 3. CO_2 replacement tests continued for about 5 hours.

In Test 1, at the end of the CO_2 replacement test, hydrate cages were occupied by 57% CO_2 and 43% CH_4. In Test 2, hydrate cages were occupied by 55% CO_2 and 45% CH_4, while in Test 3, gas inside hydrates was constituted by 24% CO_2 and 76% CH_4. In Test 4, with pure water, hydrates cages were occupied by 20% CH_4 and 80% CO_2.

Total moles of gas inside hydrates after CO_2 replacement process in Test 4, are higher than in tests with salt. Total hydrate moles after CO_2 replacement are double the total number of methane hydrate moles in table 1, for all the tests. The difference is in the CH_4 moles recovered from hydrates after the replacement.

Comparing results in Table 1 with those in Table 2, CH_4 moles recovered from hydrates are 0.008 in Test 1, 0.012 in Test 2 and 0.029 in Test 3, while 0.056 in Test 4. Especially, for Test 1 and Test 2, these amounts are much lower than the number of CO_2 moles entrapped in hydrates. This means that

CO_2 replacement occurred only to a small extent, while there was a significant formation of new CO_2 hydrates.

At this point, the exchange factor is introduced, as the ratio of the CH_4 moles produced from the hydrate phase and the CO_2 moles sequestered in the hydrate phase and it gives an extent of the replacement process. The exchange factor is equal to 0.11 in Test 1, 0.19 in Test 2 and 0.40 in Test 3. The values of the exchange factor in Test 1 and Test 2 are much lower than 1, demonstrating that the main phenomenon which took place, was the formation of new CO_2 hydrates. Further details on such aspect can be obtained from temperature profiles during CO_2 injection in Figure 5.

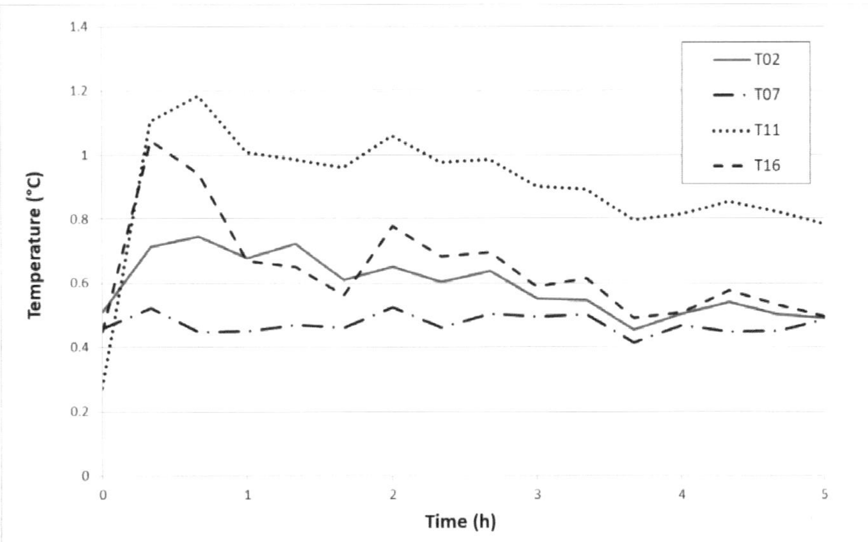

Figure 5. Temperature profiles during CO_2 replacement -Test 3.

The test starts at an average internal temperature equal to 0.4 °C. Over time, the first temperature peaks are recorded by thermocouples T11 and T16, respectively in the middle region and in the low region of the sediment.

T11, in particular, is the thermocouple located in an area where there is a significant presence of methane hydrates (see Figure 4). It is therefore very likely that in this region the preponderant phenomenon is CO2 replacement. T16 instead is located in the bottom area of the sediment near the CO2 injection point, where methane hydrate formation is marginal if compared to T11 area. Here, mainly new CO2 hydrate formation took place, also because it is an area of the reactor rich in water. T07 seems not to be affected by temperature changes. It suggests that CO2, injected at the bottom, is entrapped in the solid phase only in the central and lower part of the reactor, not being able to reach the upper region of the sediment. Test 4, without salinity, shows an exchange factor equal to 0.25, which is similar to Test 2 and lower than Test 3.

4. Conclusion

The present paper discusses the results obtained from experimental tests on methane hydrate formation and CO_2 replacement process in presence of salt water. Test were carried out in a lab-scale reactor of 1 l volume, in presence of water with a salinity equal to 37g l^{-1}. One test was carried out with pure water for sake of comparison. The initial pressure was about 60 bar. The number of hydrate moles formed with salt water is lower than that with pure water. This confirms the inhibitor effect of salt on hydrate formation. CH_4 hydrate formation starts in the upper part of the sediment and then moves downwards. The most significant temperature increase occurs at the center of the sediment.

During CO_2 injection, not only CH_4-CO_2 replacement took place, but also new CO_2 hydrate formation. The exchange factor, in fact, ranges from 0.11 to 0.4 for tests with salt.

CO_2, injected at the bottom, replaced methane mainly in the central part of the reactor, while the bottom region is interested by new CO_2 hydrate formation. So, CO_2 is entrapped in the solid phase only in the central and lower part of the reactor, not being able to reach the upper region of the sediment.

Acknowledgements

The authors gratefully acknowledge financial support derived from MIUR by the PRIN Project 2017 entitled: "Methane recovery and carbon dioxide disposal in natural gas hydrate reservoirs" Prot.20173K5L3K.

References

[1] Sloan, E. D. Fundamental principles and applications of natural gas hydrates, Nature, 2003, 426, 353-363.
[2] Makogon YF et al. Natural gas-hydrates - A potential energy source for the 21st Century, Journal of Petroleum Science and Engineering, 2007, 56, 14-31
[3] Trofimuk AA et al. Accumulation of natural gases in hydrate formation zones of the world'oceans. Doklady Akademii Nauk SSSR, 1973.
[4] Milkov AV. Global estimates of hydrates-bound gas in marine sediments: how much is really out there? Earth Science Reviews, 2004, 66, 183-197.
[5] Klauda JB, Sandler SI. Global Distribution of Methane Hydrate in Ocean Sediment. Energy & Fuels, 2005, 19 (2), 459-470.
[6] Ruppel, C., Methane hydrates and the future of natural gas, Gas hydrates project, U.S. Geological Survey, Woods Hole, MA, MITEI Natural Gas Report, Supplementary Paper on Methane Hydrates, 2011.
[7] Dornan P et al. Free energies of carbon dioxide sequestration and methane recovery in clathrate hydrates. The Journal of chemical physics, 2007, 127, p.124510.
[8] Fitzgerald GC et al. Large scale reactor details and results for the formation and decomposition of methane hydrates via thermal stimulation dissociation, Journal of Petroleum Science and Engineering, 2012, 94-95,19.
[9] Rossi F et al. Experiments on methane hydrates formation in seabed deposits and gas recovery adopting carbon dioxide replacement strategies (2019) Applied Thermal Engineering, 148, pp. 371-381.
[10] Castellani, B et al. Energy and environmental analysis of membrane-based CH4-CO2 replacement processes in natural gas hydrates (2019) Energies, 12 (5), art. no. 850.
[11] Castellani, B et al. Simulation of CO2 storage and methane gas production from gas hydrates in a large scale laboratory reactor (2016) Journal of Petroleum Science and Engineering, 147, pp. 515-527.
[12] Maekawa T et al. Pressure and temperature conditions for in sodium chloride methane hydrate solutions dissociation. Geochemical Journal, Vol. 29, pp. 325 to 329, 1995.
[13] Fitzgerald GC, Castaldi MJ, Zhou Y. Large scale reactor details and results for the formation and decomposition of methane hydrates via thermal stimulation dissociation. J Petroleum Sci Eng 2012;94-95:19-27.
[14] P.D. Dholabhai, P. Englezos, N. Kalogerakis, P.R. Bishnoi. Equilibrium conditions for methane hydrate formation in aqueous mixed electrolyte solutions Can J Chem Eng, 69 (1991), pp. 800-805
[15] M.D. Jager, E.D. Sloan. The effect of pressure on methane hydration in pure water and sodium chloride solutions Fluid Phase Equilib, 185 (2001), pp. 89-99
[16] P. Englezos, P.R. Bishnoi. Prediction of gas hydrate formation conditions in aqueous electrolyte solutions AIChE J, 34 (1988), pp. 1718-1721

Microgrid application of liquefied air energy storage (LAES) systems

Beatrice Castellani[a*], Parminder Kaur[a], Adriano Sciacovelli[b], Federico Rossi[a], Yulong Ding[b]

[a]CIRIAF, University of Perugia, Engineering Department, Via G. Duranti 67, 06125 Perugia, Italy
[b] Birmingham Centre for Energy Storage, University of Birmingham, Edgbaston, Birmingham, B15 2TT, United Kingdom
*corresponding author: beatrice.castellani@unipg.it

Abstract. An energy and economic analysis on small-scale LAES (liquefied air energy storage) system is presented. The LAES operative parameters were analyzed via MATLAB simulations. The optimal case is given with a flowrate of 1000 kg/h and 4 turbine expansions, resulting in a net electric power output in the discharging process of 260 kW and the optimal values are 450 K for inlet temperature and 150 bar for the discharge pressure. For the charging process the specific consumption is significantly affected by the storage pressure. Also the round trip efficiency is influenced by the storage pressure: for a storage pressure of 8 bar, it is about 25% compared to the 12% in case of a storage pressure equal to 1.10 bar. The LCOS obtained is between 1.2 €/kWh and 1.8 €/kWh. This values are higher than other results in literature due to the scale of the system.

1. Introduction

The evolution of the energy transition based on renewable energy sources (RES) reached an important milestone with the recognition from the European Union (EU) of the importance of energy communities. In fact, with two Directives [1-2] the EU has given the definitions of the energy communities and asked the Member States to introduce them into their national legislations. Energy communities are emerging in the huge transformation of the electrical systems. The paradigm based on centralized generation is being replaced by a system based on distributed generation. In the energy community scenario, energy storage systems (ESS) has also become one of the key building blocks providing distribution grids with lots of benefits in terms of stability, reliability, quality, and control. Small-scale applications of ESS are therefore needed.

There is a wide range of ESS to store electrical energy. A common approach is classifying ESS in accordance to the form of used energy: mechanical, electrochemical, chemical, electrical and thermal energy storage systems [3].

The most common mechanical storage systems are pumped hydroelectric power plants, compressed air energy storage (CAES), flywheel energy storage and mechanical springs. Electrochemical storage systems consist of various types of batteries. Chemical energy storage focuses on hydrogen and synthetic natural gas (SNG) as secondary energy carriers and, finally, electrical storage systems include double-layer capacitors and superconducting magnetic energy storage [3-9]. Technologies used for high power ranges and energy capacities are pumped hydro storage (PHS) and CAES [10].

CAES uses the surplus electricity to compress air. The high-pressure air is saved in sealed devices (such as underground mine, large air storage tank) and is released to drive the expander generating electricity when the demand of electricity is high. It is used only in stationary applications. There are seven commissioned CAES facilities worldwide (four of these in the United States) and three large facilities are under construction in China [11]. New approaches, based on hybrid systems are being explored to potentially balance cost/performance trade-offs. For example, in the Netherlands, CAES was recently combined with Li-ion batteries for ancillary services. This approach is expected to both extend the Li-ion life and significantly improve the total lifetime cost [11]. Because of the storage characteristics of large capacity and long-term, CAES has attracted the attention and research of a large number of scholars around the world. The Huntorf power station in Germany and the Mcintosh power station in United States are two typical large-scale CAES power stations in the world. CAES technology is a research hotspot of large-scale energy storage technology at present. Analyses in literature show that CAES could potentially compete with Li-ion for about 60 GWh of the total 150-GWh projected capacity required in 2030 [12].

An upgrade of the CAES technology is Liquid air energy storage (LAES). With a working principle similar to that of CAES, LAES uses liquid air as the medium for energy storage [13]. The benefits are: i) energy storage density much higher than CAES, ii) no restriction on geographical condition. The development of LAES technology began in 2005 when the Leeds University in the UK and Highview Power completed the first full-system patent of LAES technology. In addition, the first complete independent demonstration LAES system in the world was built in Slough, UK, in 2011 [14].

At present, there have been a lot of studies on the performance, economy and dynamic simulation of large-scale LAES. The research is devoted to improve the overall efficiency of the system by improving the performance of the components or to coupling LAES with other systems to make full use of the excess energy [15-18].

At small-scale, Highview Power developed a 300 kW/2.5 MWh LAES pilot plant that has been operating in the UK since 2010. Literature studies on LAES small scale applications, applied to demand side management (i.e. load levelling, peak shaving, time shifting, backup power), are devoted to find the optimal plant configuration which minimizes the specific consumption of the air liquefaction process [19]. A preliminary analysis in [19] consists in comparing different liquefaction cycles by changing the operating parameters like pressure and recirculation fraction. From the simulation results, Claude and Kapitza cycles offer the lowest specific consumption; the performance is incremented of 25% with the two stage compression. From the study, the pressurization of the phase separator positively influences the performance, decreasing the specific consumption by 21% with a phase separator pressure of 4 bar. Also the combined effect of both pressurized phase separator and incremented operating pressure can decrement the specific consumption below 500 kWh/t. Increasing the operating pressures increases the sizes of the components, the complexity of the cycle and the costs.

The optimal configuration of an air liquefier can be a two stage compression Kapitza cycle with an operating pressure in the range of 38–45 bar and a phase separator pressure of 6 bar. This range of operating parameters provides a specific consumption range of 520–560 kWh/t which is interesting for LAES application in microgrids.

A study on a pilot plant scale LAES was carried out by Morgan et al. [20]; a round trip efficiency of 8% was obtained because of the small size of the pilot plant, and due to the fact that only 51% of the accessible cold was recovered.

In another work [21] the authors analyze the performance of a single-effect absorption chiller using a Water-Lithium Bromide solution combined with a small air liquefier with a liquid air production capacity of 0.834 t/h. In the suggested solution, the waste heat of the compression phase of the liquefaction cycle is recovered and used to run the absorption cycle, where the resulting cooling power is adopted to reduce the specific consumption and improving the exergy efficiency of the system. The results reveal a decrease of the specific consumption of about 10% (537 kWh/t to 478 kWh/t) and an increment of exergy efficiency of about 11.5%.

The paper by S. Mazzoni et al. [22] offers the use of polygeneration power plants in the smart energy scenarios as a valid option to the traditional production systems in terms of cost savings, pollutant emission reductions and for overall improvement in performance efficiency.

In this specific work the considered LAES system is a small scale one for microgrid application. In literature few studies are conducted about small scale LAES in microgrid application. "Microgrid scale" refers to a liquefaction plant with daily production between few tons and few tens of tons of liquid air. That correlates to a size of air liquefaction plant between lab and industrial scale. The modeling of the LAES system for energy and economic analyses is proposed, together with a comparison with literature data.

2. Methodology

The liquid air energy storage (LAES) operating principle includes 3 main phases: liquefaction of gaseous air when energy is available at off-peak times, storage of liquid air in insulated tanks and expansion of liquid air in turbines to generate power in the peak of demand. The research is carried out through MATLAB model and articulated in the following activities:

- Energy analysis of the LAES system with the calculation of the main operating parameters and the round trip efficiency
- Economic evaluation.

2.1. Energy analysis

Firstly the expansion process is analyzed, considering that the number of the expander units is in the range from two to four. The number of expanders in the expansion unit directly influences the energy output in the discharge phase, which also impacts the overall efficiency of the system. So, the electric power output in the expansion unit with different stages is calculated. Other two parameters of the discharging stage are evaluated: the efficiency of the cold storage and the effectiveness of the superheater. As far as the charging stage is concerned, the power consumption of the compression unit is estimated. Finally, the key parameter that describes the performance of the system - the round trip efficiency - is calculated.

2.1.1 Discharging stage. During the discharging stage, liquid air is pumped by a cryogenic pump from the storage tank and regasified to ambient temperature. The cold energy released during the regasification is stored in a Cold Storage System (CSS) in order to reuse the waste cold for the charging stage. The air expansion is a multi-stage process with superheaters (SHs). The CSS is modeled by means of its utilization factor (%), evaluated as the ratio between the effective thermal power recovered and the maximum available thermal power:

$$\eta_{HGCS} = \frac{\dot{Q}_{u,HGCS}}{\dot{Q}_{tot,HGCS}} \tag{1}$$

For the expansion unit, three different cases were analyzed: from 2 to 4 expansion stages. Air pressure and temperature entering the expansion stage are considered constant. The superheater is modeled as a two-stream heat exchanger. The total output power of the discharging stage is the net Electric power output of the system, $P_{net,d}$ and it is evaluated with the following equations, as elsewhere in literature [23-24]:

$$P_{net,d} = n_{ex} \cdot \dot{m}_{LA} \cdot cp_{ave,air} \cdot TIT \cdot (1 - \frac{1}{\alpha_e^{eta_{pol}}}) \tag{2}$$

$$eta_{iso,ex} = \frac{\alpha_e^{\frac{1}{eta_{pol}} - 1}}{\frac{1}{n} - 1} \tag{3}$$

$$\alpha_e = (\beta_{ex})^{\frac{k-1}{k}} \tag{4}$$

where n_{ex} is the number of the expansion processes, $\dot{m}_{LA}$ [kg/h] is the liquid air production at the end of charge phase, $cp_{ave,air}$ [kJ/kgK] is the average isobaric specific heat of air, k is the specific heat ratio of air, eta_{pol} is the polytropic efficiency of the expansion process, β_{ex} is the expansion ratio, TIT is turbine inlet temperature [°C] and p_{amb} [bar] is the ambient pressure.

2.1.2 Charging stage. The charging stage is formed by a compression section and a liquefaction section. A recuperative process is included in the charging stage of the LAES. The recuperative process consists of two stages of compression, with intercooling, a recuperative heat exchanger, an expander, a J-T valve, a phase separator and liquid tank. The charging stage is modelled through the evaluation of the Specific Consumption (SC). The SC can be expressed as a function of both charge pressure and recirculation fraction, as expressed in [23]. Recirculation fraction x_{RF} is the ratio of the mass flow elaborated by the J-T valve and the mass flow entering the recuperative heat exchanger.

2.1.3 Round trip efficiency. The key parameter evaluated in the energy analysis is the round trip efficiency, η_{RT} [%], defined as [23-24]:

$$\eta_{RT} = \frac{P_{net,d}}{P_{net,ch}} \tag{5}$$

Where $P_{net,ch}$ [kW] is the net electric power consumed during the LAES charge phase, $P_{net,d}$ [kW] is the net electric power produced by the power turbines. The specific values of the produced and consumed net electric power are also evaluated per unit of produced liquid air:

• Specific electric power output (SP) [kW/kg]
• Liquefaction specific consumption (SC) [kW/kg]

2.2. Economic evaluation

The economic evaluation and comparison with other storage technologies was carried out calculating the Levelized Cost of Storage (LCOS), which offers a quick comparison of the cost of electricity-to-electricity storage systems [25]. The LCOS is defined as the total cost over the entire lifetime of the plant divided by the total amount of electricity produced by the storage system, and has been used in this study to evaluate the economic potential of the small-scale LAES system. It can be expressed as:

$$LCOS = \frac{I_o + \sum_{t=1}^{t=n} \frac{TC_t}{(1+i)^t}}{\sum_{t=1}^{t=n} \frac{EOUT_t}{(1+i)^t}} \tag{6}$$

Where I_o represents the capital expenditure for investment, TC_t denotes the annual total costs at year t, $EOUT_t$ stands for the annual electricity outputs, and n is the lifetime of the plant [years]. The annual costs and the annual electricity outputs are discounted with the interest rate i.

3. Results

In this section, the results of the energy and economic evaluation are presented and discussed with respect to the state of the art.

3.1. Energy results

The discharging stage was modelled assuming the following ranges:

- Discharging pressure p_d: 50–150 bar
- Storage pressure p_s: 1.10 and 8 bar.

Three cases are studied to evaluate the net Electric power output of the system and the SP: from 2 to 4 expansions. For each case the results are analyzed for two different mass flow rates (800 kg/h and 1000 kg/h). Data of the net Electric power output are compared in Table 1. In all the three cases, the net power output range is higher for a flow rate equal to 1000 kg/h.

Table 1: Net electric power output for two values mass flow rates (800 kg/h and 1000 kg/h).

N. expansions	Flow rate 800 kg/h	Flow rate 1000 kg/h
2	80-160 kW	100-200 kW
3	100-200 kW	120-240 kW
4	100-220 kW	140-260 kW

Table 2 shows the results of the discharging stage calculations considering the best conditions, while Table 3 shows the results of the charging stage calculations.

Table 2: Discharging stage calculations.

Case	N. expansions	P_d (bar)	TIT (K)	$\dot{m}_{LA}$ (kg/h)	P_{netd} (kW)
1	2	150	450	1000 kg/h	200
2	3	150	450	1000 kg/h	240
3	4	150	450	1000 kg/h	260

Table 3: Charging stage calculations.

Case	P_s (bar)	P_{ch} (bar)	SC (kW/kg)
1	1.10	50-100	2- 3.2
2	8	50-100	0.9-1.4

The round trip efficiency η_{RT} is evaluated for $\dot{m}_{LA}$=1000 kg and two different $p_{storage}$ (1.10 bar and 8 bar). Figure 1 shows the round trip efficiency η_{RT} as a function of the SC and the SP for the two values of storage pressure.

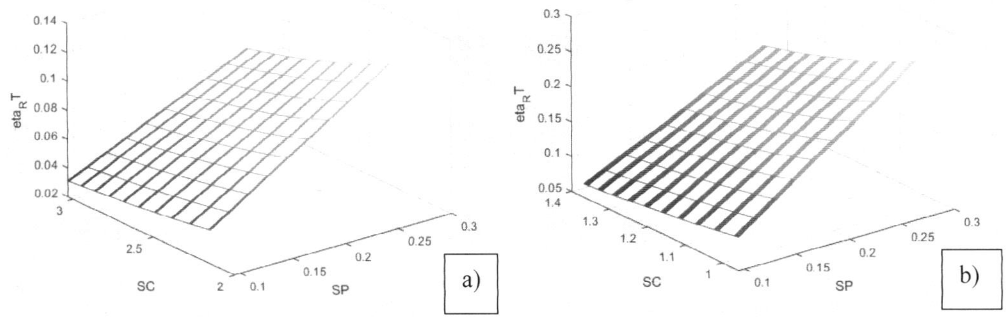

Figure 1: Performance of the $\boldsymbol{\eta_{RT}}$ at $\boldsymbol{p_{storage}}$= 1.10 bar (a) and 8 bar (b).

An increase of the SP and a decrease of the SC positively contributes to the η_{RT}. The increase of the SP is related to a higher TIT and a higher discharge pressure, the SC depends instead from the charging pressure. The two graphs allows to highlight the significant dependence of η_{RT} from the storage pressure. For a storage pressure equal to 1.1 bar, the range of the η_{RT} obtained is between 2% and 11%. For a storage pressure equal to 8 bar, the range of the η_{RT} obtained is between 5% and 25%. This results are comparable with others in literature [23-24], showing that there is a strong dependence of the round trip efficiency on the storage pressure.

3.2 Economic results

The capital expenditure for investment consists of three terms; for each one the following values are assumed:

- Capital expenditure for the liquefier: 1100 €/kW
- Capital expenditure for the turbine: 270 €/kW
- Capital expenditure for storage tank: 25.3 €/kW.

The annual total cost (TC) is calculated as the sum of OPEXE (the annual energy based maintenance expense), OPEXP (the annual power based operating expense), EC (the annual electricity purchasing cost) and IC (the insurance costs for all the devices). The OPEXE is assumed equal to 0.00253 €/kW and the OPEXP equal to 10.67 €/kW [26]. The annual electricity purchasing cost is determined by the electricity price. Since the cost of electricity may vary over time, it is assumed an average electricity price of 3 €ct/kWh. As the price is influenced by the number of full load hours per year, both should be examined in combination. For this work it is assumed that the energy storage system working with two cycles per day and so 7300 cycles over the lifetime of 10 years. The insurance costs for all the devices is considered $0.5I_o$ [27].

The P_{netd} and P_{netch} are referred to the case of 4 expansions and $\dot{m}_{LA}$=1000 kg/h; and 170 kW is the storage capacity of the tank. The interest rate i is assumed 2% and the lifetime of the plant is 20 years. The range of the LCOS obtained is between 1.2 €/kWh and 1.8 €/kWh. Such values are higher than the range presented in literature for a large scale LAES system and equal to 0.21-0.649 €/kWh [26]. This may depend from the low round trip efficiency of the small scale system, in fact the LCOS decreases with the increment of the round trip efficiency.

Finally, another study [27] shows the LCOS for different storage systems. PSH system has the lowest cost with 5–9 €ct/kWh and CAES systems have a similar LCOS in the range 7–12 €ct/kWh. Li-ion batteries offer an LCOS between 23 and 37 €ct/kWh at 365 cycles per year; this value is higher than that of Pb batteries which present an LCOS of 15–19 €ct/kWh. VRF batteries have a LCOS of 32 €ct/kWh. The LCOS of battery technologies is expected to decrement in the next decade thanks to the technological developments and decreasing I_o.

Such LCOS values are low compared to the system analysed in this work. Nevertheless, both studies in [26-27] considered large-scale storage systems, as the comparison shows that there is an increase in costs in the small scale: a small scale cycle optimization is therefore necessary.

4. Conclusion

In this work, an energy and economic analysis on small-scale LAES is presented. The LAES operative parameters were analyzed via MATLAB simulations: specific consumption, net electric power output and round trip efficiency.

To evaluate the net electric power output three different cases were analysed: the first one considers 2 expansions, the second one 3 expansions and 4 in the last one. For each case the results are analysed for two different mass flow rates (800 kg/h and 1000 kg/h). The optimal case is given with a flowrate of 1000 kg/h and 4 turbine expansions, resulting in a net electric power output in the discharging process of 260 kW and the optimal values are 450 K for inlet temperature, 1000 kg/h for the mass flow rate and 150 bar for the discharge pressure.

For the charging process the specific consumption is significantly affected by the storage pressure. With its increase, a lower specific consumption is obtained considering the same range of the charge pressure and recirculation fraction, resulting in a maximum value of the specific consumption in the charging process of 1.4 kW and minimum value of 0.9 kW for a storage pressure of 8 bar compared to the 2-3.2 kW for a storage pressure of 1.10 bar. Also the round trip efficiency is influenced by the storage pressure.

Its increase allows to have a higher efficiency: for a storage pressure of 8 bar the round trip efficiency is about 25% compared to the 12% in case of a storage pressure equal to 1.10 bar.
For the economic feasibility, the range of the LCOS obtained is between 1.2 €/kWh and 1.8 €/kWh. These values are higher than any other technologies studied in literature in fact the LCOS varies between about 0.10 and 0. 60 €/kWh for the CAES, PSH, Li-ion, Pb and vanadium redox flow (VRF) batteries. This may be caused by the small scale of the system.

References

[1] Directive (EU) 2018/2001 of the European Parliament and of the Council of 11 December 2018 on the Promotion of the Use of Energy from Renewable Sources. 2018. Available online: https://eur-lex.europa.eu/legal-content/EN/TXT/?uri=CELEX%3A32018L2001 (accessed on 3 June 2021). (permalink).

[2] Directive (EU) 2019/944 of the European Parliament and of the Council of 5 June 2019 on Common Rules for the Internal Market for Electricity and Amending Directive 2012/27/EU. 2019. Available online: https://eur-lex.europa.eu/legal-content/EN/TXT/?uri=CELEX%3A32019L0944 (accessed on 3 June 2021). (permalink).

[3] Zatti M, Moncecchi M, Gabba M, Chiesa A, Bovera F, Merlo M. Energy Communities Design Optimization in the Italian Framework. Applied Sciences. 2021; 11(11):5218.

[4] S. Koohi-Fayegh, M.A. Rosen, A review of energy storage types, applications and recent developments, Journal of Energy Storage, Volume 27, 2020, 101047

[5] G. Fuchs, B. Lunz, M. Leuthold, D.Uwe Sauer. Technology Overview on Electricity Storage Overview on the potential and on the deployment perspectives of electricity storage technologies On behalf of Smart Energy for Europe Platform GmbH (SEFEP), 2012.

[6] O. Palizban, K. Kauhaniemi, Energy storage systems in modern grids—Matrix of technologies and applications, Journal of Energy Storage, Volume 6, 2016, 248-259.

[7] MC. Argyrou, P Christodoulides, SA. Kalogirou, Energy storage for electricity generation and related processes: Technologies appraisal and grid scale applications. Renewable and Sustainable Energy Reviews, 94, 2018, 804-821.

[8] H. Lopes Ferreira, R. Garde, G. Fulli, W. Kling, J.Pecas Lopes. Characterisation of electrical energy storage technologies, Energy, 53, 2013, 288-298.

[9] F. Díaz-González, A. Sumper, O. Gomis-Bellmunt, R. Villafáfila-Robles. A review of energy storage technologies for wind power applications. Renewable and Sustainable Energy Reviews, 16, 4, 2012, 2154-2171.

[10] Castellani B, Presciutti A, Filipponi M, Nicolini A, Rossi F. Experimental investigation on the effect of phase change materials on compressed air expansion in CAES plants. Sustainability, 7, 2015, 9773-9786.

[11] L. Chen, T. Zheng, S. Mei, X. Xue, B. Liu, Q. Lu. Review and prospect of compressed air energy storage system J. Mod. Power Syst. Clean Energy, 4, 2016, 529-541

[12] Energy Storage Grand Challenge: Energy Storage Market Report. U.S. Department of Energy Technical Report NREL/TP-5400-78461 DOE/GO-102020-5497, December 2020 (visited online on December 17,2021 at: https://www.energy.gov/sites/prod/files/2020/12/f81/Energy%20Storage%20Market%20Report%202020_0.pdf

[13] Y. Xie, X. Xue. Thermodynamic analysis on an integrated liquefied air energy storage and electricity generation system, Energies, 11, 2018, 1-12.

[14] H. Wang, M. Su. Application and prospect of liquid air energy storage technology, Sino-global Energy, 26, 2021, 90-95

[15] H. Lars, R. Span. Influence of the heat capacity of the storage material on the efficiency of thermal regenerators in liquid air energy storage systems. Energy, 74, 2019, 235-45.

[16] H. Lars, R. Span, M. Pascal, S. Viktor. Investigation of a liquid air energy storage (LAES) system with different cryogenic heat storage devices. Energy Procedia, 158, 2019, 4410-5.

[17] N. Mohammad Hossein, M. Zeynalian, A. Reza Razmi, A. Arabkoohsar, Soltani M. Energy, exergy, and economic analyses of an innovative energy storage system; liquid air energy storage (LAES) combined with high-temperature thermal energy storage (HTES). Energy Convers Manag, 226, 2020.
[18] Vecchi A., Naughton J., Li Y., Mancarella P., Sciacovelli A. Multi-mode operation of a Liquid Air Energy Storage (LAES) plant providing energy arbitrage and reserve services e analysis of optimal scheduling and sizing through MILP modelling with integrated thermodynamic performance. Energy, 200, 2020.
[19] E. Borri, A. Tafone, A. Romagnoli, G. Comodi, A preliminary study on the optimal configuration and operating range of a "microgrid scale" air liquefaction plant for Liquid Air Energy Storage, Energy Conversion and Management, 143, 2017, 275-285.
[20] R. Morgan, S. Nelmes, E. Gibson, G. Brett, Liquid air energy storage – analysis and first results from a pilot scale demonstration plant, Appl Energy, 137, 2015, 845-853.
[21] E. Borri, A. Tafone, G. Comodi, A. Romagnoli, Improving liquefaction process of microgrid scale Liquid Air Energy Storage (LAES) through waste heat recovery (WHR) and absorption chiller, Energy Procedia, 143, 2017, 699-704.
[22] S. Mazzoni, S Ooi, A. Tafone, E. Borri, G. Comodi, A. Romagnoli, Liquid Air Energy Storage as a polygeneration system to solve the unit commitment and economic dispatch problems in micro-grids applications, Energy Procedia, 158, 2019, 5026-5033.
[23] A. Tafone, A. Romagnoli, E. Borri, G. Comodi, New parametric performance maps for a novel sizing and selection methodology of a Liquid Air Energy Storage system, Applied Energy, 250, 2019, 1641-1656.
[24] A. Tafone, E. Borri, G. Comodi, A. Romagnoli, Parametric performance maps for design and selection of Liquid Air Energy Storage system for mini to micro-grid scale applications, Energy Procedia, 158, 2019, 5053-5060.
[25] R.F. Cascone, P. Sonti, LCOS – A Key Metric for Cost of Energy Storage, 2019.
[26] C. Xie, Y. Li, Y. Ding, J. Radcliffe, Evaluating Levelized Cost of Storage (LCOS) Based on Price Arbitrage Operations: with Liquid Air Energy Storage (LAES) as an Example, Energy Procedia, 158, 2019, 4852-4860.
[27] V. Jülch, Comparison of electricity storage options using levelized cost of storage (LCOS) method, Applied Energy, 183, 2016, 1594-1606.

Reverse flow in a pressure suppression system due to condensation instabilities

G. Giambartolomei[1*], A. Pesetti[1], L. Berti[1], M. Raucci[1], R. Lazzeri[1], D. Aquaro[1]

[1] Department of Civil and Industrial Engineering (DICI), University of Pisa, Largo L. Lazzarino 2, 56122 Pisa (Italy)

*Corresponding author e-mail: guglielmo.giambartolomei@ing.unipi.it

Abstract. The Vacuum Vessel Pressure Suppression System (VVPSS) is designed to protect the ITER reactor (Vacuum Vessel, VV) from an overpressurization accident. The VVPSS is mainly composed of four Vapour Suppression Tanks (VSTs) partially filled with water inside which venting pipes (spargers) are immersed to direct the steam in the subcooled pool to condense it. Since the maximum pressure allowed inside the VV is only 0.15 MPa, a sub-atmospheric pressure must be maintained inside the VSTs. During a series of experimental campaigns funded by ITER Organization to study and validate the correct operability of the VVPSS and conducted in a large-scale plant built at the B. Guerrini laboratory of the University of Pisa, pressure instability and reverse flow through the sparger holes have been observed. The analysis of acceleration signals recorded during this pressure inversion, showed large amplitude oscillations at low frequency due to the Condensation Induced Water Hammer (CIWH).

1 Introduction

The International Thermonuclear Experimental Reactor (ITER) experiments will take place inside the Vacuum Vessel (VV), a hermetically sealed steel doughnut-shaped container that houses the fusion reactions and acts as a first safety containment barrier. The fusion phenomena occur inside the plasma and to be able to form this state of matter, very low pressures (high vacuum) are required. The VV is therefore designed to withstand a maximum internal pressure of 0.15 MPa during a postulated accidental scenario of a Loss of Coolant Event (LOCA) or an Ingress of Coolant Event (ICE) both produced by a damage to the hot plasma facing components caused by a plasma stability disruption. To protect the VV from an internal over pressurization caused by these in-vessel coolant postulated leakages, a key safety system called Vacuum Vessel Pressure Suppression System (VVPSS) has been designed downstream of the reactor. The design of the VVPSS is conceptually similar to the design of suppression systems of Boiling Water Reactors (BWRs) with the substantial difference that in the former, the suppression tanks are required to operate at sub-atmospheric pressure since, as previously mentioned, the ITER VV operates at extremely low vacuum conditions. The VVPSS is mainly constituted by 4 Vapour Suppression Tanks (VSTs) measuring 4.7 m in height, with a diameter of 6.2 m and an inner volume of 100 m^3 each partially filled with water, 3 with 60 m^3 and one with 40 m^3. The 3 VSTs containing 60 m^3 of water, also called Large LOCA Tanks (LLTs), are designed to manage bigger LOCA events (cat. III and IV), while the one containing 40 m^3, also called Small

LOCA Tank (SLT), is designed to manage smaller LOCA events (cat. II). Each tank is crossed by a sparger immersed in water to release the steam. Figure 1 illustrates the four tanks located at the underground level -2 of the ITER facility building.

Figure 1: Section view of the VVPSS

The fundamental phenomenon that governs this security system is the Direct Contact Condensation (DCC) of steam at sub-atmospheric pressure conditions, a physical condition that is poorly studied in the scientific literature. For this reason, the reliability, and the efficiency of this kind of fusion reactor safety system must be experimentally investigated and validated. In this new scenario, where the pressure values are sub-atmospheric, the situation can be very different from the atmospheric one. For example, for a given superheated steam temperature, in the DCC of steam at sub-atmospheric conditions, the subcooling temperature needed to condense the steam is always lower with respect to the atmospheric condition, because the saturation temperature is lower. Pressure oscillations deriving from unstable condensation regimes such as chugging and condensation oscillations, occur at different values of the controlling parameters G_S/P_W (ratio between the steam mass flux and the downstream pressure) and T_W (water temperature). These oscillations can cause damage to the tank, especially if their frequency is close to the natural frequency of the surrounding structures. The potential danger related to this issue has been quite extensively studied in the case of DCC of steam when several parameters vary but always at atmospheric pressure conditions [1–8] with rather conflicting results. During the postulated accidental event in the VV, the steam can also react with hot surfaces [9] and some non-condensable gases (hydrogen and oxygen gases due to the water radiolysis or thermolysis) may be mixed in it. The presence of non-condensable gases can drastically modify the phenomenon of condensation from the oscillation point of view [10]. Some scholars have focused attention on another phenomenon linked to oscillations and equally dangerous as it occurs at low frequency and with large amplitudes, namely the so-called Condensation Induced Water Hammer (CIWH) [11–14]. So far only a few experimental studies on the oscillations resulting from DCC of steam at sub-atmospheric pressures have been carried out in a small-scale plant [15,16].

The aim of this paper is to describe a violent pressure fluctuation phenomenon that trigger heavy oscillations on the sparger observed during some condensation tests. The tests were conducted in a large-scale test facility of the University of Pisa [17,18].

2 Experimental activity

2.1 Experimental apparatus

Recently a Large-Scale Test Facility (LSTF) was built at the laboratory "B. Guerrini" of the University of Pisa to verifies the safe and correct operability of the VVPSS at the foreseen parameters and at different operating conditions (Figure 2). The main component of the LSTF is the Experimental Test Tank (ETT), a 92 m^3 (scaling factor 92/100 with respect to one ITER VST) steel vessel filled with 55 m^3 of softened (1 °f) water.

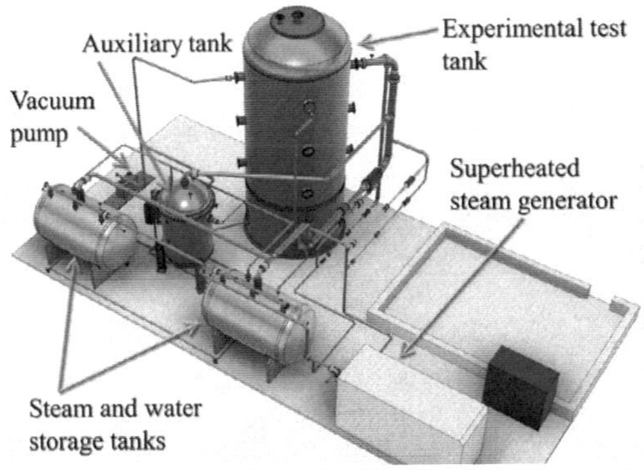

Figure 2: Large Scale Test Facility (LSTF) layout rendering (left) and aerial picture of the facility (right)

Before each test, a 45 kW vacuum pump creates the sub-atmospheric pressure conditions inside the ETT with a maximum volumetric flow rate of 1800 m^3/h, to a pressure value corresponding to the saturation pressure at the water pool initial temperature. The superheated steam to be condensed inside the ETT is provided by a 1750 kW electrical steam generator. The steam generator is able to produce up to 520 g/s of superheated steam at a maximum temperature of 250°C and a maximum pressure of 25 bar abs. The maximum steam flow rate provided by the steam generator is only 1/10 of the maximum steam mass flow rate expected during a postulated ITER cat. IV LOCA event, that is not expected to last more than 500 s. To overcome this technical limitation, the LSTF can undergo an upgrade to the Full-Scale Test Facility (FSTF) thanks to the ability to store saturated water at a maximum pressure of 25 bars abs and 220 °C of temperature in two horizontal tanks of 15 m^3 of inner volume. This hot water undergoes a lamination process through two globe valves and becomes superheated steam to be injected into the ETT with a maximum steam mass flow rate of 5 kg/s for a maximum period of time of 1000 s. Three steam control lines are equipped with vortex flow meters and control valves according to the required steam mass flow rate. These lines end in the ETT with a sparger system immersed in the subcooled pool. The research program foresaw the use of two kind of spargers: the first one called SPS-A is a 100 holes sparger with a diameter similar to a DN125 pipe, while the SPS-B is a 1000 holes sparger with a DN450 pipe diameter. In both spargers the holes' diameter is 10 mm and the distance between the first row of holes and the last one is 266 mm, as in the actual VST sparger. The ETT is heavily equipped with temperature and pressure sensors aimed to characterize the temperature and pressure distribution in the inner volume in the vertical and radial direction. The position of each sensor was defined in order to have a complete and well-defined measurement of the temperature and the pressure field in radial and axial direction inside the ETT as well as in order to minimize their interference on the physical phenomena (i.e., water convective motions, steam condensation, etc.). The distribution of sensors in the axial and radial directions permitted also to investigate the thermal stratification and the condensing steam jet plume extension (enabling the experimental determination of the heat transfer coefficient at the steam-water interface). A total number of 59 TE and 16 PE are positioned in the ETT inner space, distributed on 16 horizontal surfaces and in 14 layers. In addition to all these probes, to characterise the steam properties before the injection, the steam temperature and pressure measurement were recorded inside the sparger (TE1001 and PE1001 label, respectively). The acquisition frequency of the temperature and pressure sensors is 5 and 50 Hz, and the accuracy is ± 0.1 °C and 0.25% of full scale, respectively. In the vicinity of the holes' area of the spargers, the density of the sensors is greater than in the rest of the tank volume. The sparger acceleration is measured by means of a stainless-steel waterproof sensitive low noise monoaxial accelerometer (Hansford Sensors, HS-100S) stud mounted 450 mm above the flange of the pipe with a wide and flat pass-band from 1 to 10000 Hz throughout a measurement range from 0 to 50 g. The acquisition of the accelerometric signals is made with a HBM Quantum MX 840 and with the acquisition software Catman easy. The accelerometer is screw mounted on a metal plate welded to the structure (DN300 pipe of the sparger System), as depicted in Figure 3. Six video cameras are present inside the ETT to record images of the condensation tests from different views.

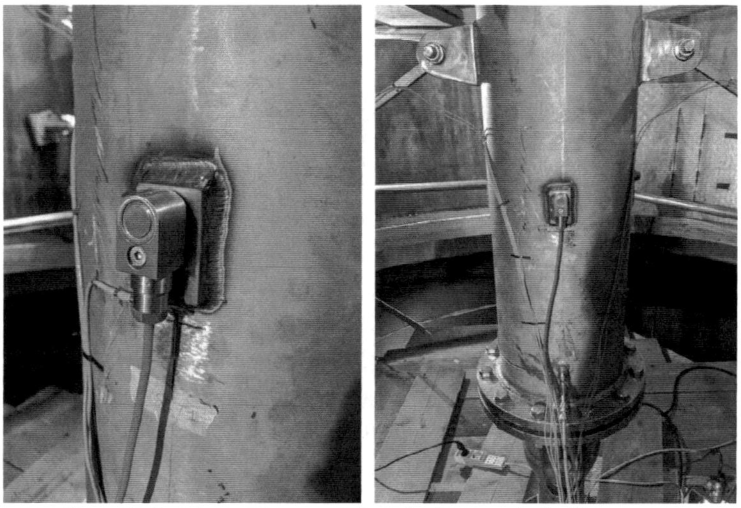

Figure 3: Accelerometer mounted on the large-scale facility sparger

2.2 Experimental procedures

The experimental activity performed with the LSTF consisted in 174 tests subdivided in four experimental campaigns, the first three with the sparger A and the last one with the sparger B. The tests were steady state with pure steam or transient tests both with pure steam and with steam mixed with non-condensable gas (air).

The objectives of the experimental tests have been to essentially investigate the unstable condensation regimes, i.e., situations where the ratio between the mass flow rate per hole and the downstream pressure is relatively low. In this paper we will focus on a phenomenon that was observed in 6 tests of the fourth experimental campaign, that is the one conducted with the 1000 holes sparger. These tests were conducted with pure steam injected in the subcooled water pool. During these tests the steam mass flow was quite constant with an average value of 496 g/s, close to the maximum allowed by the steam generator. A period of 120 s, representative of this particular phenomenon, will be analyzed below.

3 Discussion of the results

During the fourth experimental campaign, performed with the sparger B (SPS-B), in 6 tests at relatively low water pool average temperatures significant steam jets instabilities and reverse flow were registered. During the DCC of steam, the water from the outside enters the sparger. This phenomenon causes the condensation of steam to take place inside the sparger with the consequent rise of great vibrations due to the collapse of steam bubbles that affect the structures inside the ETT.

Below are some diagrams of pressure and acceleration recorded signals showing the onset of the Condensation Induced Water Hammer (CIWH) phenomenon during the test. Figure 4 shows that the pressure difference across the sparger surface (inside, PE1001 and outside, PE1207) are often negative in a 120 s long test period. In other words, the pressure outside of the sparger is often greater than the pressure inside and because of this a reverse flow occurred.

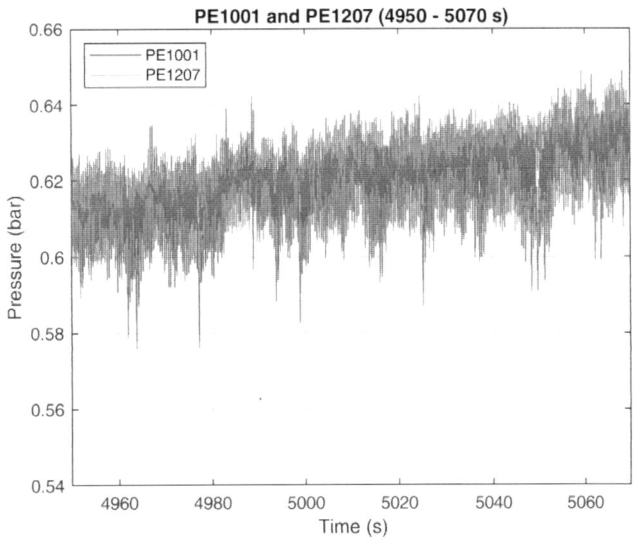

Figure 4: Pressure difference inversion

The total average temperature of the water pool inside the ETT is 39.9 °C while the temperatures in front of the sparger holes, and in the upper layers close to the water surface are about 70 °C; there is a considerable thermal stratification inside the ETT. The downstream pressure ranges between 0.61 and 0.64 bar.

The corresponding accelerometric signal acquired with a sampling frequency of 4800 Hz is shown in Figure 5 (left). A maximum acceleration peak of 5.1461 g occurs in this time interval. The amplitude of the acceleration ranges between ± 0.9 g. The Power Spectral Density (PSD) of the accelerometric signal using Welch's overlapped segment averaging estimator shown in Figure 5 (right) shows a peak around 2 Hz but smaller frequency contents are also present. During this test strong vibrations occurred with big noise heard also at a considerable distance from the ETT.

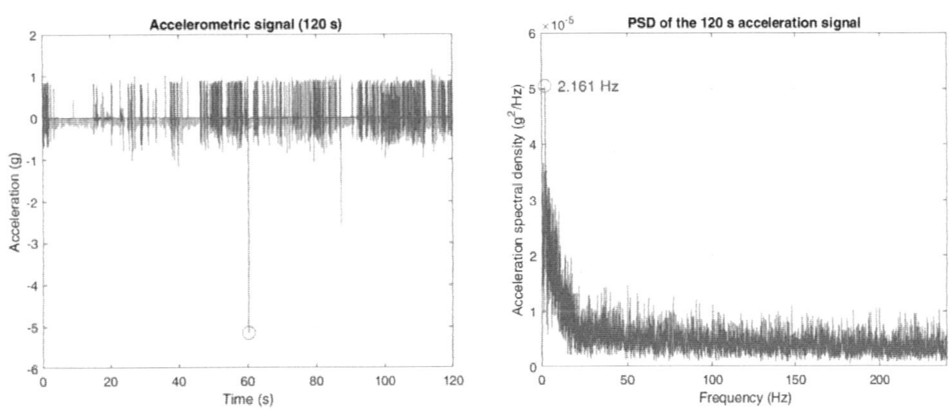

Figure 5: Time domain and frequency domain of the 120 s long acceleration signal

Some video camera images related to this analysed test period have been observed and analysed. The images are acquired at 50 fps, therefore the time distance between each frame is 0.02 s. Since the maximum power of the acceleration signal is found at a frequency of 2.161 Hz, the time period of this periodic phenomenon present in the signal is about 0.46 s. As can be seen from Figure 6 there are at least two cyclic situations. Starting from a condition where no steam is exiting from the sparger holes (first frame), the same situation happens after 0.44 s (last frame) which corresponds to a frequency of about 2 Hz. From the same figure it can also be seen that the same initial condition (first frame, no steam coming out from the sparger holes) occurs after a period of 0.14 s which corresponds to a frequency of 2 Hz.

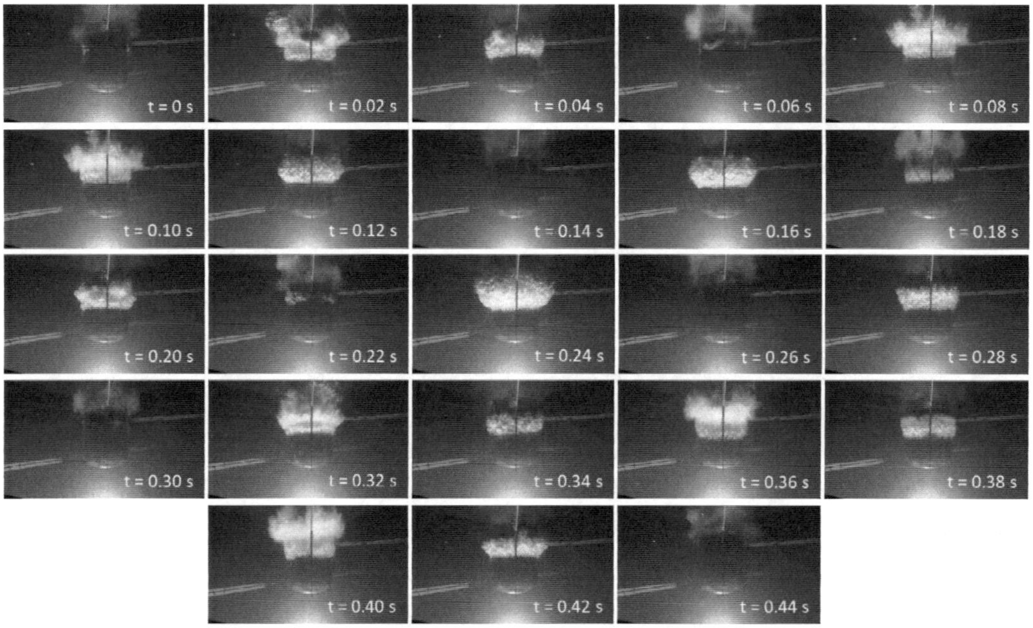

Figure 6: 23 frames extracted from the video of the 120 s acceleration signal

In this thermohydraulic situation (Q = 496 g/s, P_{VS} = 0.3801 bar, T_w = 39.9 °C) the steam exits from about 8 upper rows of holes, and it coalesces creating an almost axial symmetrical shape. This shape is very similar to the one observed by Tanskanen et al. [19]. The steam does not condense immediately in front of the holes but in the upper colder zone. At certain times the steam condenses inside the sparger determining a reverse flow and causing large vibrations on the structures. This particular and potentially dangerous condition is the phenomenon of Condensation Induced Water Hammer (CIWH) mentioned in [10,12,14,20–22]. It must be stressed out that even if the average temperature of the water pool is relatively low (about 40 °C), the temperature of the water in the proximity of the sparger holes is close to 70 °C that is close to the saturation temperature. Being the subcooling temperature in front of the holes almost zero, the steam jet plumes must expand towards the upper layers to increase the heat exchange surface and therefore be able to condense. In this case, the most important water temperature value is the temperature close to the sparger's holes, where condensation takes place, not the average temperature of the entire pool. Kukita et al. [2] and Gregu et al. [6] found maximum loads and highest pressure peaks at relatively low pool water temperature, 37 °C and 25 °C respectively. This is consistent with what has been observed in our test because the reverse flow entrained water at about 20 °C of temperature from the bottom. In the experimental test described before (see the instants t = 0.12 s, t = 0.16 s and t = 0.24 s in Figure 6), the steam jet plumes exit from about 600 holes (the upper 12 circumferential rows of 50 holes each only 13.3 mm distant from the hole centres) and increase their toroidal shape like volume. After this, the steam volume moves upward, coalesces forming a donut shape. The steam disappears because water is sucked up inside the sparger by the steam bubble collapse. What we observed during our experimental campaigns is also consistent with what has been observed by Park et al. [4]. The FFT analysis showed that several dominant frequencies over a broad frequency range exist during the direct contact steam condensation process using a multi-hole sparger. Another analogy with literature resides in the chugging frequencies of the steam bubbles observed. The highest frequency peak observed in the PSD plot of Figure 5 is 2.161 Hz; the frequencies observed by Hujala et al. [23] ranged from 1 to 3 Hz. Also in the experiments conducted by Li et al. [14] the water was sucked up inside the sparger during the chugging regime.

4 Conclusions and future perspective

The experimental activity of DCC of steam at sub-atmospheric pressure conditions showed significant steam jet instabilities and reverse flow in the sparger. During steam condensation strong vibrations and loud noises were experienced. An acceleration signal of the sparger showed a broad frequency content.

Two periodic phenomena have been detected: the steam jets coalesce to form a toroidal shape around the sparger surface (first phenomenon) and the steam condenses inside the sparger creating strong vibrations on the structures (second phenomenon). The steam coalescence is due to the high water pool temperature (subcooling temperature close to zero) in the vicinity of the sparger holes. The higher the temperature of the water, the more the steam jets need to expand in volume to condense (they need greater heat exchange surfaces). As the volume of the jets is larger, they are more likely to meet and coalesce.

When the steam condenses inside the sparger due to a reverse flow between the inside and the outside of the sparger surface, the phenomenon of Condensation Induced Water Hammer (CIHW) occurs.

The relatively small steam mass flow rate (chugging regime) due to a large (DN450) sparger with a large number of holes, are parameters from which the phenomenon of CIWH could arise.

Future experimental campaigns will focus on the CIWH phenomenon, for determining the conditions in which it happens.

Declaration of competing interest

The authors declare that they have no known competing financial interests or personal relationships that could have appeared to influence the work reported in this paper.

Acknoledgements

This work was founded by ITER Organization under the service contract IO/17/CT/4300001491. The support of IO is gratefully acknowledged. The authors wish to thank all the UNIPI DICI's technicians involved in the implementation, commissioning, conditioning and operation of the facility.

References

[1] Giencke, E., 1981, "Pressure Distribution Due to a Steam Bubble Collapse in a BWR Pressure Suppression Pool," Nucl. Eng. Des., 65(2), pp. 175–196.

[2] Kukita, Y., Namatame, K., Takeshita, I., and Shiba, M., 1987, "LOCA Steam Condensation Loads in BWR Mark II Pressure Suppression Containment System," Nucl. Eng. Des., 102(2), pp. 225–228.

[3] Cho, S., Song, C., Park, C., Yang, S., and Chung, M., 1999, "Characteristics of Pressure Oscillation Induced by Direct Contact Condensation of Steam Discharged through Sparger in a Pool of Subcooled Water," *Proceeding of the Korean Nuclear Society Spring Meeting*.

[4] Park, C. K., Song, C. H., and Jun, H. G., 2007, "Experimental Investigation of the Steam Condensation Phenomena Due to a Multi-Hole Sparger," J. Nucl. Sci. Technol., 44(4), pp. 548–557.

[5] Zhao, Q., Chen, W., Yuan, F., Wang, W., Chong, D., and Yan, J., 2015, "Pressure Oscillation and Steam Cavity during the Condensation of a Submerged Steam Jet," Ann. Nucl. Energy, 85, pp. 512–522.

[6] Gregu, G., Takahashi, M., Pellegrini, M., and Mereu, R., 2017, "Experimental Study on Steam Chugging Phenomenon in a Vertical Sparger," Int. J. Multiph. Flow, 88, pp. 87–98.

[7] Xu, Q., Liu, W., Li, W., Yao, T., Chu, X., and Guo, L., 2019, "Experimental Investigation on Interfacial Behavior and Its Associated Pressure Oscillation in Steam Jet Condensation in Subcooled Water Flow," Int. J. Heat Mass Transf., 145, p. 118779.

[8] Xu, Q., Guo, L., and Chang, L., 2017, "Mechanisms of Pressure Oscillation in Steam Jet Condensation in Water Flow in a Vertical Pipe," Int. J. Heat Mass Transf., 110, pp. 643–656.

[9] Gaeta, M. J., Merrill, B. J., Bartels, H. W., Laval, C. R., and Topilski, L., 1997, "Short-Term Hydrogen Production Issues for ITER," Fusion Technol., 32(1), pp. 23–34.

[10] Yang, X., Liu, J., Fu, P., Chen, N., and Wei, J., 2019, "Experimental and Theoretical Study of Pressure Oscillation of Unstable Steam-Air Jet Condensation in Water in a Rectangular Channel," Int. J. Multiph. Flow, 119(July), pp. 14–27.

[11] Urban, C., and Schlüter, M., 2014, "Investigations on the Stochastic Nature of Condensation Induced Water Hammer," Int. J. Multiph. Flow, 67, pp. 1–9.

[12] Wang, L., Yue, X., Chong, D., Chen, W., and Yan, J., 2018, "Experimental Investigation on the Phenomenon of Steam Condensation Induced Water Hammer in a Horizontal Pipe," Exp. Therm. Fluid Sci., 91, pp. 451–458.

[13] Barna, I. F., Imre, A. R., Baranyai, G., and Ézsöl, G., 2010, "Experimental and Theoretical Study of Steam Condensation Induced Water Hammer Phenomena," Nucl. Eng. Des., 240(1), pp. 146–150.

[14] Li, S. Q., Lu, T., Wang, L., and Chen, H. S., 2017, "Experiment Study on Steam-Water Direct Contact Condensation in Water Flow in a Tee Junction," Appl. Therm. Eng., 120, pp. 99–106.

[15] Lo Frano, R., Mazed, D., Olcese, M., Aquaro, D., Del Serra, D., Sekachev, I., and Giambartolomei, G., 2018, "Investigation of Vibrations Caused by the Steam Condensation at Sub-Atmospheric Condition in VVPSS," Fusion Eng. Des., 136(September 2017), pp. 1433–1437.

[16] Mazed, D., Lo Frano, R., Aquaro, D., Del Serra, D., Sekachev, I., and Olcese, M., 2018, "Experimental Investigation of Steam Condensation in Water Tank at Sub-Atmospheric Pressure," Nucl. Eng. Des., 335(January), pp. 241–254.

[17] Pesetti, A., Marini, A., Raucci, M., Giambartolomei, G., Olcese, M., Sarkar, B., and Aquaro, D., 2021, "Large Scale Experimental Facility for Performance Assessment of the Vacuum Vessel Pressure Suppression System of ITER," Fusion Eng. Des., 171, p. 112523.

[18] Pesetti, A., Raucci, M., Giambartolomei, G., Marini, A., Sarkar, B., Olcese, M., and Aquaro, D., 2021, "Full Scale Test Facility for Operability Verification of Tokamak Pressure Suppression System at Sub-Atmospheric Conditions," Fusion Eng. Des., 168.

[19] Tanskanen, V., Jordan, A., Puustinen, M., and Kyrki-Rajamäki, R., 2014, "CFD Simulation and Pattern Recognition Analysis of the Chugging Condensation Regime," Ann. Nucl. Energy, 66, pp. 133–143.

[20] Easom, B. H., 1992, *The Effect of Noncondensible Gas on Steam Bubble Collapse Induced Water Hammer Signature Redacted Signature Redacted Signature Redacted.*

[21] Gregu, G., Takahashi, M., Pellegrini, M., and Mereu, R., 2017, "Experimental Study on Steam Chugging Phenomenon in a Vertical Sparger," Int. J. Multiph. Flow, 88, pp. 87–98.

[22] Zhao, Q., and Hibiki, T., 2018, "Review: Condensation Regime Maps of Steam Submerged Jet Condensation," Prog. Nucl. Energy, 107(November 2017), pp. 31–47.

[23] Hujala, E., Tanskanen, V., and Hyvärinen, J., 2018, "Frequency Analysis of Chugging Condensation in Pressure Suppression Pool System with Pattern Recognition," Nucl. Eng. Des., 339, pp. 244–252.

Hidden Grid Background Oriented Schlieren in studying convective flows

Dario Ambrosini[1], **Tullio de Rubeis**[1], **Domenica Paoletti**[1], **Jagadesh Ramaiah**[2], **Rajshekhar Gannavarpu**[2]

[1]DIIIE, University of L'Aquila, P.le Pontieri 1, 67100 L'Aquila, Italy
[2]Dept. of Electrical Engineering, Indian Institute of Technology Kanpur, Kanpur-208016, India

dario.ambrosini@univaq.it

Abstract. In this paper, we present a new BOS (Background Oriented Schlieren), based on a hidden grid, for studying heat flows. In the setup, we record a grid-based intensity pattern whose phase map carries information about the temperature gradient. The background (undistorted) pattern is hidden in the light source. Quantitative analysis is obtained by a windowed Fourier transform approach. Some experimental results are given to demonstrate the feasibility of the technique.

1. Introduction

Optical methods have an established fluid flow and heat and mass transfer studies tradition. This success is based on the typical features of optical techniques, such as high spatial and temporal resolution, good accuracy, and non-intrusiveness (i.e., the phenomena are not affected by the measurement system itself). Although optical methods in flow visualization are rather traditional, they are experiencing a flourishing season, thanks to the advances in hardware and software [1].

An evident trend is towards simplifying measurement techniques accompanied by increasingly efficient data processing. An example of this trend is the widely used technique of Background Oriented Schlieren (BOS) [2, 3]. BOS is based on the idea of mapping refractive index fields by studying the distortion of a background pattern. Considering that the refractive index can be modified by different physical phenomena (heat transfer, mass transfer...), information can be obtained from refractive index variations about the corresponding phenomena.

Examples of topics addressed in literature are quantitative measurements [4], uncertainty [5], and 3D [6, 7] applications. Data processing in BOS is usually done by cross-correlation [2], but also phase extraction has been introduced [8]; phase estimation allows a multi-scale approach by wavelet transform [9].

A variety of background patterns have been proposed: grids and lines [10, 11], dots [12, 13], gradient patterns [14], speckle [15-17] and natural background [18].

The main feature of BOS is the great experimental simplicity; it requires only a background pattern, a light source to illuminate it, and a recording device. A basic setup is shown in Figure 1.

A test object, in which density gradients (due, for instance, to a heat transfer phenomenon) induce refractive index gradients, is placed between the background and the camera.

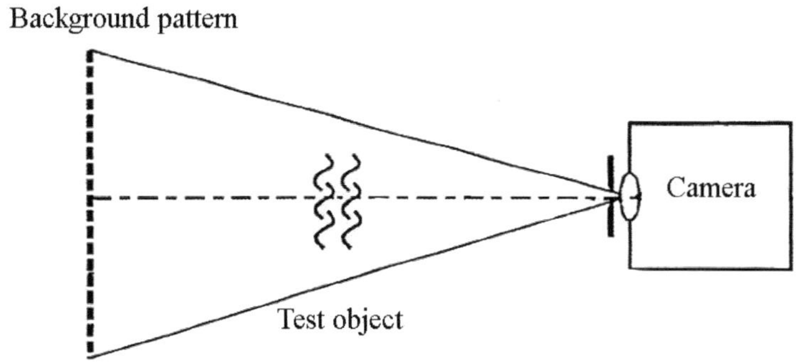

Figure 1. Diagram of Background Oriented Schlieren (light source not shown).

When refractive index gradients are present, the background, seen through the test object, is distorted. We can obtain information about the phenomenon that has induced the refractive index variation from this deformation.

The second advantage of BOS, and one of the reasons for its success, lies in an efficient data processing. The distortion of the background can be considered as an apparent displacement and measured by Digital Image Correlation (DIC) [19] or using algorithms developed for Particle Image Velocimetry (PIV) [2, 20]. It can be also measured by optical flow [21, 22] or obtained as a phase change [8, 9, 23].

The apparent displacement is proportional to the refractive index gradients or, considering a heat transfer phenomenon, proportional to the temperature gradients [2, 4].

This paper introduces a new BOS technique, based on a hidden grid coupled to the robust data processing provided by windowed Fourier transform, for studying heat flows.

Some experimental results are given to demonstrate the feasibility of hidden grid technique for quantitative visualization of convective heat transfer. The features of the new arrangement are highlighted in the Conclusions.

2. Hidden Grid BOS

There are many optical techniques based on grids and lines, typically fringe projection [24] and moiré [25, 26]. Grids and lines were also used in BOS and strictly related techniques [11, 23, 27]. All these grids and lines are visually evident, e.g., projected or stamped on a background.

Implicit or hidden grids were proposed in moiré metrology [28]. Typical implicit grids are those induced by the structure of charge-coupled device (CCD) detectors or liquid-crystal display (LCD) projector.

In this paper we introduce a new variation of BOS, based on a background pattern implicit in the light source.

The sketch of the experimental setup for Hidden Grid BOS is shown in Figure 2.

A light source, a video projector, sends a white image on a ground glass that acts as the background screen. A "hidden" grid, due to the pixel structure of the projector, is thus present on the ground glass.

The recording device is a CMOS TV Camera (Silicon Video 9T001C) with a 2048×1536 pixels resolution and a pixel size of $3.2\ \mu m \times 3.2\ \mu m$.

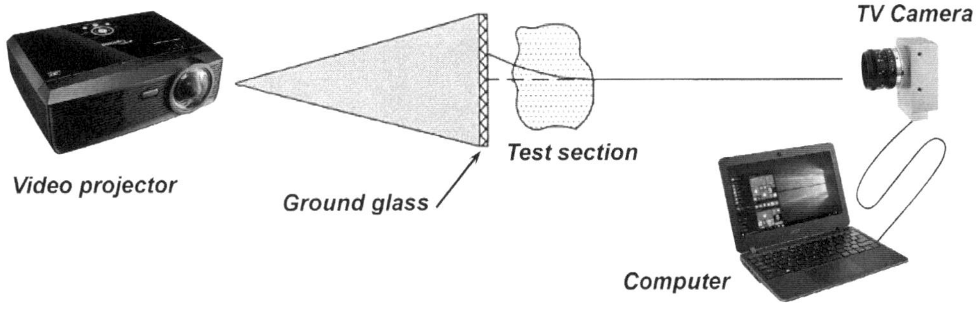

Figure 2. Hidden Grid BOS experimental setup.

Background images are captured using a TEC-55 55 mm F2/8 Telecentric Computar Lens; this lens exhibits good resolution and contrast, with low viewing angle and magnification errors.
The feasibility of Hidden Grid BOS is demonstrated by a preliminary study of natural convective flow, usually a demanding task because of the comparatively small heat transfer.
The test section consisted of two vertical channels, asymmetrically heated in air; more details are given in [29]. The central aluminium plate (thickness 0.012 m, height 0.175 m, length 0.3 m), whose dimensions are chosen to favour a two-dimensional thermal field, is heated by electrical plane resistances, and instrumented with chromel-alumel thermocouples (± 0.1 K). Images are recorded with the heated plate at constant temperature. The central plate is shrouded by two unheated vertical walls, having the same height and length of the heated plate. Figure 3 shows a photograph of the test section and a typical Hidden Grid BOS image, in which the grid on the ground glass can be appreciated.

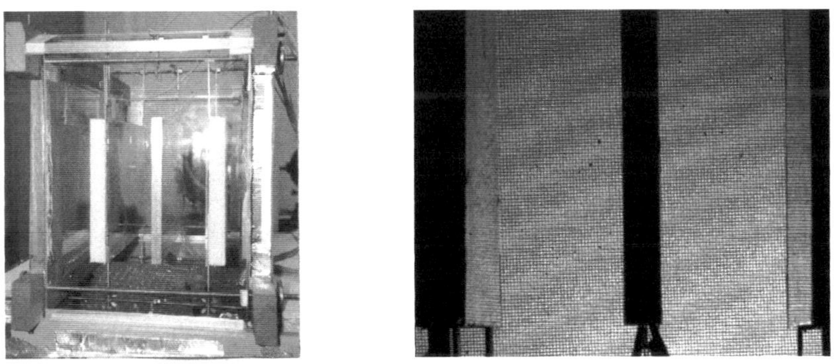

Figure 3. A photograph of the test section (left, courtesy G. Tanda) and a Hidden Grid BOS image (right).

3. Preliminary experimental results

For processing the hidden grid images, we used the Windowed Fourier Transform (WFT) [30, 31] method in our analysis.
Windowed Fourier transform is a robust phase estimation method based on local processing of the given data within a chosen spatial window. Figure 4 shows first experimental results obtained using our hidden grid background oriented schlieren setup.

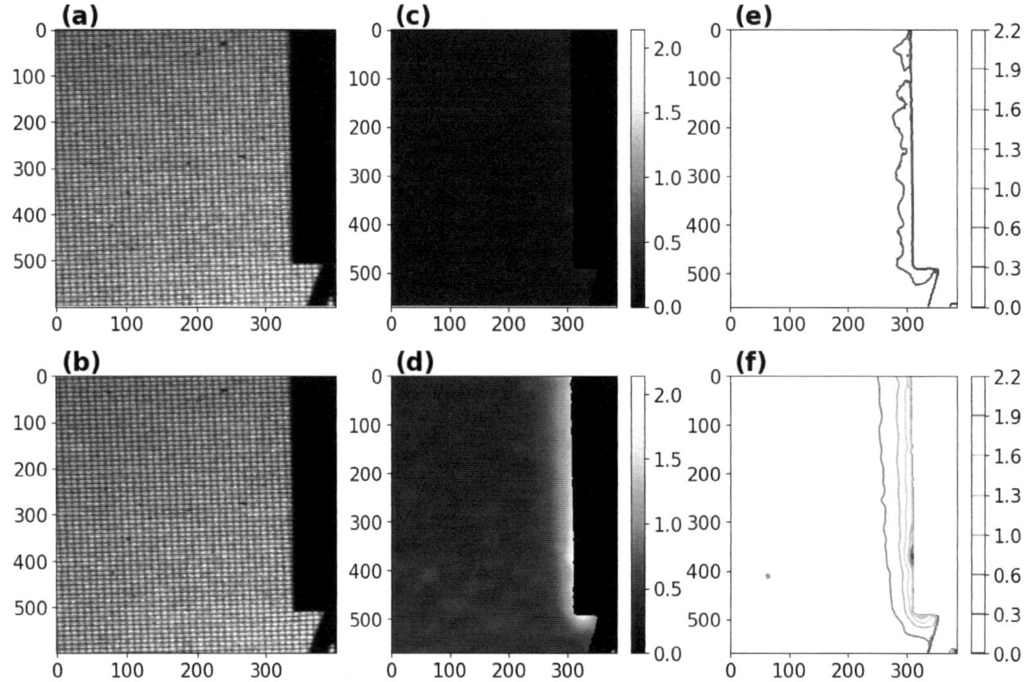

Figure 4. (a-b) Experimentally recorded intensity patterns: (a) 296 K, (b) 320 K. (c-d) Phase maps (in radians) estimated using the windowed Fourier transform method. (e-f) Contour plots of the estimated phase.

Prior to recording these images, an initial reference pattern is taken, with the plate at room temperature (293 K), this pattern is used to remove carrier and any background noise. In our analysis, all the phase maps are measured with respect to the reference pattern.

Figures 4 (a,b) show the intensity patterns recorded at different heating states, namely 296 K (temperature difference with respect to room temperature 3 K) and 320 K (temperature difference with respect to room temperature 27 K). The relative phase maps estimated using WFT are shown in Figures 4 (c,d). As the phase is proportional to the temperature gradients, the boundary layer can be visualized; the layer is particularly evident in Figure 4 (d).

The phase gradually decreases as we move away from the plate. Figures 4 (e,f) show the contour plots of the corresponding phase maps obtained using the WFT method. These iso-phase lines are also lines of constant temperature gradients.

In addition, we also show the line profiles (along row 400) for the two estimated phase maps corresponding to temperatures 296 K and 320 K in parts (a,b) of Figure 5. These plots also help in quantitative visualization of spatial phase variation.

Regarding computational cost, the execution time required by the windowed Fourier transform method for processing a recorded intensity pattern was about 10 seconds on a workstation with Intel Xeon processor (E5-1660), 3.20 Gigahertz clock frequency and 160 Gigabyte memory.

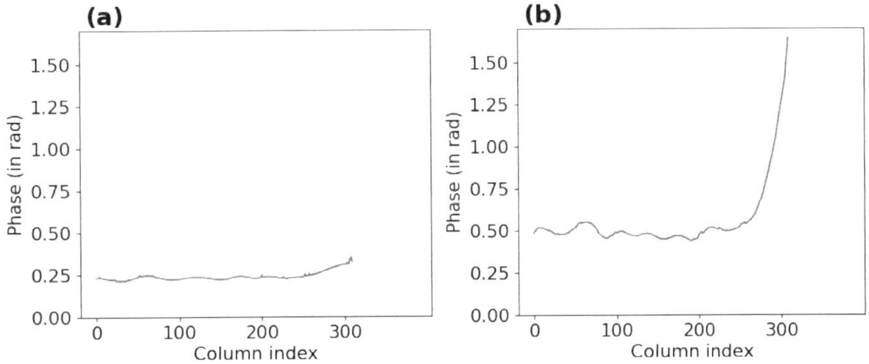

Figure 5. (a) Line profile of estimated phase map corresponding to intensity pattern recorded at 296 K for row number 400. (b) Line profile of estimated phase map corresponding to intensity pattern recorded at 320 K for row number 400.

Further insight can be gained by a comparison with the iso-displacement lines (that are also lines of constant temperature gradients) obtained by schlieren [32, 33].
Figure 6 shows lines at equal light shift obtained by focal filament schlieren [33], in the same test section but with a temperature difference between heated plate and ambient air of 45 K.

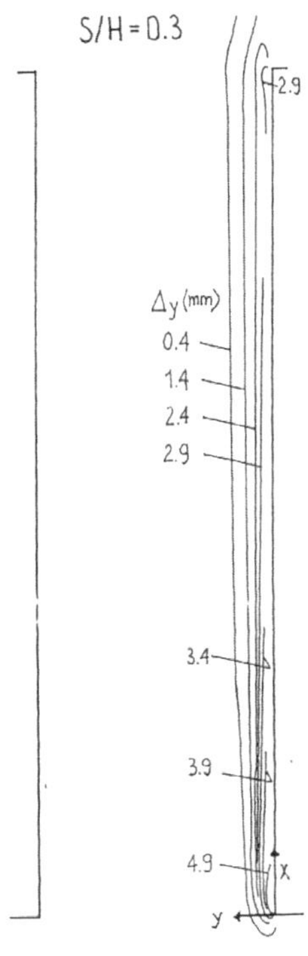

Figure 6. Lines of equal light shift (i.e., lines of constant temperature gradient) obtained by focal filament schlieren [33].

The results are in good qualitative agreement, considering that Hidden Grid BOS has a much simpler and cheaper setup with respect to the traditional knife and mirror schlieren system of [33].

4. Conclusions

In this paper we introduced Hidden Grid BOS, a new type of Background Oriented Schlieren in which the grid is implicit in the light source. The setup has several advantages: with respect to BOS using a laser source, it is less noisy. Furthermore, illumination is very efficient because a white blank image is projected. This implies the possibility of using high f-numbers on the lens, thus obtaining a high depth of field, useful to keep in focus both the background and the test section. The setup is also relatively flexible, thanks to the LCD projector, and can be adapted to different test sections. The coupling with data processing by WFT is also very effective. A possible shortcoming is the high computational cost. However, this cost should reduce significantly in the next years.

Examples of current trends in BOS research include 3D applications [6, 7, 34-36], quantifications and/or improvement of BOS features [5, 23, 37-39], the use of smartphones [40], and the visualization of convective indoor air flows [41].

In several of these trends, hidden grid BOS can represent a useful tool thanks to its features.

References

[1] Prenel J-P and Ambrosini D 2012 Flow visualization and beyond *Opt Lasers Eng* **50** 1–7.

[2] Raffel M 2015 Background-oriented schlieren (BOS) techniques *Exp Fluids* **56** 60.

[3] Settles G S and Hargather M J 2017 A review of recent developments in schlieren and shadowgraph techniques *Meas Sci Technol* **28** 042001.

[4] Hargather M J and Settles G S 2012 A comparison of three quantitative schlieren techniques *Opt Lasers Eng* **50** 8–17.

[5] Rajendran L K, Zhang J, Bhattacharya S, Bane S P and Vlachos P P 2020 Uncertainty quantification in density estimation from background-oriented Schlieren measurements *Meas Sci Technol* **31** 054002.

[6] Grauer S J, Unterberger A, Rittler A, Daun K J, Kempf A M and Mohri K 2018 Instantaneous 3D flame imaging by background-oriented schlieren tomography *Combust Flame* **196** 284-99.

[7] Cai S, Wang Z, Fuest F, Jeon Y, Gray C and Karniadakis G 2021 Flow over an espresso cup: Inferring 3-D velocity and pressure fields from tomographic background oriented Schlieren via physics-informed neural networks *J Fluid Mech* **915** A102.

[8] Ambrosini D, Paoletti D, Di Biase R, Rastogi P K and Siva Gorthi S 2012 Role of data processing in measuring temperature gradients with DOE Schardin's schlieren #2 *Opt Lasers Eng* **50** 1069–74.

[9] Rajshekhar G and Ambrosini D 2018 Multi-scale approach for analyzing convective heat transfer flow in background-oriented Schlieren technique *Opt Lasers Eng* **110** 415–9.

[10 Sutherland B R, Dalziel S B, Hughes G O and Linden P F 1999 Visualization and measurement of internal waves by 'synthetic schlieren'. Part 1. Vertically oscillating cylinder *J Fluid Mech* **390** 93–126.

[11] Dalziel S B, Hughes G O and Sutherland B R 2000 Whole-field density measurements by "synthetic schlieren" *Exp Fluids* **28** 322–35.

[12] Raffel M, Richard H and Meier G E A 2000 On the applicability of background oriented optical tomography for large scale aerodynamic investigations *Exp Fluids* **28** 477–81.

[13] Sourgen F, Leopold F and Klatt D 2012 Reconstruction of the density field using the Colored Background Oriented Schlieren Technique (CBOS) *Opt Lasers Eng* **50** 29–38.

[14] Mier F A and Hargather M J 2016 Color gradient background-oriented schlieren imaging *Exp Fluids* **57** 95.

[15] Ambrosini D, Paoletti D and Di Biase R 2010 White-light speckle photography technique applied for free convection heat transfer problems *Flow Meas Instrum* **21** 98–104.

[16] Meier A H and Roesgen T 2013 Improved background oriented schlieren imaging using laser speckle illumination *Exp Fluids* **54** 1549.

[17] Nakamura Y, Suzuki T, Kinefuchi K and Sasoh A 2021 Speckle beam-oriented schlieren technique *Exp Fluids* **62** 13.
[18] Hargather M J and Settles G S 2010 Natural-background-oriented schlieren imaging *Exp Fluids* **48** 59–68.
[19] International Digital Image Correlation Society, Bigger R, Blaysat B, Boo C, Grewer M, Hu J, Jones A, Klein M, Raghavan K, Reu P, Schmidt T, Siebert T, Simenson M, Turner D, Vieira A and Weikert T 2017 *A Good Practices Guide for Digital Image Correlation* ed E Jones and M Iadicola (International Digital Image Correlation Society), https://www.idics.org/guide/.
[20] Raffel M, Willert C E, Scarano F, Kahler C J, Wereley S T and Kompenhans J 2018 *Particle Image Velocimetry – A Practical Guide* 3rd ed. (Cham; Switzerland: Springer International Publishing).
[21] Liu T and Shen L 2008 Fluid flow and optical flow *J Fluid Mech* **614** 253–91.
[22] Liu T, Merat A, Makhmalbaf M H M, Fajardo C and Merati P 2015 Comparison between optical flow and cross-correlation methods for extraction of velocity fields from particle images *Exp Fluids* **56** 166.
[23] Shimazaki T, Ichihara S and Tagawa Y 2022 Background oriented schlieren technique with fast Fourier demodulation for measuring large density-gradient fields of fluids *Exp Therm Fluid Sci* **134** 110598.
[24] Xu J and Zhang S 2020 Status, challenges, and future perspectives of fringe projection profilometry *Opt Lasers Eng* **135** 106193.
[25] Amidror I 2009 *The Theory of the Moiré Phenomenon, Volume I: Periodic Layers* (London: Springer).
[26] Wang Q and Ri S 2022 Sampling Moiré method for full-field deformation measurement: A brief review *Theor App Mech Lett* **12** 100327.
[27] Ambrosini D and Paoletti D 2007 Heat transfer measurement by a diffractive optical element fringe projection *Opt Eng* **46** 093606.
[28] Garavaglia M and Laquidara A P 2001 Optoelectronic hidden grids and moiré patterns: basics and applications in distance measurement *Opt Eng* **40** 2510.
[29] Ambrosini D and Tanda G 2006 Comparative measurements of natural convection heat transfer in channels by holographic interferometry and schlieren *Eur J Phys* **27** 159–72.
[30] Kemao Q 2007 Two-dimensional windowed Fourier transform for fringe pattern analysis: Principles, applications and implementations *Opt Lasers Eng* **45** 304–17.
[31] Dai X, Xie H and Wang Q 2014 Geometric phase analysis based on the windowed Fourier transform for the deformation field measurement *Opt Laser Technol* **58** 119–27.
[32] Devia F, Milano G and Tanda G 1994 Evaluation of thermal field in buoyancy-induced flows by a Schlieren method *Exp Therm Fluid Sci* **8** 1–9.
[33] Ambrosini D, Paoletti D and Tanda G 2004 Investigation of natural convection in vertical channels by Schlieren and Holographic Interferometry *J Flow Vis Image Proc* **11** 323–34.
[34] Grauer S J and Steinberg A M 2020 Fast and robust volumetric refractive index measurement by unified background-oriented schlieren tomography *Exp Fluids* **61** 1-17.
[35] Unterberger A and Mohri K 2022 Evolutionary background-oriented schlieren tomography with self-adaptive parameter heuristics *Opt Express* **30** 8592-8614.
[36] Winter K O and Hargather M J 2019 Three-dimensional shock wave reconstruction using multiple high-speed digital cameras and background-oriented schlieren imaging *Exp Fluids* **60** 1-13.
[37] Sugisaki H, Lee C, Ozawa Y, Nakai K, Saito Y, Nonomura T, Asai K and Matsuda Y 2022 Single-pixel correlation applied to background-oriented schlieren measurement *Exp Fluids* **63** 1-7.
[38] Reichenzer F, Schneider M and Herkommer A 2021 Improvement in systematic error in background-oriented schlieren results by using dynamic backgrounds *Exp Fluids* **62** 1-18.
[39] Ramaiah J, Ajithaprasad S, Gannavarpu R and Ambrosini D 2020 Fast and robust method for flow analysis using GPU assisted diffractive optical element based background oriented schlieren (BOS) *Opt Lasers Eng* **126** 105908.
[40] Settles G S 2018 Smartphone schlieren and shadowgraph imaging *Opt Lasers Eng* **104** 9-21.
[41] Becher L, Voelker C, Rodehorst V and Kuhne M 2020 Background-oriented schlieren technique for two-dimensional visualization of convective indoor air flows *Opt Lasers Eng* **134** 106282.

Experimental Characterization of the Airflow within a Car Cabin

Authors: M. Bertone[1,*], A. Sciacchitano[2], F. Arpino[1], C. Canale[1], G. Cortellessa[1], G. Grossi[1], L. Moretti[1]

[1] Department of Civil and Mechanical Engineering, University of Cassino and Southern Lazio, via G. Di Biasio 43, Cassino, Italy
[2] Faculty of Aerospace Engineering, Delft University of Technology, Kluyverweg 1, Delft, The Netherlands

[*] Corresponding author's e-mail address: michele.bertone@unicas.it

Abstract. The ventilation flow in a car cabin has been experimentally investigated. The study has been carried out in a car commercially available, by testing one ventilation mode (panel-vent mode) at one fan strength (level 3 of the 4 available) with fresh air intake (without any re-circulation). The flow velocity at the exit of the vents has been measured using a 5-hole pressure probe. The flow velocity fields inside the car cabin have been measured by particle image velocimetry (PIV) in three longitudinal sections: (i) the car centre plane, including both the front and rear area; (ii) the driver's seat centre plane, only in the front area; (iii) the passenger's seat centre plane, only in the front area. At these longitudinal planes, the time-average flow velocity is presented and discussed. The experimental results provide new insights in the ventilation flow in a car cabin.

1. Introduction

The car cabin is an indoor micro-environment, where people spend a considerable amount of time [1, 2]; therefore, thermal comfort and indoor air quality play a key role and need to be considered to ensure the well-being of the occupants. To estimate and improve such parameters, accurate information about the car cabin environment is crucial, especially on the ventilation [3].

Previous studies on ventilation flow within the car cabin have used Computational Fluid Dynamics (CFD) simulations or experiments carried out with scaled or simplified models. Lee and Yoon [3] used a 1/10-scale car model to visualise the airflow inside the cabin and to measure the variation in temperature fields. Yang et al. [4] employed a car model made of acrylic material and an heating, ventilation and air-conditioning (HVAC) device to simulate ventilation conditions; the authors performed Particle Image Velocimetry (PIV) measurements to characterize the air flow inside the acrylic cabin and compared their results with those obtained by CFD simulations. Inside the cabins of actual cars, PIV measurements are often hindered by the limited access for the illumination and imaging systems. As a consequence, researchers had to perform geometrical modifications to the cars to guarantee the optical access. For instance, Lee et al. [5] replaced both the roof and the door passenger's seat part replaced with tempered glass to perform PIV in longitudinal planes. Similarly, Huera-Huarte et al. [6] removed the two doors of the co-pilot side and replaced them with two acrylic sheets to carry out PIV measurements in longitudinal and horizontal planes. Ullrich et al. [7] performed PIV measurements in a car with the HVAC unit without the filter and compared the results with a CFD study. The flow within the car cabin is hard to study due to its dependence on several parameters, such as the position and shape of the vents, mode of ventilation, fan strength (air flow rate), geometry and size of the cabin [8, 9]. In all experimental studies reported in the literature, the car model used was scale-down or simplified, reducing or replacing interior vehicle elements. Therefore, the previous studies analyse the ventilation flow in the cabin in experimental conditions that are quite different from the operating conditions of a real vehicle. To overcome this limitation and to ensure a trustable reproduction of actual operating conditions, an experiment with a real car is needed.

In the present study, a real car was used to study the effect of ventilation on the flow field in the car cabin. The velocity fields of the ventilation flow inside the cabin were measured using the PIV measurement technique at three characteristic longitudinal planes; a 5-hole pressure probe was employed to measure the flow velocity at the outlet of the vents. The aim of this paper is to present a comprehensive benchmark for one ventilation mode (panel-vent mode) at a fixed fan speed of the car ventilation system useful for the validation of Computational Fluid Dynamics (CFD) numerical simulations. Measurements at all four fan speeds has been also investigated and will be presented in future papers. Validated numerical model can be used to investigate Indoor Air Quality (IAQ) and comfort of the occupants for the analyzed or any other car cabin.

2. Methodology

The experiments were carried out in a real car, a Peugeot 206; the measures of the car are 1430 mm (width) x 1430 mm (height) x 3835 mm (length). A schematic of the car with the main measures is shown in Figure 1.

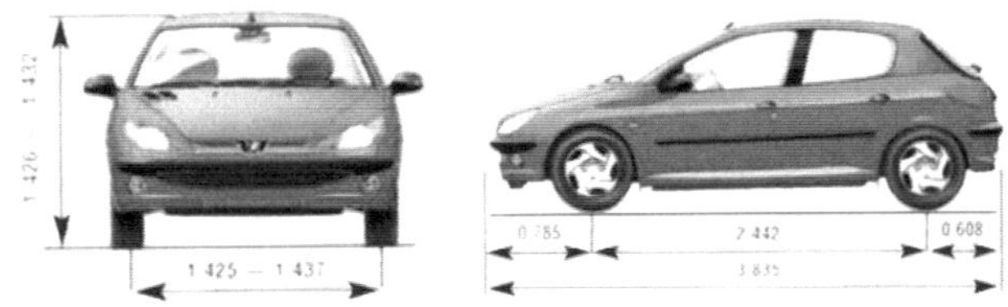

Figure 1 - View of the car with the main measures in metres.

The panel-vent mode was tested under one fan strength, at level 3 out of the 4 available for mass flow rate regulations and using fresh air intake (without any re-circulation). The panel-vent mode is such that each of the four inlet vents in the dashboard panel, indicated with numbers 3 and 4 in Figure 2, are open, while all the other vents are closed. The fans grille has been positioned with a neutral tilt and fixed with clips, according to previous studies by Adhikari et al. [10] and Piovano et al. [11], showing that an accurate fixing position of the grille is decisive for a repeatable examination of the air distribution in a car cabin.

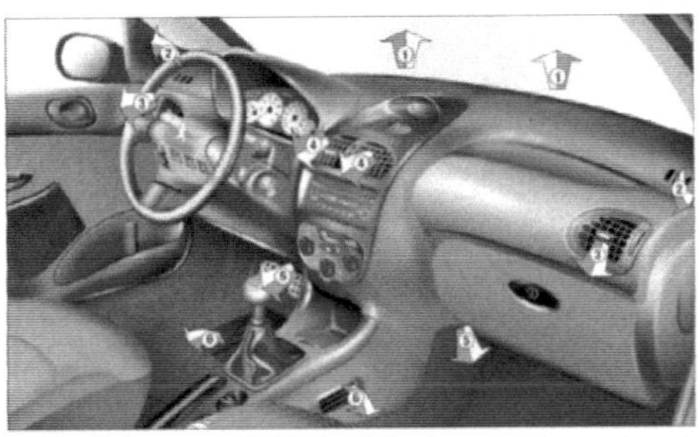

Figure 2 – View of the cabin car with all the vents highlighted. In the current experiments, only the vents indicated with 3 and 4 were open, whereas all the others were kept closed.

No modifications have been made to the interior parts of the car cabin, including the seats, side doors, ceiling, and floor. Only the reflecting surfaces inside the cabin have been covered with non-reflecting material to avoid interferences in the processing of the collected data.
The flow velocity for each opening vent has been individually measured using a 5-hole pressure probe to allow investigation of the flow structures associated with each vent to give a proper definition of such boundary conditions. Despite the ventilation inducing a complex 3-dimensional flow in the car cabin, the planar PIV method has been used, since the lateral flow motion is quite weak in comparison to the vertical and longitudinal velocity components [12].
A preliminary study on the identification of the exit sections was performed; the ventilation system was seeded with water-glycol droplets of 1 μm median diameter produced by the SAFEX seeding generator. The flow patterns have been observed within the cabin, identifying by visual inspections the sections where the airflow exited. From this preliminary investigation, it was noticed that the air exited the cabin from several locations, being the vehicle not hermetic; however, a larger amount of air exited from the rear area of the passenger compartment, which is consistent with what reported in the study of Ullrich et al. [7].

2.1. Airflow at vents outlet, 5-hole pressure probe
A multi-hole pressure probe, the ProCap Compact 5-hole digital probe by Streamwise, has been used to obtain a 3D measurement of the airflow velocity exits of the vents. The 5-hole probe allows obtaining a 3D measurement of the flow velocity, determining the velocity, yaw angle, and pitch angle from measurements of the differential pressure [13] between pairs of holes located near the tip of the sensors [14]; a detailed description of the 5-hole probe is available in Paul et al. [15].
For each vent, a control volume has been defined and the measurements were carried out with an acquisition frequency of 60 Hz, with a sampling of 20 minutes for each vent. In post-processing, to show the velocity field, a surface parallel to the plane of the vents has been selected at a distance of approximately 1 cm from the latter.

2.2. PIV experimental setup
Planar PIV measurements at three longitudinal sections have been performed: (i) the car centre plane, including both the front and rear area (shown in Figure 3, detail a); (ii) the driver's seat centre plane, only at the front area (shown in Figure 3, detail b); (iii) the passenger's seat centre plane, only at the front area (shown in Figure 3, detail c). Each of these three different setups has required specific positioning of the cameras and laser head and specific system calibration.

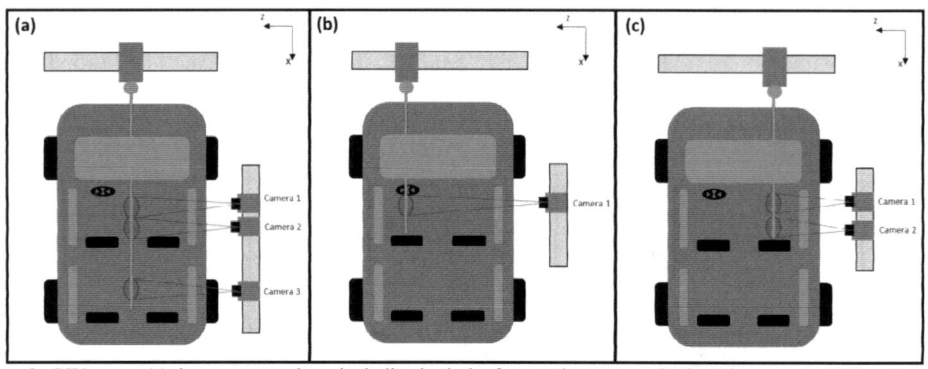

Figure 3 - PIV setup: (a) the car centre plane, including both the front and rear area; (b) the driver's seat centre plane (only the front area); (c) the passenger's seat centre plane (only the front area).

The experiment has been conducted in the backyard of the Aerodynamics Laboratories of the Delft University of Technology. To protect the equipment (seeding generator, laser, cameras and computer) in case of rain, and to confine the laser light, the car and the equipment have been placed inside a tent of size 3 m (width) x 2.6 m (height) x 6 m (length). The temperature and humidity of the environment inside the tent have been checked during all the measurements, as well as the internal temperature of the cabin (summarized in Table 1). The seeding was generated using a SAFEX seeding generator, which produces water-glycol droplets of 1 μm median diameter. During the experiment, for all measurements performed, the seeding has been directly introduced inside the car cabin placing the SAFEX seeding generator inside. The particles were illuminated by a dual-cavity Evergreen Laser (Nd: YLF, pulse energy of 200 mJ at 15 Hz, wavelength of 532 nm) and images were recorded with three LaVision CMOS cameras (6.5 μm pixel size, 2560 × 2160 pixels maximum resolution, 12 bit). Depending on the setup (see Figure 3), three, two or one cameras have been used. The cameras have been equipped with a Nikon object of 35 mm and 50 mm focal length and 4 or 5.6 as optical aperture, depending on the camera. The field of view and the optical magnification factor changed based on the setup and camera used. For setup (a) and setup (b), the cameras 1 and 2 were set in a way that the FOVs of each camera were overlapping between them, to obtain a larger image composed using the 2 images acquired by each camera (see section 3.2). In overlapping areas, arithmetic mean has been used to determine the velocity. Focal length, optical aperture, field of view and optical magnification factor are summarized in Table 1. The image acquisition has been conducted at a frequency of 15 Hz, except for the acquisition at setup (a) with camera 3, which is conducted at a frequency of 5 Hz. The time intervals between two consecutive particle images varied between 1 and 8 ms, depending on the camera used. Table 1 summarizes the main experimental parameters used in each configuration.

Table 1- summary of all parameters used during the acquisition

	Camera	f #	Lens objective (mm)	Field of view (mm)	Magnification factor	Frequency (Hz)	Δt (ms)	Temp out (°C)	Humidity (%)	Temp in (°C)
Setup a	1	5.6	35	653 x 545	0.025	15	1	30.2	28	43.6
	2	4	50	454 x 382	0.037	15	1	30.2	28	43.6
	3	4.6	35	472 x 397	0.035	5	2	28.4	25	37.8
Setup b	1	5.6	35	813 x 688	0.02	15	8	27.0	30	39.0
Setup c	1	5.6	35	524 x 437	0.032	15	5	29.1	33	39.5
	2	4	50	351 x 297	0.047	15	1.7	29.2	33	39.8

The data sets consist of 500 double-frame images for each experimental condition. The processing has been conducted via the LaVision Davis 8.6 software, using Gaussian interrogation windows of 128 × 128 pixels with 75% overlap for the initial passes and 32 × 32 pixels with 75% overlap for the final passes. The resulting instantaneous velocity fields have been ensemble-averaged to obtain the spatial distribution of the time-average velocity.

3. Results and discussions

The results from the experiments are presented in two sections; the first one discusses the results obtained from the 5-hole pressure probe for characterization of the airflow exits from the vents, whereas the second section discusses the PIV results in the longitudinal planes.

3.1. Airflow measurement at vents outlet, 5-hole pressure probe

Figure 4 shows the mean velocity fields of the outlet flow rate from the four vents opened in the panel-vent mode. The velocity fields show an asymmetry, as the velocity fields on the right (passenger's side) result in a higher magnitude than the velocity field on the left (driver's side). This is probably due to the design of the Peugeot 206 car, where the air intake duct is located on the right of the car hood. This influences the flow pattern inside the car cabin, as shown in section 3.2. Moreover, the shape of the mean velocity field from the vent at the left of the driver seat is quite different with respect to the others, which could be due to the presence of the steering wheel that has an influence on the velocity field.

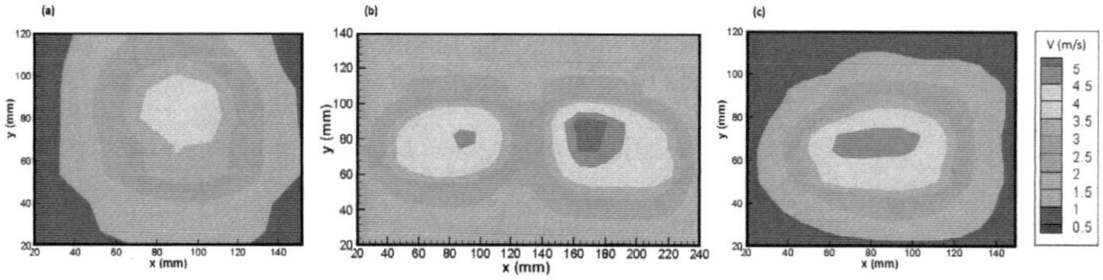

Figure 4 - mean velocity field in m/s of inlet flow rate from the vents active in panel-vent mode: (a) left vent, near the driver seat; (b) central vents; (c) right vent, near the passenger seat

3.2. Airflow measurement at longitudinal planes, PIV

Figure 5 shows the mean velocity fields at all measurement regions. The flow structure is complicated and large- and small-scale vortices coexist in the car cabin.

In Figure 5, detail a, the jet stream issued from the central dashboard vents is directly observed thanks to the measurement plane aligned at the centre of the two central panel vents. The jet stream is narrow and concentrated, probably due to the high fan strength (level 3 of the 4 available), the vents' shape and the tilt of the vents' grille. The flow issued impinges on the vehicle ceiling, leading to the formation of a large-scale clockwise vortex that characterizes both the front and rear area. From the rear area, part of the airflow is re-circulating through the vortex from the rear to the front area and part of the airflow goes to the parcel shelf, the section identified as the flow exit from the cabin. Despite the qualitative preliminary investigation has shown that the flow exits the passenger compartment in multiple locations, including the parcel shelf, the PIV measurement has confirmed that the flow of air is directed in that area, which can be considered the exit area. On average, the magnitude of the velocity vectors in the rear part is smaller than those in the front compartment. The region of the image that is affected by reflection due to the presence of the steering wheel, near the inlet section ($x = 0.15$ m, $y = 0.1$ m), has been masked.

In front of the driver's seat (Figure 5, detail b), two vortices rotating in opposite directions, one clockwise near the seat and one anti-clockwise near the windscreen, are located. Despite the fan strength imposed being very high (level 3 out of the 4 available), the magnitude of the vector in this area is small in comparison to the central plane.

In the passenger seat (Figure 5, detail c), the flow pattern is complicated and the magnitude of the velocity is larger compared with those of the driver's seat and smaller compared with those in the central plane. This may be caused by the different layout of the vehicle dashboard due to the different configurations required for the driver and passenger, such as the steering wheel at the driver's seat, which could cause an asymmetry in the flow. The asymmetry could be also due to the difference in the magnitude of the velocity flow in the outlet vents active in the panel-vent mode, greater on the vents on the right than on the vents on the left (see paragraph 3.1).

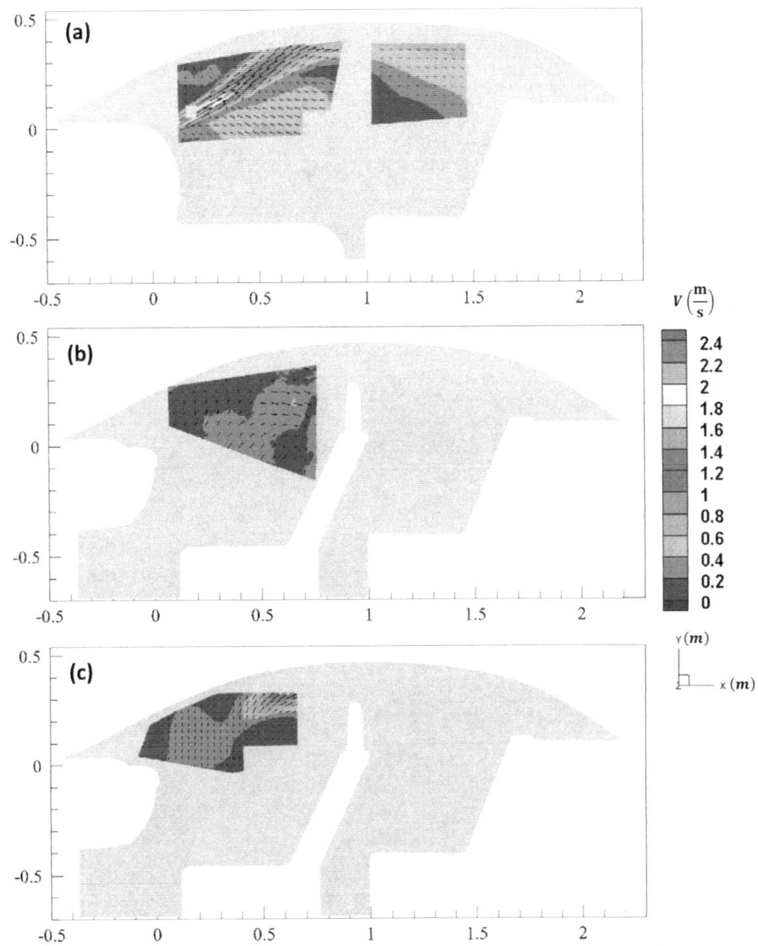

Figure 5 - mean velocity fields in the x-y plane inside the car cabin: (a) the car centre plane, including both the front and rear area; (b) the driver's seat centre plane (only the front area); (c) the passenger's seat centre plane (only the front area).

Figure 6 shows (a) mean velocity and (b) measurement uncertainty at a 95% confidence level carried out at the car centre longitudinal plane (front area). The airflow velocity reaches a maximum velocity of 2.4 m/s in the jet stream near the outlet of the fans and decreases considerably from a point around 0.4 m. The uncertainty of the jet stream is around 0.055 m/s in the core zone, and reduces to 0.04 m/s in the tail area.

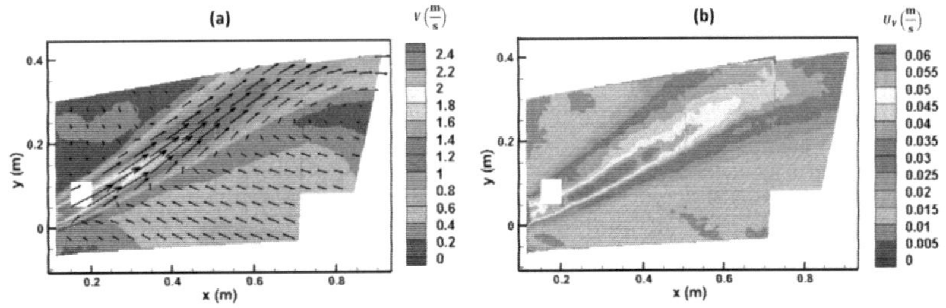

Figure 6 - (a) mean velocity fields and (b) uncertainty (both in meters per second) at the car centre plane (front area)

Figure 7 shows measurements of the (a) mean velocity and (b) measurement uncertainty at a 95% confidence level carried out in a longitudinal plane that goes through the middle of the car in the rear area. The magnitude of velocity vectors is around 0.4 m/s and the uncertainty is around 0.03 m/s.

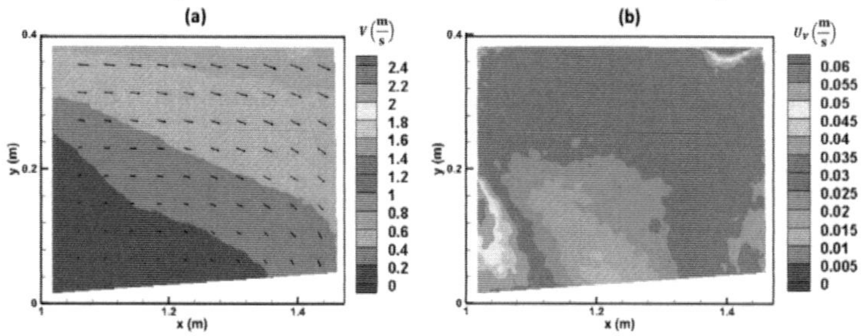

Figure 7 - (a) mean velocity fields and (b) uncertainty (both in meters per second) at the car centre plane (rear area)

Figure 8 shows (a) mean velocity and (b) measurement uncertainty at a 95% confidence level carried out at the longitudinal plane that goes through the centre of the driver seat. The maximum velocity is 0.5 m/s and the magnitude of the velocity vectors is smaller than those in the front compartment.

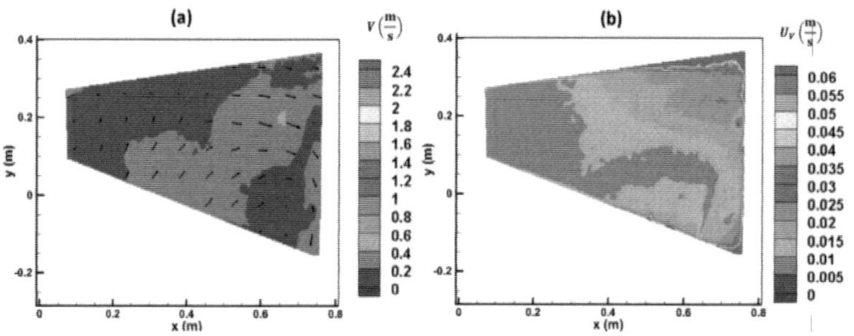

Figure 8 - (a) mean velocity fields and (b) uncertainty (both in meters per second) at the central section of the driver's seat

Figure 9 shows measurements carried out in the longitudinal plane that goes through the passenger's seat. The maximum velocity is 1 m/s.

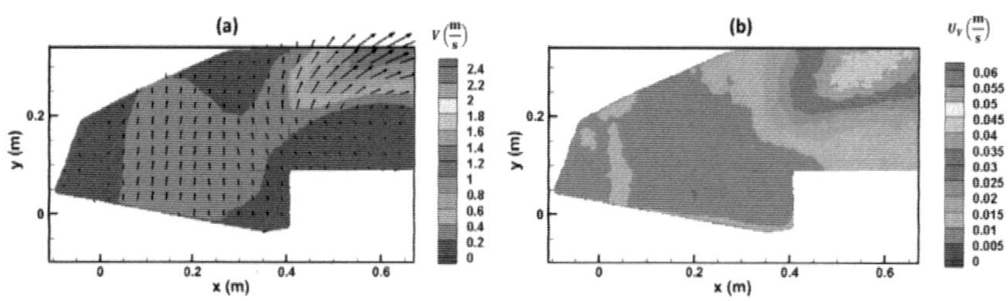

Figure 9 - (a) mean velocity fields and (b) uncertainty (both in meters per second) at the central section of the passenger seat

4. Conclusions

The ventilation flow in a car cabin has been experimentally investigated with the aim to propose a comprehensive benchmark for Computational Fluid Dynamics (CFD) numerical simulations validation. The study has been carried out in a commercial car cabin without any internal or external modification, by testing one ventilation mode (panel-vent mode) at one fan strength (level 3 of the 4 available) with fresh air intake (without any re-circulation). Obtained data represent preliminary results of an experimental campaign in which the velocity airflow field at all four fan speeds has been investigated. The results refer to the outgoing velocity field of vents active in panel-vent mode, measured using a 5-hole pressure probe and to three characteristic longitudinal planes, measured using the PIV technique.

The velocity field on the outgoing velocity field from the vents has shown an asymmetry between vents on the left and right sides, a difference which also occurs in the velocity field measured with the PIV in the driver's and passenger's plans. On the longitudinal centre plane, the jet stream outgoing from the vents is narrow and concentrated and the flow issued impinges the vehicle ceiling, leading to the formation of a large-scale clockwise vortex that characterizes both the front and rear area. The experimental results allow a better understanding of the ventilation flow in a car cabin and allows the construction of a benchmark case for the validation of CFD numerical simulations. Validated numerical model represents a useful tool for the design and optimization of the ventilation system to improve Indoor Air Quality (IAQ) and comfort of the occupants in any car cabin. Measurements at all four fan speeds has been also investigated and will be presented in future papers.

References

[1] G. Buonanno, G. Giovinco, L. Morawska, and L. Stabile, 'Tracheobronchial and alveolar dose of submicrometer particles for different population age groups in Italy', *Atmos. Environ.*, vol. 45, no. 34, pp. 6216–6224, Nov. 2011, doi: 10.1016/j.atmosenv.2011.07.066.

[2] European Communities, 'How Europeans spend their time', Luxemburg: European Communities, 2004.

[3] S.-J. Lee and J.-H. Yoon, 'Temperature field measurement of heated ventilation flow in a vehicle interior', *Int. J. Veh. Des.*, vol. Vol. 19, no. No. 2, p. pp 228-243, 1998.

[4] J. H. Yang, S. Kato, and H. Nagano, 'Measurement of airflow of air-conditioning in a car with PIV', *J. Vis.*, vol. 12, no. 2, pp. 119–130, Jun. 2009, doi: 10.1007/BF03181954.

[5] J. P. Lee, H. L. Kim, and S. J. Lee, 'Large-scale PIV measurements of ventilation flow inside the passenger compartment of a real car', *J. Vis.*, vol. 14, no. 4, pp. 321–329, Dec. 2011, doi: 10.1007/s12650-011-0095-9.

[6] F. J. Huera-Huarte, X. Cort, E. Aramburu, X. Vizcaino, and L. Casto, 'DPIV Measurements of the HVAC Aerodynamics Inside a Passenger Car', *SAE Tech. Pap.*, pp. 2014-36–0214, Sep. 2014, doi: https://doi.org/10.4271/2014-36-0214.

[7] S. Ullrich, R. Buder, N. Boughanmi, C. Friebe, and C. Wagner, 'Numerical Study of the Airflow Distribution in a Passenger Car Cabin Validated with PIV', in *New Results in Numerical and Experimental Fluid Mechanics XII*, vol. 142, A. Dillmann, G. Heller, E. Krämer, C. Wagner, C. Tropea, and S. Jakirlić, Eds. Cham: Springer International Publishing, 2020, pp. 457–467. doi: 10.1007/978-3-030-25253-3_44.

[8] F. Arpino *et al.*, 'Risk of SARS-CoV-2 in a car cabin assessed through 3D CFD simulations', *Indoor Air*, vol. 32, no. 3, Mar. 2022, doi: 10.1111/ina.13012.

[9] F. Arpino, G. Cortellessa, G. Grossi, and H. Nagano, 'A Eulerian-Lagrangian approach for the non-isothermal and transient CFD analysis of the aerosol airborne dispersion in a car cabin', *Build. Environ.*, vol. 209, p. 108648, Feb. 2022, doi: 10.1016/j.buildenv.2021.108648.

[10] V. P. Adhikari, A. Nassar, and Q. H. Nagpurwala, 'Numerical Studies on the Effect of Cooling Vent Setting and Solar Radiation on Air Flow and Temperature Distribution in a Passenger Car', *Numer. Stud. Eff. Cool.*, pp. 2009-28–0048, Dec. 2009, doi: 10.4271/2009-28-0048.

[11] A. Piovano, L. Lorefice, N. Paola, L. Miretti, and E. Ribaldone, 'Automated CFD multi-objective optimization for a passenger vehicle panel duct', 2014.

[12] J.-H. Yoon and S.-J. Lee, 'Velocity field measurements of ventilation flow in a vehicle interior', *Int. J. Veh. Des.*, vol. 31, no. 1, p. 96, 2003, doi: 10.1504/IJVD.2003.002050.

[13] R. Robinson, D. Butterfield, D. Curtis, and T. Thompson, 'Problems with Pitots. Issues with flow measurement in stacks', *International Conference on Emission Monitoring*, Milan (Italy), 2004.

[14] Denis William Bryer and R. C. Pankhurst, 'Pressure-probe Methods for Determining Wind Speed and Flow Direction', *Aeronaut. J.*, vol. 75, p. 660, 1971, doi: 10.1017/S0001924000046315.

[15] A. Paul, R. Upadhyay, and A. Jain, 'A novel calibration algorithm for five-hole pressure probe', *Int. J. Eng. Sci. Technol.*, vol. 3, no. 2, Jul. 2011, doi: 10.4314/ijest.v3i2.68136.

Experiments of convective evaporation of refrigerant R513A in a horizontal stainless-steel tube

A Arcasi[1], R Mastrullo[1], A W Mauro[1*], A F Passarelli and I Viscardi[1]

[1] Department of Industrial Engineering, Federico II University of Naples, P.le Tecchio 80, 80125 Naples (Italy)
*Corresponding author e-mail: wmauro@unina.it

Abstract. Refrigerant R513A represents an interesting solution for the retrofit of conventional high-GWP fluorinated gases, such as R134a, R401A, R401B and R409A for low and medium temperature applications. R513A is an azeotropic mixture (almost zero-temperature glide at any operating pressure) made up of R134a and R1234yf (44% and 56% in mass, respectively), allowing at the same time a very low GWP of 580 and favourable safety characteristics such as no flammability and no toxicity (A1 ASHRAE class). The boiling performance of this blend is scarcely explored and studied in scientific literature, especially in case of commercial tubes typically adopted for refrigeration purposes. For this reason, this paper presents two-phase flow boiling experiments of refrigerant R513A in a 6.00 mm horizontal stainless-steel tube. Heat is provided by means of Joule effect directly on the tube surface, and the peripheral average heat transfer coefficients are obtained by measuring the temperatures at four sides (top, bottom, left and right) of the channel. The effect of the operating conditions is experimented and discussed, by varying the mass flux between 150 and 300 kg/m^2 s, saturation temperature between 20 and 50°C and imposed heat flux between 5 and 20 kW/m^2. Also, a comparison with the boiling performance of refrigerant R134a is proposed within the same operating conditions. Finally, the assessment of well-known flow boiling prediction methods is presented and discussed.

Keywords: *R513A; flow boiling, heat transfer coefficient, correlation assessment*

1. Introduction

In the last decades, the global warming issue has become of primary importance and several regulations have been involved to limit the impact of the anthropogenic activities on Earth's temperature increase. Among them, the Kigali Amendment to the Montreal Protocol [1] promoted the phase-out of the hydrofluorocarbon refrigerants (HFCs) and the European F-GAS regulation [2] restricted the use of high-GWP (Global Warming Potential) fluorinated gases in several applications, thus leading to their replacement. In this context, R134a refrigerant (GWP value of 1430) has to be substituted with new low-GWP mixtures such as R1234yf, R1234ze and R513A for medium temperatures and other blends like R452A, R455A and R448A for low temperature applications. On this regard, Citarella et al. [3] proposed a thermo-economic analysis to explore possible design options for a 2.5 kW commercial refrigeration unit by using several mixtures (R454C, R449A, R452A and R455A), finding that R454C could be a suitable alternative as long-term replacement for these applications. Sanchéz et al [3] experimentally studied the energy performance of a cooler using R290 as an alternative of R134a and found that propane reduce energy about 27.5% compared to the one of R134a. Zheng et al [5] compared the performance of R290 and R134a in a cool storage air conditioning system and found that the system COP was comparable to the one of R134a whereas the R290 system had a higher volumetric refrigerating capacity. Mota-Babiloni et al. [6][7] experimentally analyzed the influence of using R134a and R513A refrigerants in a vapor compression system by varying evaporating and condensing conditions: it was found that R513A refrigerant leads both to a higher cooling capacity and COP values compared to the ones of R134a. The same considerations have been conducted by Zhang et al [8] and Sun et al [9]. New data on the two-phase heat transfer performance of these blends are particularly useful to test the accuracy of existing prediction methods for the correct design of condenser and evaporators employed in refrigeration systems. A literature review has pointed out numerous experimental studies on boiling of new R134a possible substitutes, either including pure fluids as R1234yf and R1234ze in conventional [10] and mini-channels [11], or non-azeotropic blends related to lower temperature applications, such as R452A [12], R455A [13] and R448A [14]. As regards instead the boiling performance of R513A, to the best of the authors' knowledge, the existing literature on this topic is scarce and limited to narrow diameter tubes. Diani et al [15][16] investigated both flow boiling and pressure drop performance of R513A in a 2.5 mm smooth and microfin tube comparing experimental results with the common predictive methods available in literature. In this study, experimental data of R513A during flow boiling in a 6.00 mm horizontal stainless-steel tube are collected. The effect of the operating parameters on the heat transfer coefficient in terms of saturation temperature, mass flux and heat flux are evaluated and discussed, by varying the saturation temperature between 20 and 50 °C, mass flux between 150 and 300 kg/m^2 s and heat flux between 5 and 20 kW/m^2. Finally, heat transfer comparison with the reference fluid R134a in the same operating conditions is investigated and the assessment of the well-known heat transfer coefficient correlations is proposed.

2. Experimental setup

2.1. Refrigerant and water loops

Figure 1 provides a photograph and a schematic representation of the experimental rig used in this paper. It consists of two different closed loops, one for the working fluid object of the study and one for the demineralized cooling water. In the refrigerant loop (black line) the employed fluid is pumped by a gear pump and goes into a Corolis mass flow meter. Then it passes into a preheating section whose heat capacity is about 3.6 kW and where it partially evaporates, achieving the desired vapor quality value. Successively it goes through the test section where the measurement of the heat transfer coefficient is attained. Finally, the saturated fluid reaches a plate condenser and a sub-cooler, before closing the loop. The demineralized cooling water loop (blue line) consists of a circulation pump and it feeds the sub-cooler and the condenser before going into the thermostatic bath, which controls its thermal level. More details of the whole experimental apparatus can be found in previous publications of the same authors [13][17]. The whole test bench is covered with several layers of synthetic foam to minimize the heat to and from the environment.

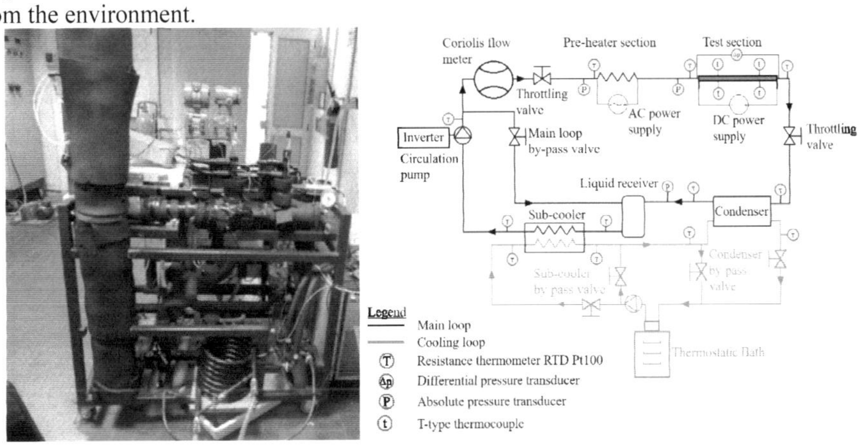

Figure 1. Photograph and representation of the experimental rig.

2.2. *Test section and measurement instrumentation*
All tests were performed in a single smooth horizontal stainless steel tube (AISI SS316) whose inner and outer diameters are 6.0 ±0.05 mm and 8.0 ±0.05 mm, respectively. All the geometrical features and characteristics of the test section are shown in Figure 2. Heat is provided by means of Joule effect directly on the tube surface with a DC power supply unit and two copper electrodes welded on the surface at a distance of 193.7 ±0.79 mm. Two wires are clamped in B and D position for the evaluation of the heat flux imposed. Four T-type thermocouples are placed on the external surface of the tube (point C) in the top, bottom, left and right position in order to evaluate the outer temperature distribution of the tube. The list of the measurement sensors and transducers employed in the experimental campaign is reported in Table 1 with their operative range and instrumental accuracy.

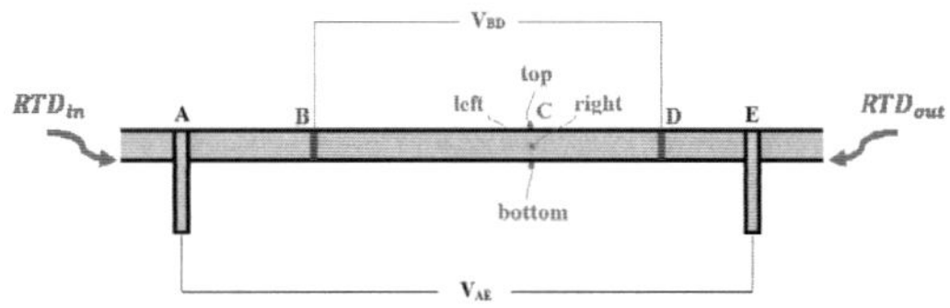

Geometrical features	
A,E	copper electrodes
B,D	voltage measurement points
C	temperature measurement point
AE	193.7 ±0.79 mm
BD	101.6 ±0.41 mm
AC	146.7 ±0.64 mm
d	6.00 ±0.05 mm (internal diameter)
D	8.00 ±0.05 mm (outer diameter)

Figure 2. Schematization and main geometrical characteristics of the test section.

Table 1. Specifications of the transducers and sensors with their operative range and accuracy.

Measurement	Range	Instrument accuracy
Temperature (4-wire Pt100 RTDs)	-80/250 °C	±0.180 °C
Wall temperature (T-type thermocouples)	10/90 °C	±0.1 °C
Inlet absolute pressure transducer (test section)	0/35 bar	±0.1% reading
Absolute pressure transducer (test rig)	0/50 bar	±0.3% reading
Differential pressure (test section)	0/14.51 kPa	±0.06 kPa
Coriolis flow meter $\dot{m}$	2.3/115.7 g/s	±1.0% reading
Electric power (pre-heater section)	0/3.6 kW	±1.0% reading
Voltage (test section) V_{BD}	0/5 V	±0.03% reading
Current (test section) I_{tube}	0/300 A	±1% reading

3. Method

3.1. *Data reduction*
The average heat transfer coefficient for each point is evaluated by the Newton equation with the assumption of uniform heat flux q and by considering the average wall temperature $T_{wall,mean}$ from the four T-type thermocouples measurements $T_{wall,o}$, as shown in Equation (1):

$$h_{mean} = \frac{q}{T_{wall,mean} - T_{sat,C}} \tag{1}$$

where the average wall temperature and the uniform heat flux are given by equations (2) and (3):

$$T_{wall,mean} = \frac{T_{wall,top} + T_{wall,bot} + T_{wall,left} + T_{wall,right}}{4} \tag{2}$$

$$q = \frac{V_{BD} \cdot I_{tube}}{\pi d \overline{BD}} \tag{3}$$

The inner $T_{wall,i}$ is obtained from the outer wall temperature $T_{wall,o}$ by considering the analytical solution to the heat transfer phenomenon with the hypothesis of 1-D heat flux in the radial direction and uniform heat generation inside the tube, as shown in Equation (4):

$$T_{wall,i} = T_{wall,o,i} + \frac{V \cdot I}{4\pi\lambda_{tube}L_h} \cdot \frac{\left(\frac{D}{d}\right)^2 \cdot \left(1-\log\left(\left(\frac{D}{d}\right)^2\right)\right)-1}{\left(\frac{D}{d}\right)^2-1} \tag{4}$$

The saturation temperature $T_{sat,C}$ is a function of the local pressure value P_C and the latter is a function of the pressure drop ΔP_{tot} across the test section.

$$T_{sat} = f(P_C) = f\left(P_{in} - \frac{\overline{PC}}{L_h}\Delta P_{tot}\right) \tag{5}$$

The local vapor quality value is instead a function both of the local pressure value and the local enthalpy i_C:

$$x_C = f\ (P_C, i_C) \tag{6}$$

Finally, the local enthalpy is obtained by an energy balance considering the test section (subscript *TS*) and the preheater section (subscript *preh*), thus leading Equation (7):

$$i_C = i_{preh} + \frac{\dot{Q}_{preh}}{\dot{m}} + \frac{\dot{Q}_{TS} \cdot \left(\frac{\overline{AC}}{\overline{AE}}\right)}{\dot{m}} \tag{7}$$

All the transport and thermodynamic properties of the working fluid are calculated by the software Refprop 9.1 [18] whereas the data reduction is executed with a MATLAB [19] code.

3.2. Uncertainty analysis

The instrumental uncertainty was combined with the standard deviation of the measurement in the recording time. The law of propagation of error [20] was implemented for the derived parameters uncertainty using a coverage factor of 2, thus achieving a confidence level higher than 95%.

Table 2 reports the maximum uncertainty of the parameters of interest. For most of the database, the operating parameters' uncertainty is very low, whereas it becomes significantly higher if the dry-out occurrence is taken into account.

Table 2. Maximum uncertainty of parameters both considering and excluding dry-out conditions.

Parameter	Maximum uncertainty for 90% of the database	Maximum uncertainty for the whole database (including dry-out conditions)
Saturation temperature T_{sat}	±0.04 °C	±0.06 °C
Mass flux G	±7.6 %	±15 %
Heat flux q	±0.76 %	±1.3 %
Vapor quality x	±0.06	±0.07
Average heat transfer coefficient h_{mean}	±5.02 %	±30.4 %

4. Experimental results

4.1. Operating conditions and experimental procedure

All the experimental tests were carried-out in steady state conditions, with a record frequency of 1 Hz and in a recording time of 90 s. Specifically, the saturation temperature, mass flux and heat flux values were set by adjusting the demineralized water temperature through the thermostatic bath, by fixing the inverter frequency and regulating the electrical voltage of the DC power supply unit, respectively. Finally, the desired vapor quality value was regulated through the heat rate in the preheater section. In the present study, the azeotropic mixture R513A has been studied by varying the saturation temperature from 20 to 50°C, mass flux from 150 and 300 kg/m^{2}s and imposed heat flux between 5 and 20 kW/m^2 in order to evaluate the effect of the operating parameters on the heat transfer coefficient.

4.2. Effect of the saturation temperature on the heat transfer coefficient

Figure 3 shows the effect of the saturation temperature on the average heat transfer coefficient as a function of the vapor quality for a fixed mass flux and heat flux values of 400 kg/m^{2}s and 20 kW/m^2, respectively. It can be noticed that the increase of the saturation temperature positively affects the heat transfer coefficient. In fact, the heat transfer coefficient increases from 7.8 kW/m^{2}K to 8.9 kW/m^{2}K (+14%) at a vapor quality of 0.3 and from 8.0 kW/m^{2}K to 9.4 kW/m^{2}K (+17.5%) at a vapor quality of 0.6 when the saturation temperature increases from 40°C to 50 °C. This can be explained by the increase of the nucleate boiling contribution probably due to the lower surface tension and consequently to the higher bubble growth.

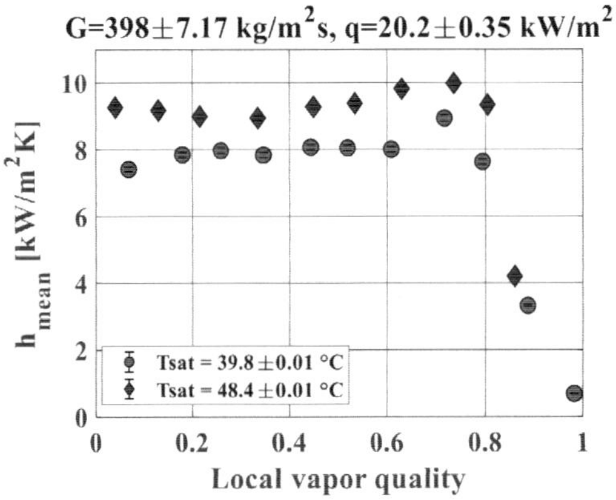

Figure 3. Heat transfer coefficient as a function of vapor quality. Effect of the saturation temperature for a mass flux of 400 kg/m^{2}s and a heat flux of 20 kW/m^2.

4.3. Effect of the mass flux on the heat transfer coefficient

The effect of the mass flux value on the heat transfer coefficient as a function of the vapor quality is reported in Figure 4 for a saturation temperature of 40 °C and a heat flux value of 10 kW/m^2 (Figure 4a) and a saturation temperature of 40°C and an imposed heat flux of 20 kW/m^2 (Figure 4b). The increase of the mass flux value positively affects the heat transfer coefficient. Particularly, for low mass flux value and relatively low imposed heat flux such as 150 kg/m^{2}s and 10 kW/m^2 the decreasing trend in the heat transfer coefficient is probably due to an anticipated dry-out phenomenon. As the mass flux value increases, the heat transfer coefficient is rather constant with the vapor quality due to the non-negligible convective heat transfer (see Figure 4a). For higher imposed heat flux values such as 20 kW/m^2, instead, the heat transfer coefficient seems to be not affected by the mass flux effect (see Figure 4b), implying that the nucleative boiling contribution plays a major role.

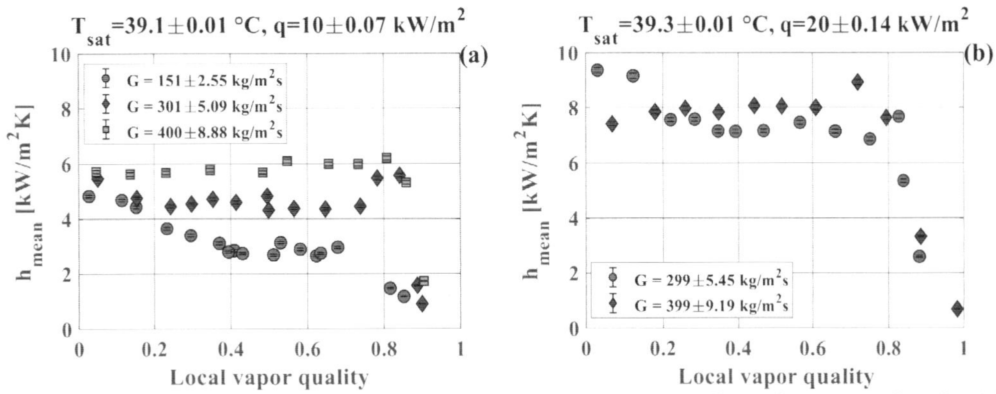

Figure 4. Heat transfer coefficient as a function of vapor quality. Effect of the mass flux for (a) a saturation temperature of 40 °C and a heat flux value of 10 kW/m^2 and (b) saturation temperature of 40 °C and heat flux value of 20 kW/m^2.

4.4. Effect of the heat flux on the heat transfer coefficient

Figure 5 exhibits the effect of the imposed heat flux as a function of the vapor quality for a fixed mass flux of 300 kg/m^{2}s and a saturation temperature of 40 °C. Particularly, at low vapor qualities the nucleate boiling contribution is stronger and the increasing in the heat flux significantly affects the heat transfer coefficient. In fact, it raises from 4.7 kW/m^{2}K for an imposed heat flux of 10 kW/m^2 to 9.1 kW/m^{2}K (+94%) for 20 kW/m^2 at a vapor quality value of 0.2. This behavior becomes less intense as the vapor quality increases: at a vapor quality of 0.6, the heat transfer coefficient passes from 4.3 kW/m^{2}K at a heat flux of 10 kW/m^2 to 7.1 kW/m^{2}K at 20 kW/m^2 (+65%). Finally, the vapor quality for dry-out inception is approximately equal to 0.83 and seems not to be greatly affected by the heat flux variation.

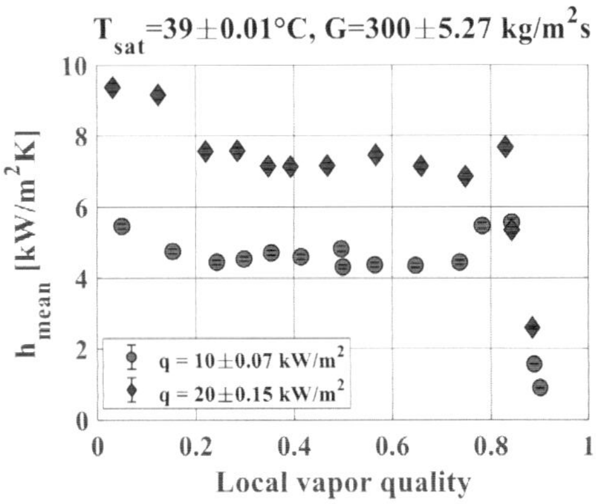

Figure 5. Heat transfer coefficient as a function of vapor quality. Effect of the heat flux for a saturation temperature of 40 °C and a mass flux value of 300 kg/m^{2}s.

4.5. Comparison of the heat transfer performance with R134a

The heat transfer performance of R513A has been compared with the one of pure R134a in the same operating conditions. Figures 6a-b show the effect of the working fluid for different mass flux and heat flux values. It is worth noting that, for low mass flux values, the effect of the working fluid is negligible (Figure 6a), probably due to the earlier dry-out phenomenon occurring in both cases, whereas R134a heat transfer coefficient is significantly higher as the mass and heat flux values increase (Figure 6b). In fact, for a fixed vapor quality value of 0.5, the heat transfer coefficient of R513A is 7.1 kW/m^{2}K whereas the one of R134a is 11.2 kW/m^{2}K (+58%).

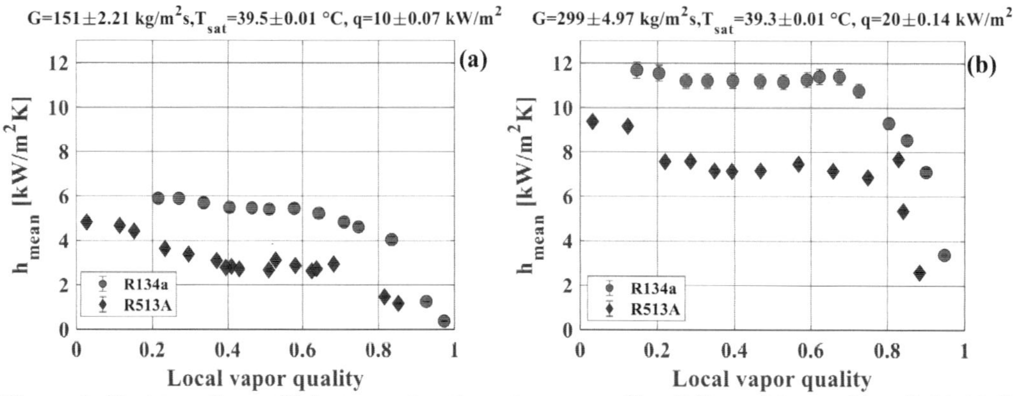

Figure 6. Heat transfer coefficient as a function of vapor quality. Effect of the working fluid (a) for mass flux value of 150 kg/m^{2}s, saturation temperature of 40 °C and heat flux of 10 kW/m^2 and (b) for mass flux of 300 kg/m^{2}s, saturation temperature of 40 °C and heat flux value of 20 kW/m^2.

5. Assessment of the existing prediction methods

Experimental data for both refrigerants R513A and R134a were compared to the predicted values from well-known correlations taken from literature in Figure 7a-b. The statistical analysis includes the Mean Absolute Percentage Error MAPE, the Mean Relative Percentage Error MRPE and the percentage of data points falling into an error range of ±30% ($\delta_{\pm 30\%}$). Two correlations have been chosen from the available ones thanks to their good agreement. Gungor-Winterton (1986) [21] method, developed for water and halogenated refrigerants, slightly underestimates R513A and R134a database (MRPE of -11.7% and -33.5%, respectively), with most of 60% of the R513A dataset falling into a range of error of ±30%. The modification of Gungor-Winterton [21] correlation proposed by Del Col (2010) [22] provides a slightly overestimation of R513A dataset (MRPE of 6%) and it shows a good agreement for the entire dataset, with most of 74% and 75% of the R513A and R134a points falling into an error range of ±30%, respectively.

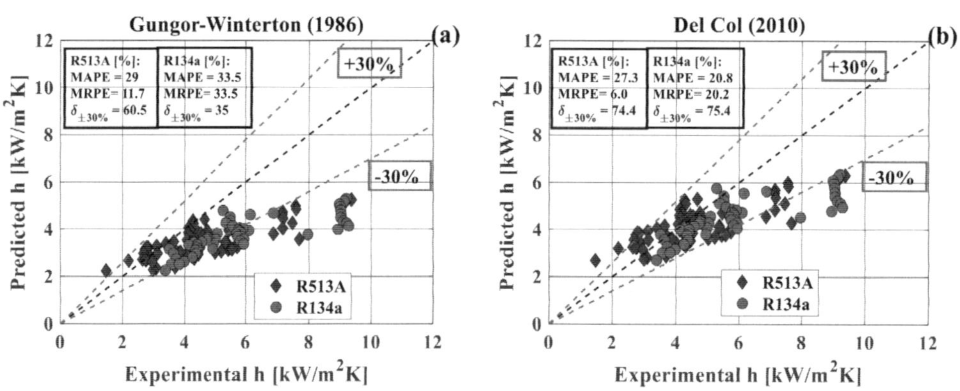

Figure 7. Predicted vs experimental heat transfer coefficient. (a) Gungor-Winterton (1986) method [21]; (b) Del Col (2010) correlation [22].

6. Conclusions

An experimental campaign for heat transfer coefficient data during flow boiling of R513A in a 6.00 mm single horizontal stainless-steel tube was carried-out. The main results are reported as follows:

- The mass flux and the heat flux positively affect the heat transfer coefficient trend. In particular, for low mass flux value such as 150 kg/m^{2}s and low imposed heat flux such as 10 kW/m^2, an anticipated dry-out occurs. As the mass flux increases, the heat transfer coefficient is relatively constant with the vapor quality value, probably due to the rise of the convective heat transfer. On the other hand, when the imposed heat flux is relatively high such as 20 kW/m^2, the heat transfer coefficient seems to be not affected by the increase of the mass flux value.
- The saturation temperature positively affects the heat transfer performance of R513A as well. For a mass flux value of 400 kg/m^{2}s and an imposed heat flux value of 20 kW/m^2, the percentage difference in the heat transfer coefficient reaches about 20% when the saturation temperature rises from 40 °C to 50 °C. This behaviour is probably due to the lower surface tension and consequently to the higher bubble formation, involving a higher nucleate boiling contribution.
- When compared to R134a refrigerant in the same operating conditions, the heat transfer coefficients of R513A are similar to those obtained with R134a during anticipated dry-out. The (negative) difference between them is significantly higher with the increasing of mass flux and heat flux values: at a mass flux of 300 kg/m^{2}s, saturation temperature of 40 °C and heat flux of 20 kW/m^2, the percentage difference reaches 60%.
- The assessment of two-phase heat transfer coefficient predictive methods has shown that Del Col correlation [22] provides the best agreement with the experimental data with more of 74% of the whole database falling into an error band of ±30% and a MAPE index equal to 27.3% (R513A) and 20.8% (R134a), respectively.

References

[1] Heath E 2017 *Int. Legal Mater* **56** 193-205
[2] The European Parliament and the Council of the European Union 2014 *Off. J. Eur. Union* **150** 195-230
[3] Citarella B, Viscito L, Mochizuki K, Mauro A W 2022 *En. Conv. Manag.* **253** 115152
[4] Sànchez D, Andreu-Nàcher A, Calleja-Anta D, Llopis R, Cabello R, 2022 *En. Conv. Manag.* **256** 115388
[5] Zheng H, Tian G, Zhao Y, Jin C, Ju F, Wang C 2022 *Case Stud. Th. Eng.* **36** 102203
[6] Mota-Babiloni A, Belman-Flores J M, Makhnatch P, Navarro-Esbrì J, Barroso-Maldonado J M 2018 *Energy* **162** 99-110
[7] Mota-Babiloni A, Navarro-Esbrì J, Pascual-Miralles V, Barragàn-Cervera A, Maiorino A 2019 *Appl. Therm. Eng* **147** 482-491
[8] Yang M, Zhang H, Meng Z, Qin Y 2019 *Appl. Therm. Eng* **146** 540-547
[9] Sun J, Li W, Cui B 2020 *Int. J. Refrigeration* **112** 348-356
[10] Longo G A, Mancin S, Righetti G, Zilio C 2019 *Int. J. Refrigeration* **108** 336-346
[11] Del Col D, Bortolin S, Torresin D, Cavallini A 2013 *Int. J. Refrigeration* **36** 353-362
[12] Mastrullo R, Mauro A W, Viscito L 2019 *Exp Th Fluid Sc* **105** 247-260
[13] Mauro A W, Napoli G, Pelella F, Viscito L 2020 *Int. J. Refrigeration* **119** 195-205
[14] Lillo G, Mastrullo R, Mauro A W, Pelella F, Viscito L 2019 *Appl Th Eng* 114146
[15] Diani A, Rossetto L 2019 *Int. J. Refrigeration* **107** 301-314
[16] Diani A, Rossetto L 2020 *Int. J. Heat Mass Transfer* **162** 120402
[17] Arcasi A, Mauro A W, Napoli G, Viscito L 2022 *Int. J. Heat Mass Transfer* **188** 122599
[18] Lemmon E W Mc Linden M O, Huber M L, 2009 *REFPROP NIST* Database **23**
[19] MATLAB 2019a Release, *Mathworks*
[20] Moffat J R 1985 *J. Fluids Eng.* **107** 173-178
[21] Gungor K, Winterton R 1986 *Int. J. Heat Mass Transfer* **29** 351-358
[22] Del Col D 2010 *Exp. Therm. Fluid Sci.* **34** 234-245

Journal of Physics: Conference Series **2509** (2023) 012026 doi:10.1088/1742-6596/2509/1/012026

Experimental analysis of drop size distribution and nucleation site density during dropwise condensation from humid air flow

M Mirafiori[1], M Tancon[1], S Bortolin[1*], A Martucci[1] and D Del Col[1]

[1] University of Padova, Department of Industrial Engineering
Via Venezia 1, 35131 Padova, Italy

*Corresponding author e-mail: stefano.bortolin@unipd.it

Abstract. Condensation of the water vapor present in the air is a heat and mass transfer process encountered in many applications as humid air dehumidification and water harvesting. Depending on the wettability characteristics of the surface, condensation can take place in filmwise mode or in dropwise mode with the formation of discrete liquid droplets over the condensing surface. While dropwise condensation (DWC) of pure steam was found to promote a considerable enhancement of the heat transfer compared to filmwise condensation, when dealing with humid air DWC more investigation is needed. Modeling of DWC from humid air requires the calculation of the heat flow rate through a single droplet and the determination of the drop-size distribution. The heat exchanged through a single droplet depends on the heat and mass transfer resistances, while the drop-size distribution is also affected by nucleation site density and droplets mobility. Therefore, to better understand the DWC phenomenon with humid air and for the validation of the models, it is necessary to measure the heat flux (total and latent), droplet population and nucleation site density. In the present work, condensation tests from humid air are performed over two square (40 mm x 40 mm) aluminum samples that display different wettability. The experimental apparatus consists of a closed air loop with two main components: the environmental chamber and the test chamber. The air is conditioned in the environmental chamber and then it flows inside the test section where the vapor present in the humid air is condensed over the vertical metallic sample. Two variable speed fans are used to circulate the air. The test section is designed for heat and mass transfer measurements and for simultaneous visualization of the condensation process. As a peculiar characteristic of the present experimental technique, all the test section assembly is suspended on a high precision balance allowing a precise measurement of the mass of condensate. The effect of surface wettability on the heat and mass transfer during DWC is investigated. Time-lapse videos of the condensation process are acquired at different magnifications. By using a homemade MATLAB® program for droplet detection, recorded images are analysed allowing the determination of both the drop size density distribution (small and large droplet population) and the nucleation sites density.

1. Introduction

Commonly, heat exchangers are made of hydrophilic metals and, therefore, their surfaces have a high affinity with water (high wettability). Therefore, a condensing vapor tends to form a continuous liquid film over the surfaces of heat transfer devices, which contributes to the overall thermal resistance. The condensate film can be removed by applying suitable coatings on the heat transfer surfaces to reduce their wettability. On a low-wettability surface, condensation occurs with the formation of discrete droplets that, in a quasi-cyclical way, grow, coalesce, and are removed by external forces (e.g., gravity or vapor drag force). This condensation mode is called dropwise condensation (DWC). During condensation of pure steam, promoting DWC can enhance the heat transfer performance by an order of magnitude compared to filmwise condensation (FWC) [1,2]. On the other hand, it should be noted that the possible advantage given by DWC during condensation in presence of non-condensable gases

(NCGs) is not well established. In fact, during condensation from humid air, the large amount of NCGs strongly penalizes the overall heat transfer compared to the case of pure vapor condensation, making vapor diffusion through the NCGs layer the dominant mechanism [3–5].

During condensation from flowing humid air, since the sensible contribution to the total heat transfer can be comparable with the latent term, it is important to measure directly not only the overall heat flux but also the latent term (e.g., by using a high-precision scale). However, the works available in the literature and dealing with DWC in presence of humid air flow are usually limited to the measurement of the overall heat transfer [3,4]. These data show that surfaces with modified wettability can enhance the heat transfer even in the presence of NCGs due to the reduction of the thermal resistance associated with the liquid film and the perturbation of the diffusion boundary layer caused by the presence of the drops [3]. In addition to surface wettability, the velocity of the air/vapor mixture can also affect the condensation performance [4]. However, the magnitude of the heat transfer enhancement as well as the effects of the different operating conditions has not definitely established.

A key parameter to model DWC even in the presence of NCGs is the drop-size density distribution. Modern droplet population models classify droplets into small and large droplets based on the growth mechanism [6]. The small droplet population $n(r)$ grows by direct condensation, while the large droplet population $N(r)$ grows mainly by coalescence. The threshold radius between the two populations is called the interaction radius r_e. This value is related to the spatial distribution of the nucleation sites which, on a smooth and chemically homogeneous surface, should follow a Poisson distribution [6]. The presence of NCG should reduce the number of active nucleation sites compared to the case of pure vapor condensation, thus shifting the interaction radius r_e to higher values. However, due to the difficulty of measurement, the studies in the literature on DWC of flowing moist air provide neither measurements of the nucleation site density nor the small droplet population (down to a few micrometers).

In the present work, these shortcomings in the literature are addressed by using a new experimental apparatus for the measurement of both latent and total heat flux and the simultaneous visualization of the droplet population during condensation from flowing humid air. As a peculiar feature of the apparatus, the entire test section (figure 1a) is suspended on a weight scale, thus making possible the measure of the condensation heat flux in presence of air flow. The new experimental apparatus [4] has been used to investigate condensation from humid air on two aluminum samples with different wettability: a hydrophilic baseline surface and a sol-gel coated surface with reduced contact angle hysteresis [7]. Aluminum has been selected as material because it is commonly used in air heat exchangers. Air relative humidity and air velocity have been varied to evaluate the heat and mass transfer enhancement when promoting DWC in the presence of humid air. During the DWC experiments, microscope images have been recorded with the aim of measuring the drop-size distribution (of both small and large drops) and the nucleation sites density.

2. Experimental method

2.1. Experimental apparatus

A sketch of the test rig is shown in figure 1a. It consists of two main components: an environmental chamber, where the air is conditioned in terms of humidity, temperature and velocity, and a test section, where heat and mass measurements are performed on the metal specimen.

The environmental chamber allows controlling the air temperature through an air-air thermoelectric unit (Laird® Thermal Systems), the humidity through four ultrasonic humidifiers, and the velocity through two variable speed fans. A finned-coil heat exchanger connected to a thermostatic bath is used to control both the temperature and the humidity by dehumidifying/cooling or heating the air in the environmental chamber. To reduce heat losses to the ambient and thus ensure more stable conditions during the experiments, the environmental chamber is thermally insulated. The air temperature is measured by means of 5 kΩ calibrated NTC thermistors and by Pt100 platinum resistance thermometer. The humidity and the velocity of the humid air entering the test section are measured respectively by a cooled-mirror hygrometer (Michell Instruments® DEWMET) and by a hot wire anemometer. The conditioned airflow from the environmental chamber flows through an insulated channel to the test section. Before entering the test section, the airflow passes through a honeycomb grid. Then, the airflow returns through an insulated tube to the environmental chamber, closing the loop.

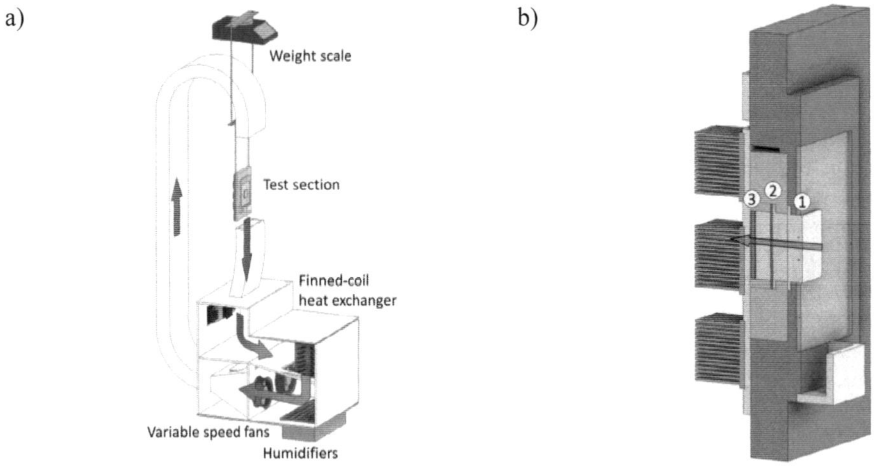

Figure 1. a) Sketch of the test rig showing the main components of the experimental apparatus. b) Cross section view of the test section assembly: 1) sample, 2) heat flux sensor, 3) Peltier module.

The test section (figure 1b) is designed for heat and mass transfer measurements during condensation of flowing moist air over a vertical specimen and for the simultaneous visualization of the phase-change phenomenon. The test section consists of a multi-component assembly designed to measure the total and latent heat flux exchanged during condensation over the metallic specimen. As shown in figure 1b, the sample is installed on the front of the test section where the humid air flows (figure 1b-1), while the aluminum heat sinks are installed on the opposite side. A Peltier module (figure 1b-3), a first aluminum block, the heat flux sensor (figure 1b-2), and a second aluminum block are respectively installed between the heat sink and the sample. All the components have a surface size of 40 mm × 40 mm to ensure one-dimensional heat flow from the sample surface to the Peltier module. The components are sealed together inside a PVC block and heat conductive paste is used to minimize thermal contact resistances. As shown in figure 1a, the test section is suspended on a high precision balance (Mettler Toledo® model PM4600). In this way, it is possible to weigh all the water produced during condensation, which total mass is the sum of the mass of liquid on the sample surface plus the condensate collected in a small tank placed just below the sample. The test section is enclosed in a transparent polycarbonate chamber and a high-speed camera is used to record both macroscopic and microscopic videos of the condensation phenomenon. It is important to note that the test section is instrumented with several sensors. In particular, the surface temperature of the sample T_w is directly obtained as the average of the readings of two Pt100 probes placed inside the sample, while the dew point temperature is given by the cooled-mirror hygrometer. Further details about the experimental apparatus can be found in [4].

2.2. Visualization technique

The visualization of the phenomenon is performed using a high-speed camera (Photron® FASTCAM Mini UX100) alternatively coupled with macro lens or micro lens. As described in [4,8], a system consisting of annular LED strips is used as a lighting source to improve droplets' detection. Videos were recorded at low framerate (0.5 fps) for the entire duration of the condensation test. The recorded images were analysed using two homemade programs written in MATLAB® code. With the aim of evaluating the nucleation sites density, the first program counts the number of drops that appears in a certain area during the initial stage of nucleation. The process is repeated on different areas cropped from the original image to get the average density of nucleation sites. The second MATLAB® program is used to evaluate the drop-size density distribution, identifying the radius and position of the drops over a certain selected area of the original image acquired by the camera. Accurate droplet detection is achieved once the program has been properly trained to determine the external radius of the light annulus reflected over each droplet. The program analyses the entire sequence of frames in order to map the droplet population over the entire video. Additional information can be found in [9].

2.3. Data reduction

As mentioned before, two different techniques have been implemented in the present apparatus for the evaluation of the heat exchanged during condensation. Using the heat flux sensor, both the convective and latent (Q_{tot}) components of the heat exchanged during condensation are measured, whereas by weighing the condensate only the latent portion is evaluated. The total heat flow rate transferred through the sample during condensation can be calculated as:

$$Q_{tot} = Q_{hfs} - Q_{par} \tag{1}$$

where Q_{hfs} is heat flow rate measured by the heat flux sensor and Q_{par} is a correction term that takes into account the parasitic heat flow rate entering the heat flux sensor from the surrounding PVC case. The latter term was estimated by performing 3D numerical simulations of the temperature field inside the test section assembly using ANSYS® Fluent 19.2. The latent heat flow rate (Q_{lat}) is evaluated from the high-precision balance as follows:

$$Q_{lat} = \dot{m} \cdot r \tag{2}$$

where r is the latent heat and $\dot{m}$ is the condensate mass flow rate. The condensate flow rate $\dot{m}$ is obtained by a linear regression of the condensate weight measured over time.

The thermodynamic and transport properties are evaluated by NIST Refprop version 10.0 [10]. Each experimental point reported in this work is obtained as the average of 20 data sets taken over a time interval of 5 minutes. Each dataset consists of 30 readings collected at 0.5 Hz by the acquisition system. The uncertainty bars are obtained considering a coverage factor k = 2 as reported in [4,11]. The combined uncertainty obtained for the total heat flow rate Q_{tot} is on average 15% while the one obtained for the latent heat flow rate Q_{lat} is around 2%. The latent heat flow rate uncertainty is due to the measurement of the mass flow rate $\dot{m}$, whose uncertainty in turn is obtained by Monte Carlo method. More details about the uncertainty analysis can be found in [4].

3. Experimental results and discussion

3.1. Heat transfer measurements

Condensation tests were performed on two aluminum samples: a mirror-polished aluminum substrate (baseline sample) and a sol-gel coated aluminum sample [5,7]. For the characterization of the two surfaces, dynamic contact angles were measured by sessile drop method. The untreated aluminum surface was hydrophilic (θ_a = 56° and θ_r = 10°), whereas the sol-gel coated surface was nearly hydrophobic (θ_a = 87° and θ_r = 72°). Due the reduced contact angle hysteresis (from 46° to 15°), the treated aluminum sample was able to promote DWC with regular shaped drops. Instead, the condensation pattern on the baseline aluminum consisted of irregular-shaped flat drops with low mobility but characterized by the same mechanisms (nucleation, coalescence, sliding) that are typical of regular DWC.

The effect of the relative humidity (RH = 50-70%) and air velocity (v_a = 1-4 m s^{-1}) on the condensation heat flux when varying the wall subcooling ($\Delta T_{d\text{-}w}$ = 7-13 K) are shown in figure 2a. When increasing the relative humidity, the amount of vapor in the air flow increases enhancing the driving force for the mass transfer mechanism. Therefore, higher condensation heat fluxes are obtained on both the samples. The coated surface shows higher condensation performance compared to the hydrophilic surface and this advantage is more pronounced at low RH. In fact, for an average increase of the heat flux equal to 18%, the condensation performance on the coated surface at RH = 50% is enhanced by about 40% compared to the baseline sample. The present results show that, due to the presence of NCGs, the condensation heat flux during condensation from humid air is at least 3 orders of magnitude lower compared to pure steam condensation [4,12,13].

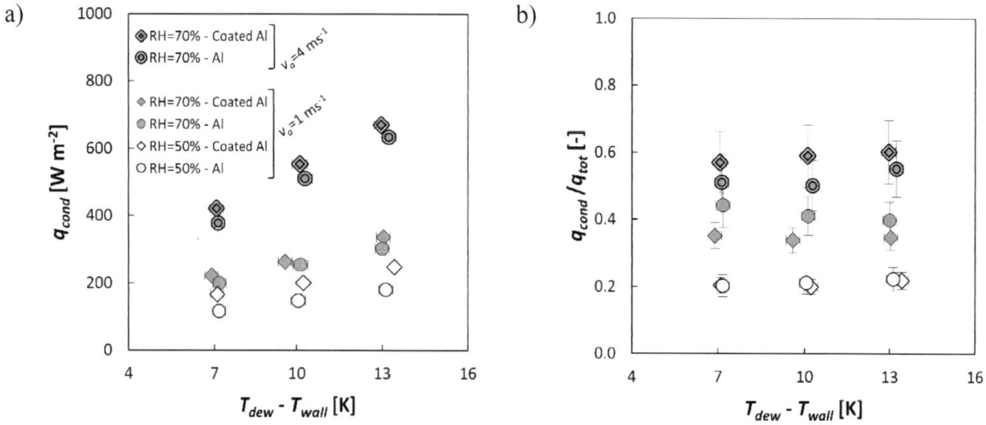

Figure 2. a) Condensation heat flux measurements (q_{cond}) at varying relative humidity and air velocity performed on the aluminum untreated and coated samples (t_a = 28 °C, RH = 50-70%, v_a = 1-4 m s^{-1}). b) Ratio between the condensation heat flux and the total heat flux (q_{cond}/q_{tot}) for the data reported in (a).

The condensation enhancement measured on the coated sample can be explained considering the formed condensate. Firstly, due to the higher droplets' mobility (lower contact angle hysteresis), the coated surface is more frequently renewed by the sliding droplets compared to the untreated surface, thus improving the heat transfer. Secondly, the structure formed by the droplets could disturb the concentration and the velocity boundary layers, enhancing the convective mass transfer of vapor molecules from the bulk to the cold surface [3]. This effect is expected to be more pronounced on the coated sample, where the regular hemispherical droplets form a structure with a greater height than the one formed by the irregular flat droplets that grow on the hydrophilic surface (θ_a = 56° and θ_r = 10°).

When the air velocity increases from 1 to 4 m s^{-1}, at fixed wall subcooling, the condensation heat flux increases by ~100% on both the samples. Due to the increased air velocity, the convective mass transfer is increased. In addition, an increase of air velocity could improve the mobility of the condensate, which is crucial for re-nucleation. As shown in figure 2a, the heat flux measured on the coated surface remains on average 18% higher compared to the hydrophilic sample, regardless of the air velocity. For comparison, as shown by Tancon et al. [2,14], the effect of the flow velocity during DWC in pure steam is considerably lower. In their velocity range (3.1-12 m s^{-1}) the HTC increased by around 20%.

On the other hand, figure 2b shows that, for the present data, depending on the experimental conditions, the latent contribution can vary from 20% to 60% of the total heat flux. It should be considered that the experimental uncertainty on the total heat flux is higher than the experimental uncertainty on the latent heat flux due to the different measurement technique (heat flux sensor for the total heat flux and weight of the condensate for the latent heat flux). Therefore, in order to accurately evaluate the effect of surface wettability on the condensation phenomenon, the measure obtained from the weight scale should be preferred.

3.2. Nucleation site density and droplet population

As a second step of the present work, the nucleation sites density N_s was investigated on the coated aluminum sample at varying experimental conditions (t_a = 28 °C, v_a = 1 m s^{-1}, RH = 50-70%, $\Delta T_{d\text{-}w}$ = 4-

10 K). Once the wall temperature drops below the dew point temperature, the water vapor contained in the humid air starts to condense. Initially, the small droplets that appear in the video have almost the same dimension (~ 1 μm), indicating that they are growing at the same rate with no coalescence events. Since the videos show the initial stage of condensation before the first coalescence occurs, for the present investigation of N_s, it has been assumed the presence of only one nucleation site under each droplet. Therefore, the evaluation of N_s relies on measuring the number of droplets on the surface before the onset of coalescence. For this task, the videos of the initial stage of the nucleation process were analysed using a homemade software developed in MATLAB® code which can detect the droplets number over the investigated area. Additional information can be found in [4].

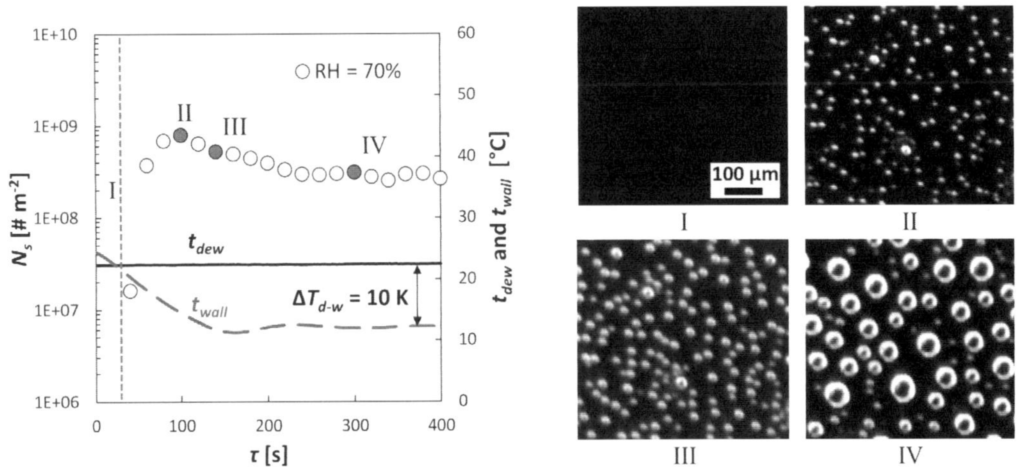

Figure 3. Drops density vs time measured on the coated sample in the early stages of condensation and, on the right, selected images showing the evolution of the process from the beginning of droplets nucleation (t_a = 28 °C, RH = 70%, v_a = 1 m s^{-1}, $\Delta T_{d\text{-}w}$ = 10 K).

As an example of the measuring technique, figure 3 shows the measured number of droplets per unit area at the early stage of condensation plotted versus time for a certain operating condition ($\Delta T_{d\text{-}w}$ = 10 K, RH = 70%). The dew point temperature (black solid line) and the surface temperature (blue dashed line) are also reported over time. On the right, the images of the main steps of the process are reported. From the instant the wall temperature falls below the dew temperature (time-I), some time passes before the first drops are detected. In fact, in addition to the limited resolution of the video, the delay can be associated to the time required for the formation of a stable droplet. After the appearance of the first droplets on the surface, the drops density increases until it reaches the maximum value after about 70 s (time-II). Hence, before the drops become large enough to merge, their number remains approximately constant over time. After the onset of coalescence, the drops density decreases as the large droplets can occupy more than one nucleation site (time-III). As discussed above, assuming only one nucleation site under each isolated drop, the maximum of the measured drop density function identifies the nucleation site density (N_s). For the case reported in figure 3, the nucleation site density resulted almost equal to 6×10^8 m^{-2}. Considering the whole database acquired at different experimental conditions, it was observed that N_s is influenced by the relative humidity, whereas the wall subcooling does not affect the nucleation sites density. In particular, when RH increased from 50% to 70%, the average nucleation sites density was found to increase from 3.3×10^8 to 6.1×10^8 m^{-2}.

For the case of nucleation sites distributed following a Poisson distribution [6] (i.e., the present case), the droplet interaction radius r_e can be calculated as half of the average distance between droplets ($r_e = 1/(4\sqrt{N_s})$). The present nucleation sites density measurements would suggest values of r_e in the range 10-20 μm. Therefore, using the image processing technique presented in section 2, which can detect droplet radii down to some microns, we are able to investigate part of the drop-size distribution of small droplets thus detecting the boundary between the two droplet populations.

Finally, the drop-size density distribution (which indicates the number of drops per unit of radius per unit of area) during condensation from humid air was investigated at fixed operating conditions (v_a = 1 m s^{-1}, t_a = 28 °C, RH = 50%, $\Delta T_{d\text{-}w}$ = 10 K). The results are reported in figure 4a. As described in section 2, the DWC images are processed by a MATLAB® program for the detection of droplets' dimension and position. Some DWC images recorded by the high-speed camera are shown in figure 4 together with the resulting images (in red and white) obtained from the droplet recognition MATLAB® code (and superimposed to the original images). The experimental drop-size distribution is computed by dividing the entire range of radii into multiple intervals, where the number of intervals is chosen in order to reveal the correct shape of the underlying distribution. For each recognized range of radii, it is possible to count the number of drops whose size falls in a specific radii range by considering the whole radii population found by analysing the video. Then, it is divided by the investigated area and by the width of the interval, obtaining the average drop-size distribution for a given condition.

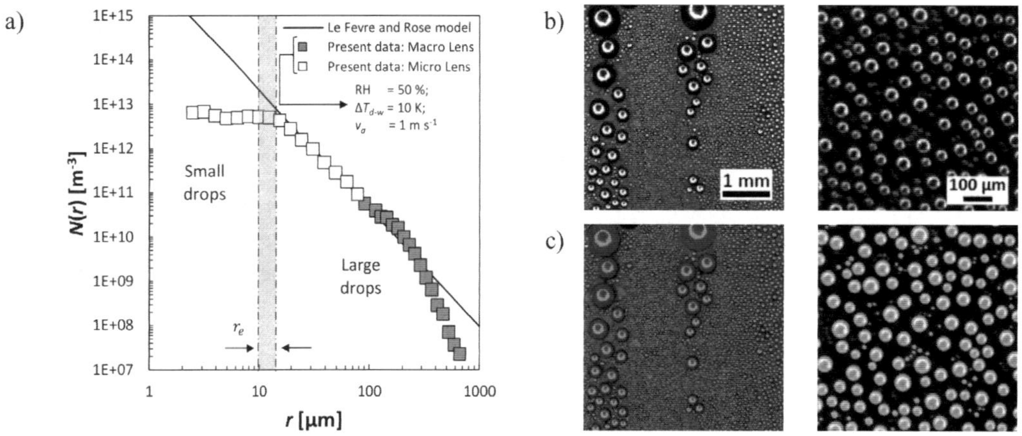

Figure 4. a) Drop-size distribution evaluated on the coated surface (t_a = 28 °C, v_a = 1 m s^{-1}, RH = 50%). b) Images of DWC taken by the high-speed camera at two different magnifications. c) Reconstructed images (in red and white) obtained by the MATLAB® code superimposed to the original images (b).

Figure 4a shows that the experimental drop-size density distribution decreases with the droplet radius. Therefore, for a given surface area, the number of small droplets is considerably higher than the number of large droplets. Furthermore, it can be observed that the distribution is characterized by two regions with different slope. The interaction radius (r_e) identifies the boundary between the small droplets (which grow by direct condensation) and the large drops (which grow mainly by coalescence) [6]. The results show that the radius at which the slope changes is around 20 µm and this agrees with the values of r_e inferred from the nucleation sites density measurements (10-20 µm).

In addition, figure 4a presents a comparison between the measurements and the predictions obtained by the Le Fevre and Rose [15] model for the distribution of large drops $N(r)$, which is given by:

$$N(r) = \frac{1}{3\,\pi\, r_{max} r^2}\left(\frac{r}{r_{max}}\right)^{-2/3} \tag{3}$$

where r_{max} is the droplet departing radius. The comparison shows that the distribution proposed by Le Fevre and Rose [15] is in good agreement with the experimental drop-size distribution for large drops in the range of radii between 20 µm and 300 µm. While, in the range of radii between 300 µm and r_{max} due to the sweeping effect of the sliding droplets, the experimental distribution was found to deviate from the power-law trend predicted by the Le Fevre and Rose [15] model.

4. Conclusions

In the present work, the heat transfer and the droplet population during condensation in presence of humid air were investigated over two different aluminum specimens: a hydrophilic sample (θ_a = 56° and θ_r = 10°) and a coated sample (θ_a = 87° and θ_r = 72°). Tests at varying experimental conditions (t_a = 28 °C, v_a = 1-4 m s^{-1}, RH = 50-70%, $\Delta T_{d\text{-}w}$ = 4-13 K) were performed using a novel technique that allows to measure, during condensation of flowing humid air, both the total heat flux (by a heat flux sensor) and the latent heat flux (by weighing the condensate) with the simultaneous visualization of the phenomenon.

The coated surface exhibited improved condensation performance compared to the hydrophilic sample, regardless of air velocity and relative humidity. Depending on the operating conditions, the increase of condensation heat flux provided by the coated surface ranges between 5% and 40%.

Starting from the images recorded during condensation, the nucleation sites density and the drop-size density distribution were evaluated on the coated aluminum sample using two homemade MATLAB® codes. In the investigated range of operating conditions, the nucleation sites density increased with the relative humidity from 3.3 × 10^8 sites/m^2 (at RH = 50%) to 6.1 × 10^8 sites/m^2 (at RH = 70%), regardless of wall subcooling. Furthermore, it was possible to partially map the small droplet population, characterized by a different slope as compared to the distribution of large drops. With respect to 50% of relative humidity, the threshold radius between the two droplet populations was measured to be approximately 20 μm.

Acknowledgements

The authors acknowledge the financial support of the Department of Industrial Engineering, University of Padova through the BIRD187743 project. The support of the European Space Agency through the MAP Condensation program ENCOM (AO-2004-096) is also acknowledged.

References

[1] Bortolin S, Tancon M and Del Col D 2022 *The Surface Wettability Effect on Phase Change* (Cham) Phase Change. Springer 29–67
[2] Tancon M, Mirafiori M, Bortolin S, Basso M, Colusso E and Del Col D 2022 *Appl. Therm. Eng.* **216** 119021
[3] Zheng S, Eimann F, Philipp C, Fieback T and Gross U 2020 *Int. J. Heat Mass Transf.* **151** 119349
[4] Tancon M, Mirafiori M, Bortolin S, Parin R, Colusso E, Martucci A and Del Col D 2022 *Exp. Therm. Fluid Sci.* **136** 110677
[5] Basso M, Colusso E, Sacco A, Tancon M, Bortolin S, Mirafiori M, Guglielmi M and Martucci A 2022 *J. Sol-Gel Sci. Technol.* **102** 466–77
[6] Miljkovic N, Enright R and Wang E N 2013 *J. Heat Transfer* **135** 111004
[7] Parin R, Rigon M, Bortolin S, Martucci A and Del Col D 2020 *Materials* **13** 878
[8] Parin R, Tancon M, Mirafiori M, Bortolin S, Moro L, Zago L, Carraro F, Martucci A and Del Col D 2020 *Appl. Therm. Eng.* **179** 115718
[9] Mirafiori M, Parin R, Bortolin S and Del Col D 2020 *J. Phys. Conf. Ser.* **1599** 012011
[10] Lemmon E W, Bell I H, Huber M L and McLinden M O 2018
[11] Bisetto A, Bortolin S and Del Col D 2015 *Exp. Therm. Fluid Sci.* **68** 216–27
[12] Shakeri Bonab M, Kempers R and Amirfazli A 2021 *Int. J. Heat Mass Transf.* **173** 121278
[13] Mirafiori M, Tancon M, Bortolin S, Martucci A and Del Col D 2022 *J. Phys. Conf. Ser.* **2177** 012046
[14] Tancon M, Parin R, Bortolin S, Martucci A and Del Col D 2021 *Int. J. Heat Mass Transf.* **165** 120624
[15] Le Fevre E J and Rose J W 1966 *Proc. Int. Heat Transf. Conf. 3nd* (Connecticut) Begellhouse 362–75

Convective condensation of R134a and R1234ze(E) inside microfin tube

A Lucchini[1], I M Carraretto[1], T N Phan[2,3], P G Pittoni[4], L Molinaroli[1] and L P M Colombo[1]

[1] Politecnico di Milano, Department of Energy, via Lambruschini 4, 20156 Milano, Italy
[2] Ho Chi Minh City University of Technology (HCMUT), 268 Ly Thuong Kiet Street, District 10, Ho Chi Minh City, Vietnam
[3] Vietnam National University Ho Chi Minh City, Linh Trung Ward, Thu Duc District, Ho Chi Minh City, Vietnam
[4] Iowa State University, 2529 Union Drive, Ames, IA 50011, United States

andrea.lucchini@polimi.it

Abstract. Environmental concerns are forcing the replacement of the commonly used refrigerants and finding new fluids is a top priority. The hydro-fluoro-olefin (HFO) R1234ze(E), because of its smaller global warming potential (GWP) and shorter atmospheric lifetime, replaced R134a. Accordingly, for HVAC systems design, a detailed knowledge of the thermo-fluid-dynamic characteristics of the fluids and reliable predictive models are required. To improve the understanding, R134a and R1234ze(E) were employed in convective condensation experiments (saturation temperature T_{sat} = 35°C, mean quality x_m = 0.1~0.9, quality changes Δx = 0.05~0.6, mass flux G = 43~444 kg·m^{-2}s^{-1}) inside a microfin tube (outer diameter D = 9.52 mm, fin number n = 60, fin height H = 0.2 mm). The results were used for two goals: the former is the comparison of the heat transfer features of the two fluids, while the latter aims at testing the performance of prediction models available in the open literature. At the saturation temperature T = 35°C, the two fluids show small differences in the thermal properties so that, as expected, the experiments highlighted a very similar behavior in the typical operating conditions of HVAC systems. In fact, for all the operating conditions marginal differences were observed in the pressure drop, the heat transfer coefficient and the flow pattern maps. The issue of prediction reliability, however, is still open. Actually, not all the models achieving good results for R134a show the same performance for R1234ze(E), especially for the pressure drop.

1. Introduction

The European Parliament developed a strategy to shift to a competitive low carbon economy. Accordingly, the European Regulation (EU) No. 29 517/2014 [1] prescribes the phase-out steps for the HFC. To properly design the next generation HVAC devices the search for new refrigerants is a top priority. The hydro-fluoro-olefin (HFO) R1234ze(E) seems a good replacement for R134a because of its environmental features: short atmospheric lifetime (17 days for R1234ze(E), 13.8 years for R134a [2]) and small global warming potential (less than 1 for R1234ze(E), 1430 for R134a [2]). Nevertheless, a negative impact on the overall performance, mostly due to vapor compression efficiency is reported as drawbacks [3]. The performance variation and its assessment were the main

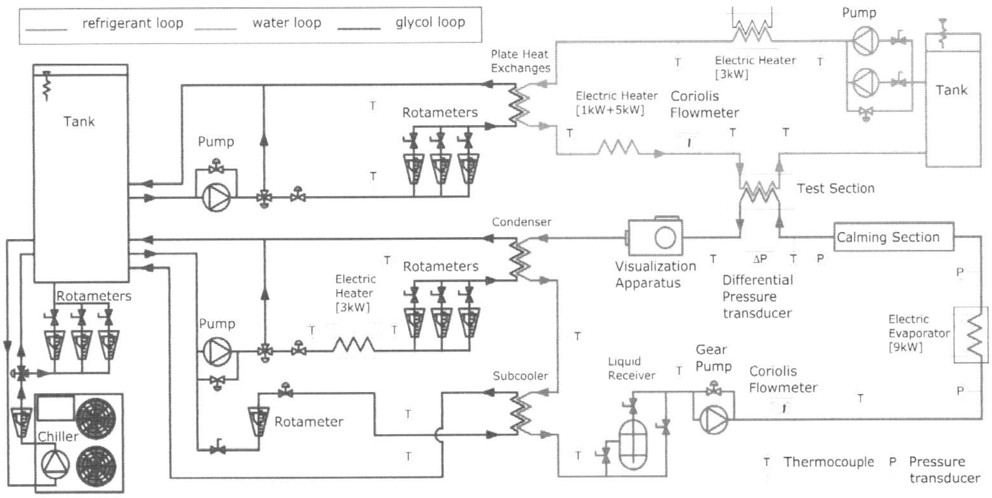

Figure 1. Experimental apparatus

concerns of the researcher in the last few years. The topics were approached at two different levels: the process (thermodynamics, transport phenomena), and the energy system. An example about the latter topics is provided by Censi and Padovan [3]: they examined a Microfin Shell-and Tube Evaporator, even though the overall heat transfer coefficient was almost unchanged, it was reported a 25% lower heat transfer rate for R1234ze(E) due to the same reduction in the volume flow rate. Colombo et al. [4] provided a further case, reporting a reduction up to 12.3% in the COP of a water-to-water heat pump compared to R134a. Zhang [5] highlighted, for a plate heat exchanger, that the HFOs refrigerants, compared to their HFC counterparts, show higher heat transfer coefficients and pressure drops when operating under the same conditions.

To correctly design HVAC components and systems, it is of paramount importance the availability of both experimental data, reporting the heat transfer characteristics of the new fluids, and reliable correlations capable for their predictions. This work compares the heat transfer performance, namely the pressure drop per unit length and the heat transfer coefficient, of R134a (reference fluid) and R1234ze(E) (drop-in replacement) during convective condensation in horizontal microfin tubes. Experimental data were compared with the prediction provided by well-known correlations developed to estimate the pressure drop per unit length [6 – 13] and the heat transfer coefficient [14 – 16].

2. Experimental apparatus

The experimental apparatus (Figure 1) is made of three main circuits (their detailed description is reported in [17]) named: the refrigerant loop (filled initially with R134a and then with R1234ze(e)), the water loop (filled with demineralized water) and glycol loop (filled with a mixture of water and ethylene glycol, 30% volume concentration).

2.1. The glycol loop

The glycol loop (blue line in Figure 1) has two main tasks: to set the operating temperature in the test section (fixing the pressure in the condenser) and to chill both refrigerant and water. A commercial chiller cools the mixture to -10°C and then it is stored in a 0.75 m^3 tank. Two independent loops, one for the water and the other for the refrigerant, are connected to the tank.

The former loop cools the water entering in the test section, which heats up because of the viscous dissipation. The other loop is in charge to set the temperature in the test section and to prevent cavitation in the refrigerant pump. The former operation is accomplished setting the mass flow rate and the temperature at the condenser inlet, using a P.I.D. driven electric heater, such that the

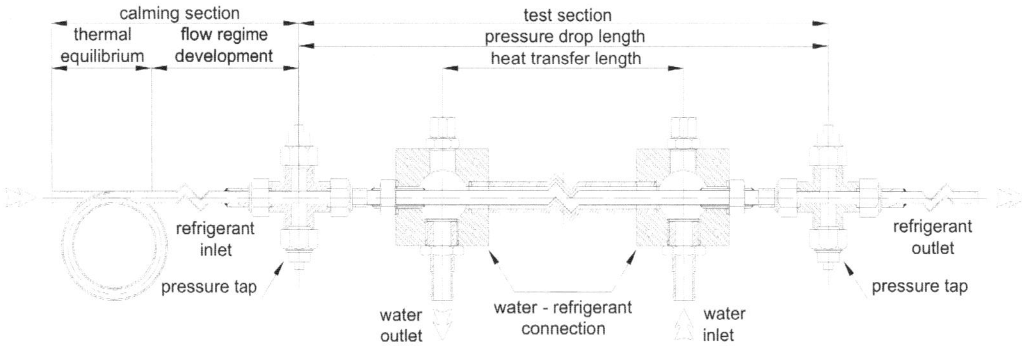

Figure 2. Test section

refrigerant pressure at the test section inlet be the saturation pressure corresponding to the test temperature. For the latter operation a bypass drains part of the mixture headed to the condenser to subcool the refrigerant entering the pump.

2.2. *The water loop*

The water loop (green line in Figure 1) is designed to exchange the thermal power required for the refrigerant phase change. The heat transfer takes place in a tube in tube heat exchanger (water in the annulus, refrigerant in the inner duct), named test section (Figure 2), thermally insulated from the surroundings. A pump drains the water from the tank, the mass is measured by a Coriolis flowmeter (range [0;400] kg/h, uncertainty 0.15% of the reading). Afterwards the flow enters in a plate heat exchanger, then a P.I.D. driven electric heater sets the inlet temperature in the test section such that the power transfer causes a -2°C temperature reduction of the water flow, two groups of 3 K-type thermocouples connected in series (uncertainty 0.1K) measure the inlet and outlet temperature.

2.3. *The refrigerant loop*

The main goal of the refrigerant loop (red line in Figure 1) is to provide, at the inlet of the test section (where the measures take place), a two-phase flow of refrigerant at specified operating conditions, they are defined by: mass flow rate, inlet quality and the inlet temperature.

A saturated liquid flow of refrigerant leaves the condenser and enters in a plate heat exchanger (subcooler) to prevent cavitation in the pump (gear type with inverter driven engine for the mass flow rate tuning). A Coriolis flowmeter (range [0;400]kg/h, uncertainty 0.15% of the reading) records the mass flow rate while a pressure transducer (relative, range [-1;30] bar, uncertainty ±1% of full scale) and a thermocouple (K-type, uncertainty 0.1K) check the thermodynamic state of the refrigerant as it enters in the evaporator, which provides the power to vaporize the amount of refrigerant specified by the test section inlet quality. A calming section (Figure 2) follows, it is made of two parts: a wrapped tube (to get the thermal equilibrium between liquid and vapor) and a straight duct (to get fully developed flow regime). Then the refrigerant enters in the test section and returns to the condenser.

The test section (Figure 2) has a heat transfer length L=1.11 m and is thermally insulated with 10 cm thick rubber foam shell. The geometrical features of the cross-section are listed in Table 1.

A differential pressure transducer (range [-103.4;103.4] kPa, uncertainty ±0.1% of the full scale), connected to two pressure taps separated by the distance l=1.3 m, records the pressure drop. While a relative pressure transducer (range [-100;1600] kPa, uncertainty ±0.25% of the full scale), connected to the tap at the inlet of the test section, reads the refrigerant pressure. The refrigerant inlet and outlet temperatures are the saturation temperatures given by the pressure readings. Two groups (one at the entrance and one at the exit) of three thermocouples are glued inside grooves (length 50 mm, depth 0.15 mm, width 0.4 mm) on the outside of the tube (top, side and bottom position), to measure the wall temperatures. The reference junction of each thermocouple (K-type, uncertainty 0.1K) is inserted in a Dewar flask filled with melting ice.

Table 1. Microfin tube J60 geometrical features

tube				cross-section			
Outer diameter		[mm]	9.52	Wet perimeter		[mm]	44.9
Fin-root diameter	D_R	[mm]	8.96	Cross section area	A	[mm^2]	62.2
Fin number	n	[-]	60	Hydraulic diameter		[mm]	5.28
Height	H	[mm]	0.2	Exchanging area ratio		[-]	1.68
Apex angle	α	[°]	40				
Helix angle	β	[°]	18				

3. Data reduction

Data processing aims at computing the quantities that identify the experimental operating conditions (saturation temperature at the test section inlet T_{ri}, mass flux G, quality change in the test section Δx, average quality in the test section x_m), the pressure drop per unit length Z and the heat transfer coefficient h. A datapoint is picked for each experimental condition, it is the average of 12 acquisitions and each of them is the mean value of 181 samples (1 Hz sampling for 3 minutes).

Preliminary checks on the thermal insulation, performed using single phase vapor flow, showed that the power transferred, computed for both refrigerant side and water side, matched within 5%.

The inlet and outlet refrigerant temperatures are the saturation temperatures corresponding to the inlet and the outlet pressure, the former is measured by a relative pressure transducer while the latter is computed subtracting to the inlet pressure the pressure drop on the test section provided by a differential pressure transducer.

$$T_{\mathrm{ri}} = T_{\mathrm{sat}}(p_{\mathrm{ri}}) \tag{1}$$

$$T_{\mathrm{ro}} = T_{\mathrm{sat}}(p_{\mathrm{ri}} - \Delta p) \tag{2}$$

Referring to the nominal geometrical features of the microfin tube, the net cross section area is:

$$A_{\mathrm{c}} = \frac{\pi D_{\mathrm{r}}^2}{4} - \frac{nH^2}{\cos\beta}\tan\left(\frac{\alpha}{2}\right) \tag{3}$$

From the Coriolis flowmeter reading, the refrigerant mass flux is computed as follows:

$$G = \frac{\dot{m}_{\mathrm{r}}}{A_{\mathrm{c}}} \tag{4}$$

Assuming steady state and negligible thermal dispersion in the test section, the power transfer takes place only between water and refrigerant:

$$\dot{Q} = \dot{m}_{\mathrm{a}} c_{\mathrm{pa}} (T_{\mathrm{ai}} - T_{\mathrm{ao}}) \tag{5}$$

The quality change can be computed as:

$$\Delta x = (x_{\mathrm{o}} - x_{\mathrm{i}}) = \frac{\dot{Q}}{\dot{m}_{\mathrm{r}} h_{\mathrm{lv}}(p_{\mathrm{ri}})} \tag{6}$$

The inlet quality can be determined performing an energy balance on the evaporator, assuming steady state and negligible thermal dispersion:

$$x_{\mathrm{i}} = \frac{\dot{Q}_{\mathrm{e}} - \dot{m}_{\mathrm{r}} c_{\mathrm{pr}}[T_{\mathrm{sat}}(p_{\mathrm{rei}}) - T_{\mathrm{rei}}]}{\dot{m}_{\mathrm{r}} h_{\mathrm{lv}}(p_{\mathrm{ri}})} \tag{7}$$

In the end the mean quality in the test section is:

$$x_{\mathrm{m}} = x_{\mathrm{i}} + \frac{\Delta x}{2} \tag{8}$$

The total pressure drop per unit length is computed as the ratio of the total pressure drop Δp to the distance l between the pressure taps (measured at the end of the building process).

$$Z = \frac{\Delta p}{l} = \frac{p_{ri} - p_{ro}}{l} \tag{9}$$

The average heat transfer coefficient h is based on the logarithmic mean temperature between wall and refrigerant temperatures. The former is the mean of the readings provided by the three thermocouples on the outside of the tube, such that logarithmic mean temperature difference is

Table 2. Experimental operating conditions

quantity	T_{ri}	G	x_m	Δx	q
	[°C]	[kg·m⁻²·s⁻¹]	[-]	[-]	[kW·m⁻²]
range	35±0.1	[43;442]	[0.10;0.91]	[-0.63;-0.05]	[-45.7;-3.5]

$$T_{\mathrm{w}} = \frac{T_{\mathrm{t}} + T_{\mathrm{s}} + T_{\mathrm{b}}}{3} \tag{10}$$

$$\Delta T_{\mathrm{ml}} = \frac{(T_{\mathrm{wo}} - T_{\mathrm{ro}}) - (T_{\mathrm{wi}} - T_{\mathrm{ri}})}{\ln \frac{T_{\mathrm{wo}} - T_{\mathrm{ro}}}{T_{\mathrm{wi}} - T_{\mathrm{ri}}}} \tag{11}$$

The heat transfer area of the refrigerant duct refers to the fin root diameter, hence the heat flux is:

$$q = \frac{Q}{\pi D_{\mathrm{i}} L} \tag{12}$$

Finally, the average heat transfer coefficient in the test section is given by the following equation:

$$h = \frac{q}{\Delta T_{\mathrm{ml}}} \tag{13}$$

The propagation uncertainty analysis (further details can be found in [17]) showed that, for each datapoint, the uncertainty related to the parameters identifying the operating conditions, the pressure drop per unit length and the heat transfer coefficient is lower than 5%.

4. Experimental results

The experiments aimed to compare the heat transfer features, during convective condensation (operating condition in Table 2), of R134a, the reference fluid, and its drop-in replacement R1234ze(E). Some preliminary remarks concerning the main differences in the thermal properties of the fluids, at the saturation temperature, are useful to understand the results (Table 3 and Table 4).

A. R1234ze(E) has 5.4% smaller phase change enthalpy than R134a, it means that, for the same quality change, the R1234ze(E) flow requires a smaller thermal power. As the experiments were performed in the same test section, the same statement holds for the heat flux too. Conversely, for a fixed heat flux, the quality change for the R1234ze(E) flow is larger.
B. R1234ze(E) has 19% smaller vapor density than R134a, it follows that, at the same mass flux, the R1234ze(E) flow has a higher mean velocity than the R134a flow.
C. R1234ze(E) shows a liquid thermal conductivity about 8% smaller than R134a. Accordingly, for the same heat flux in shear dominated flows, the logarithmic mean temperature difference for R1234ze(E) is larger than the one for R134a because, the difference between the bulk temperature and the wall temperature is larger for the R1234ze(E) flow.
D. Comparable frictional pressure drop could be expected, between R134a and R1234ze(E), because the differences in the liquid and vapor dynamic viscosities are smaller than 6%.

4.1. Complete condensation

During convective condensation the total pressure drop is the algebraic sum of a positive term related to viscous dissipation, the frictional pressure drop, and a negative term connected to the bulk velocity reduction, the accelerative pressure drop. According to Figure 3, R1234ze(E) and R134a have

Table 3. Liquid and vapor thermal properties of R134a (reference fluid) and R1234ze(E) at saturation temperature T_{ri}=35°C (source: RefProp 10)

		R134a		**R1234ze(E)**	
$p_{sat}(T_{ri})$	[MPa]	0.887		0.667	
h_{lv}	[kJ·kg⁻¹]	$1.682\cdot10^{2}$		$1.590\cdot10^{2}$	
		Liquid	**Vapor**	**Liquid**	**Vapor**
ρ	[kg·m⁻³]	$1.168\cdot10^{3}$	$4.342\cdot10^{1}$	$1.129\cdot10^{3}$	$3.527\cdot10^{1}$
c_p	[kJ·kg⁻¹K⁻¹]	1.471	1.103	1.422	1.023
k	[W·m⁻¹·K⁻¹]	$7.685\cdot10^{-2}$	$1.490\cdot10^{-2}$	$7.085\cdot10^{-2}$	$1.450\cdot10^{-2}$
M	[Pa·s]	$1.720\cdot10^{-4}$	$1.210\cdot10^{-5}$	$1.680\cdot10^{-4}$	$1.280\cdot10^{-5}$

Table 4. Variation of the thermal properties.

	$\Delta_{\%}=g/g_{ref}-1$	
$\Delta p_{sat\%}$	-24.8%	
$\Delta h_{lv\%}$	-5.4%	
	Liquid	**Vapor**
$\Delta\rho_{\%}$	-3.3%	-18.8%
$\Delta c_{p\%}$	3.3%	-7.2%
$\Delta k_{\%}$	-7.8%	-2.7%
$\Delta\mu_{\%}$	2.3%	5.9%

comparable frictional pressure drop per unit length, as suggested by remark D. The ratio of the accelerative term (void fraction computed using Zivi's correlation) to the frictional one is, for R1234ze(E), between 30% and 40% and between 25% and 30% for R134a because of the explanation given in remark B. The heat transfer coefficients (Figure 4) are similar too. The larger values recorded for R134a could be related to its larger enthalpy of phase change and liquid thermal conductivity, as indicated in remarks A and C.

4.2. *Partial condensation*

To remove the effect on the heat transfer coefficient of the difference in the phase change enthalpy, partial condensation tests were performed fixing the heat flux instead of the quality change. For the mass fluxes taken into account, according to [18], shear dominated flow regimes onsets for both the fluids. As highlighted by remark D, R134a and R1234ze(E) share similar frictional pressured drop (Figure 5) for a wide range of mass fluxes, the differences are smaller than 10%. In these operating conditions, the accelerative pressure drop provides a small contribution to the total pressure drop: in absolute terms, its ratio to the frictional pressure drop is less than 15% for both the fluids.

The experiments performed fixing the heat flux (q=15.3kW·m^{-2}) displayed that the two fluid have

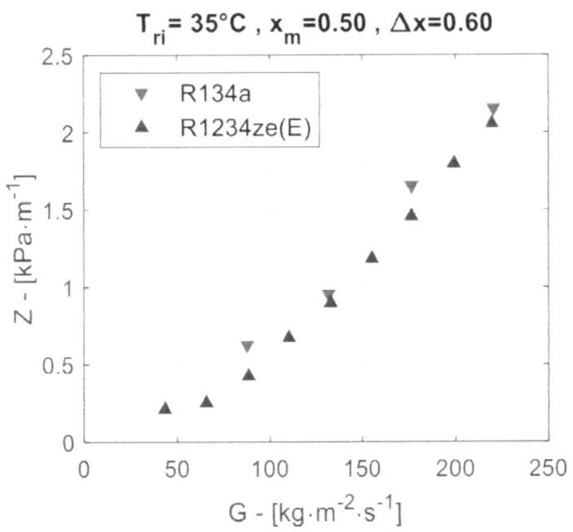

Figure 3. Frictional pressure drop per unit length as a function of mass flux.

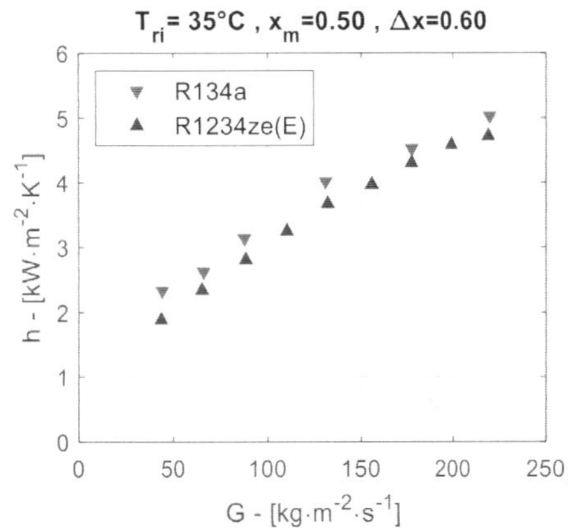

Figure 4. heat transfer coefficient as a function of mass flux.

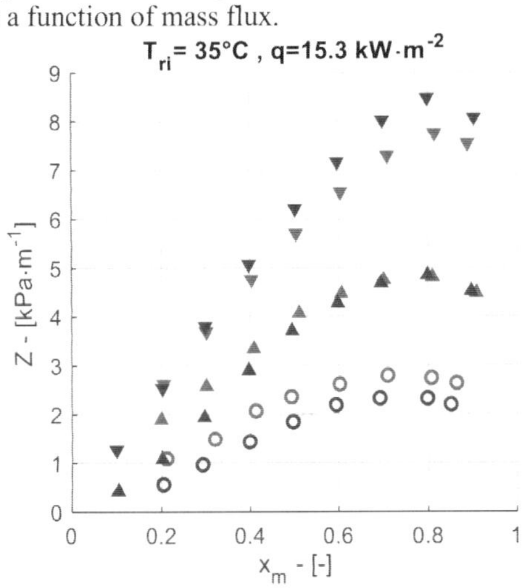

Figure 5. Frictional pressure drop per unit length as a function of mean quality (G_1=220 kg·m^{-2}·s^{-1}, G_2=330 kg·m^{-2}·s^{-1}, G_3=440 kg·m^{-2}·s^{-1}).

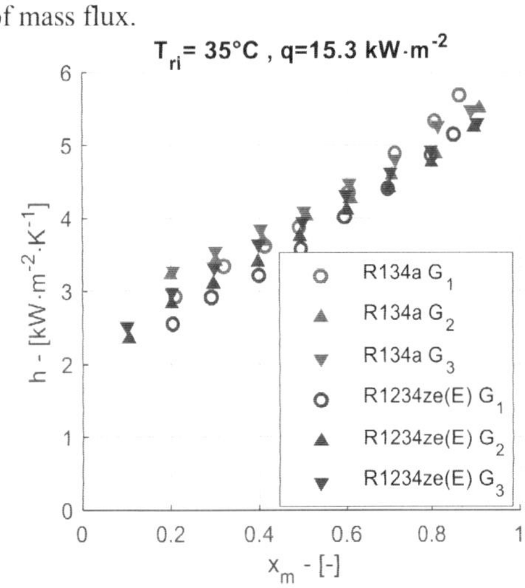

Figure 6. Heat transfer coefficient as a function of mean quality (G_1=220 kg·m^{-2}·s^{-1}, G_2=330 kg·m^{-2}·s^{-1}, G_3=440 kg·m^{-2}·s^{-1}).

Table 5. Comparison between data and the best performing correlations ($E_{\%} \leq 30\%$).

	R1234ze(E)			R134a		
Pressure drop	**$E_{\%}$**	**$E_{A\%}$**	**s**	**$E_{\%}$**	**$E_{A\%}$**	**s**
Domanski [6]	10.5%	30.9%	42.9%	-23.2%	30.9%	33.2%
Kedzierski&Goncalves [13]	-0.4%	27.0%	39.1%	-30.0%	33.9%	30.5%
Shannak [9]	16.2%	36.1%	50.9%	-20.6%	32.6%	36.8%
Muller [19]	10.7%	31.1%	45.9%	-25.1%	32.4%	32.9%
Heat transfer coefficient	**$E_{\%}$**	**$E_{A\%}$**	**s**	**$E_{\%}$**	**$E_{A\%}$**	**s**
Koyama [15]	22.3%	37.3%	49.2%	18.6%	35.1%	46.9%
Kumar&Mohseni [14]	-25.5%	26.1%	13.4%	-24.4%	25.2%	13.6%
Kedzierski&Goncalves [13]	16.8%	27.3%	33.3%	18.5%	26.6%	31.6%

similar heat transfer coefficient (Figure 6), nevertheless R134a shows higher heat transfer coefficients than R1234ze(E) because of its larger liquid thermal conductivity (remark C).

4.3. Comparison between experimental data and correlation

Many different operating conditions were tested, namely 102 for R134a and 122 for R1234ze(E), to properly benchmark the correlations available in the open literature. The pressure drop data were compared with the predictions of 8 correlations [6 – 13]. Only the four in Table 5 provide predictions with a percentage error lower than 30% for both the fluids, nevertheless the performances change a lot for the two fluids. It seems that the available correlations need further tuning to properly account for the effect of the thermal properties and the flow regimes. On the contrary, the predictive models for the heat transfer coefficient seem to work properly. Four correlations [14 – 16] were analyzed and three (Table 5) of them show both a mean percentage error smaller than 30% and similar performances for both the fluids.

5. Conclusions

The experiments point out that the two fluids have similar frictional pressure drop and heat transfer coefficient, the slightly higher values of the latter quantity for R134a are related to its larger thermal properties: phase change enthalpy, liquid and vapor thermal conductivity.

Correlations for the heat transfer coefficient seems to be reliable for both the fluids while the same does not hold for pressure drop. Even though there are models providing, for both the fluids, mean errors smaller than 30% are available their performances are quite different for the two fluids. The issue of prediction reliability, however, is still open, not all the models achieving good results for R134a show the same performance for R1234ze(E), especially for the pressure drop.

Acknowledgments: The authors with gratitude thank Professor Adriano Muzzio for the precious support he gave in developing the experimental apparatus and performing the experiments.

Funding: Research funded by MIUR through the program PRIN 2015, grant number 2015M8S2PA.

Nomenclautre

Latin symbols

A	cross sectional area [m^2]	l	distance between the pressure taps [m]
c_p	specific heat capacity [J·kg^{-1}·K^{-1}]	L	heat transfer length [m]
D	diameter [m]	m	mass flow rate [kg·s^{-1}]
$E_{\%}$	mean percentage error [-]	n	fin number [-]
$E_{A\%}$	mean absolute percentage error [-]	p	refrigerant pressure [Pa]
G	refrigerant mass flux [kg·m^{-2}·s^{-1}]	q	heat flux [W·m^{-2}]
g	generic quantity [-]	Q	thermal power exchanged [W]
H	fin height [m]	s	standard deviation [-]
h	heat transfer coefficient [W·m^{-2}·K^{-1}]	T	temperature [K]
h_{lv}	liquid vapor phase change enthalpy [J·kg^{-1}]	x	refrigerant quality [-]
k	thermal conductivity [W·m^{-1}·K^{-1}]	Z	pressure drop per unit length [Pa·m^{-1}]

Greek symbols

α	apex angle [°]	ΔT	temperature difference [K]
β	helix angle [°]	Δx	quality change in the test section [-]
$\Delta_{\%}$	percentage variation of a quantity [-]	μ	dynamic viscosity [kg·m^{-1}·s^{-1}]
Δp	pressure drop [Pa]	ρ	density [kg·m^{-3}]

Subscripts

a	water	*o*	outlet
b	bottom	*r*	refrigerant
e	evaporator	*R*	fin root
i	inlet	*s*	side
lm	logarithmic mean	*sat*	saturation
m	mean	*t*	top

References

[1] The European Parliament and Council of the European Union, "European Regulation (EU) No. 517/2014," *Offiicial J. Eur. Union*, pp. 195–230, 2014.
[2] Hodnebrog *et al.*, "Global warming potentials and radiative efficiencies of halocarbons and related compounds: A comprehensive review," *Rev. Geophys.*, vol. 51, no. 2, pp. 300–378, 2013.
[3] G. Censi and A. Padovan, "R1234ze(E) As Drop-In Replacement For R134a In A Micro-Fin Shell-And-Tube Evaporator : Experimental Tests And Calculation Model Evaporator : Experimental Tests And Calculation Model," in *18th International Refrigeration and Air Conditioning Conference*, 2021.
[4] L. P. M. Colombo, A. Lucchini, and L. Molinaroli, "Experimental analysis of the use of R1234yf and R1234ze(E) as drop-in alternatives of R134a in a water-to-water heat pump," *Int. J. Refrig.*, vol. 115, pp. 18–27, 2020.
[5] J. Zhang, M. R. Kaern, T. Ommen, B. Elmegaard, and F. Haglind, "Condensation heat transfer and pressure drop characteristics of R134a, R1234ze(E), R245fa and R1233zd(E) in a plate heat exchanger," *Int. J. Heat Mass Transf.*, vol. 128, pp. 136–149, 2019.
[6] J. Y. Choi, M. A. Kedzierski, and P. A. Domanski, "Generalized Pressure Drop Correlation for Evaporation and Condensation in Smooth and Micro-Fin Tubes," *IIF-IIR-Commission B1*, no. 1 986, 2001.
[7] M. Goto, N. Inoue, and N. Ishiwatari, "Condensation and evaporation heat transfer of R410A inside internally grooved horizontal tubes," *Int. J. Refrig.*, vol. 24, no. 7, pp. 628–638, 2001.
[8] A. Cavallini, D. Del Col, L. Doretti, G. A. Longo, and L. Rossetto, "Heat transfer and pressure drop during condensation of refrigerants inside horizontal enhanced tubes Transfert de chaleur et chute de pression lors de la Á nes a Á l ' inte Â rieur de tubes condensation de frigorige Á surface augmente Âe horizontaux a," vol. 23, pp. 4–25, 2000.
[9] B. A. Shannak, "Frictional pressure drop of gas liquid two-phase flow in pipes," *Nucl. Eng. Des.*, vol. 238, no. 12, pp. 3277–3284, 2008.
[10] L. Sun and K. Mishima, "Evaluation analysis of prediction methods for two-phase flow pressure drop in mini-channels," *Int. Conf. Nucl. Eng. Proceedings, ICONE*, vol. 2, no. 15, pp. 649–658, 2008.
[11] S. Nozu, H. Katayama, H. Nakata, and H. Honda, "Condensation of a refrigerant CFC11 in horizontal microfin tubes (Proposal of a correlation equation for frictional pressure gradient)," *Exp. Therm. Fluid Sci.*, vol. 18, no. 1, pp. 82–96, 1998.
[12] H. Müller-Steinhagen and K. Heck, "A simple friction pressure drop correlation for two-phase flow in pipes," *Chem. Eng. Process. Process Intensif.*, vol. 20, no. 6, pp. 297–308, 1986.
[13] M. A. Kedzierski and J. M. Goncalves, "Horizontal convective consensation alternative refrigerants within a micro-fin tube," *Jorunal Enhanc. Heat Transf.*, vol. 6, no. 161–178, 1999.
[14] M. A. Akhavan-Behabadi, R. Kumar, and S. G. Mohseni, "Condensation heat transfer of R-134a inside a microfin tube with different tube inclinations," *Int. J. Heat Mass Transf.*, vol. 50, no. 23–24, pp. 4864–4871, 2007.
[15] J. Yu and S. Koyama, "Condensation heat transfer of pure refrigerants in microfin tubes," *Int. Refrig. Air Cond. Conf.*, 1998.
[16] A. Cavallini *et al.*, "Condensation in horizontal smooth tubes: A new heat transfer model for heat exchanger design," *Heat Transf. Eng.*, vol. 27, no. 8, pp. 31–38, 2006.
[17] L. P. M. Colombo, A. Lucchini, T. Nhan Phan, L. Molinaroli, and A. Niro, "Design and assessment of an experimental facility for the characterization of flow boiling of azeotropic refrigerants in horizontal tubes," *J. Phys. Conf. Ser.*, vol. 1224, no. 1, 2019.
[18] L. P. M. Colombo, A. Lucchini, L. Molinaroli, A. Niro, T. N. Phan, and P. G. Pittoni, "Flow patterns during flow boiling and convective condensation of R1234ze(E) inside a microfin tube," *5-6th Therm. Fluids Eng. Conf.*, pp. 109–119, 2021.
[19] H. Müller-Steinhagen and K. Heck, "A simple friction pressure drop correlation for two-phase flow in pipes," *Chem. Eng. Process.*, vol. 20, no. 6, pp. 297–308, 1986.

Condensation heat transfer of superheated vapour of R1234ze(E) and R134a inside a brazed plate heat exchanger

G Cattelan, M Azzolin, A Diani, L Rossetto and D Del Col

Department of Industrial Engineering, University of Padova
Padova, 35131, Italy

marco.azzolin@unipd.it

Abstract. The recent restrictions on the use of refrigerants with high Global Warming Potential (GWP) have pushed the research to consider alternative solutions, such as the HydroFluoroOlefins, as possible replacements. Even though they are generally mildly flammable, their GWP is significantly lower compared to those of the HydroFluoroCarbons. This paper presents experimental measurements of R1234ze(E) and R134a heat transfer coefficients during condensation inside a brazed plate heat exchanger. Experimental tests during condensation heat transfer have been conducted considering different degrees of superheating, subcooling and outlet vapour quality. The condensation temperature varies between 34.6 °C and 42.3 °C and the refrigerant mass flux is between 9 kg m^{-2} s^{-1} and 49 kg m^{-2} s^{-1}. The results showed that the heat transfer coefficients measured with R134a are between 4% and 8% higher than those of R1234ze(E). Complete condensation experiments showed that an increase in the liquid subcooling degree significantly reduces the thermal performance at low refrigerant mass velocities. In some cases, the plate area occupied by liquid refrigerant reached almost 30% of the overall heat transfer area, thus decreasing by 2.7 times the average heat transfer coefficient when passing from 3 K to 8 K subcooling degree. A comparison with the predictions of some empirical models is also presented to assess which one can better predict the experimental data.

1. Introduction

In recent years, international organizations took important actions (the F-Gas Regulation No 517 of the European Union of 2014 [1], The Kigali Amendment of the United Nations of 2016 [2]) to promote the transition toward low Global Warming Potential (GWP) refrigerants and push the reduction in the use of Hydro Fluoro Carbons (HFC). These actions are intended to cut by 79% the HFC supply by 2030 compared with 2015 data of HFC sales. One of the most promising replacements for R134a, which is still one of the most adopted refrigerants, is the HydroFluoroOlefin (HFO) R1234ze(E). Its GWP has been estimated to be lower than 1 [3], it is not toxic but mildly flammable (classified as A2L by the ANSI/ASHRAE). In the literature, there are several works that studied the in-tube condensation of R1234ze(E) [4–8], while less attention has been paid to condensation inside brazed plate heat exchangers (BPHEs). Longo et al. [9] presented experimental measurements of partial and complete condensation of R1234ze(E) inside BPHEs. When studying complete condensation, they considered a maximum superheating and subcooling degree equal to 11 K and 4.9 K respectively. Zhang et al. [10,11] studied the same process considering different refrigerants, among which R1234ze(E) and R134a, and elaborated new equations to predict the refrigerant pressure drop and heat transfer coefficient, adding in their study the effects of surface tension. The study proposed by Jung et al. [12] considered an experimental investigation regarding the influence of heat and mass flux, vapour quality and saturation pressure on the condensation heat transfer and pressure drop obtained using R1234ze(E) inside two BPHEs having different chevron angles and performed a comparison between their results and those previously obtained using R1233zd(E). The authors found that R1233zd(E) has larger heat transfer coefficients and frictional pressure drops than R1234ze(E).

In this paper, a deeper analysis on condensation heat transfer of R1234ze(E) and R134a inside BPHE was experimentally performed. Partial condensation tests were conducted with R1234ze(E) and R134a at three mass fluxes (20, 30 and 40 kg m^{-2} s^{-1}), inlet vapour superheating equal to 15 K and outlet vapour quality from 0.05 to 0.67. Complete condensation experiments were also performed with R1234ze(E), to study the effects of varying the superheating (10 K, 20 K, 30 K) and subcooling (3 K, 5 K, 8 K). The experimental data were used to assess the prediction accuracy of some of the available correlations developed for the calculation of the refrigerant-side heat transfer coefficient during condensation inside brazed plate heat exchangers.

2. Experimental apparatus

The set-up of the experimental apparatus is shown in Fig. 1. It is composed of four loops: the refrigerant loop, the cooling brine loop, the pre-section water loop and the cold water loop. The refrigerant mass flow rate is controlled by a magnetically coupled gear pump equipped with an inverter and the pressure is regulated by a bladder accumulator linked to a nitrogen cylinder. After the pump, the refrigerant is sent to the pre-section heat exchanger as subcooled liquid, where it is vaporized and superheated using hot water from a thermostatic bath. The evaporator and the super-heater heat exchanger are adopted if the heat flow rate supplied by the thermostatic bath is not sufficient to completely vaporize the refrigerant mass flow rate. In the test section, the refrigerant condenses flowing in counter-current with cold water. A tube-in-tube heat exchanger fed with brine cooled by a chiller is then used to further cool down the refrigerant at the outlet of the condenser so that only liquid is sent to the pump. A drier and a mechanical filter are also installed before the pump. To maintain and control the refrigerant temperature, two tube-in-tube heat exchangers fed with brine are located before and after the pump. A throttling valve installed before the pre-section is used to stabilize and to operate small adjustments on the refrigerant mass flow rate.

All the adopted sensors are listed in Table 1, along with their accuracies.

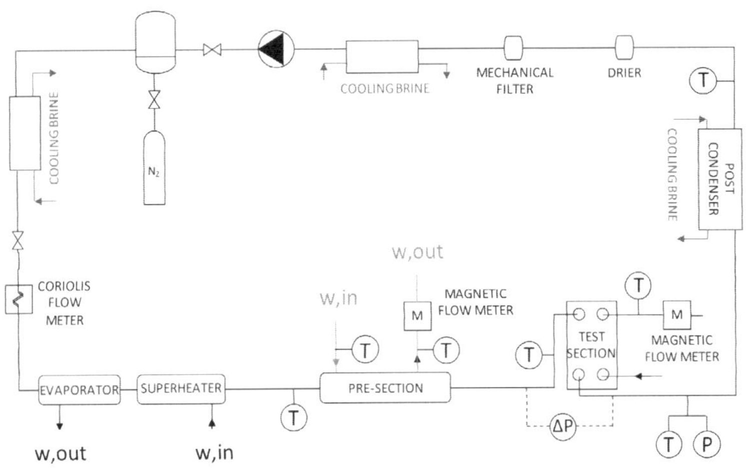

Fig. 1. Schematic of the experimental apparatus.

Table 1. Accuracies of the implemented instruments.

Sensor	Accuracy
Absolute pressure transducer	± 0.1% of the full scale (40 bar)
Differential pressure transducer	± 0.1% of the full scale (500 mbar)
Calibrated T-type thermocouples	± 0.05 K
Calibrated T-type thermopiles	± 0.03 K
Refrigerant mass flow meter	± 0.2% of the reading
Water volumetric flow meter	± 1% of the reading

3. Data reduction

3.1. Partial condensation tests

Partial condensation tests have been carried out at a fixed value of vapour superheating at the BPHE inlet and varying the outlet vapour quality. The heat flow rate exchanged in the BPHE is calculated on the water-side and it is equal to:

$$q = \dot{m}_w c_{p,w}\left(T_{w,out} - T_{w,in}\right) = \dot{m}_{ref}\left(h_{ref,in} - h_{ref,out}\right) \tag{1}$$

Eq. 1 allows to calculate the outlet refrigerant specific enthalpy and thus the vapour quality at the outlet of the BPHE as:

$$x_{out} = \frac{h_{ref,out} - h_L}{h_V - h_L} \tag{2}$$

where h_L and h_V respectively refer to the values of liquid and vapour specific enthalpies in saturation conditions, calculated at the refrigerant outlet pressure. All the water and refrigerant thermo-physical properties are evaluated using REFPROP 10 [13].

In the case that at the inlet of the BPHE the condensation temperature ($T_{sat,in}$) is higher than the wall temperature ($T_{wall,in}$), condensation process begins from superheated vapour; otherwise, in a portion of the condenser only sensible heat is exchanged until the wall temperature becomes lower than the local saturation temperature, so that condensation process can occur. Owing this, $T_{wall,in}$ needs to be calculated according to Eq. 3:

$$T_{wall,in} = K_{s-ph}\left(T_{ref,in} - T_{w,out}\right)\left(s/\lambda_p + 1/\alpha_w\right) + T_{w,out} \tag{3}$$

in which s represents the thickness of the BPHE plates, α_w the heat transfer coefficient calculated on the water-side, λ_p is the thermal conductivity of the plate and K_{s-ph} the overall heat transfer coefficient, which is evaluated at the condenser's inlet section with the following equation:

$$K_{s-ph} = \left(\frac{s}{\lambda_p} + \frac{1}{\alpha_w} + \frac{1}{\alpha_{ref,g}}\right)^{-1} \tag{4}$$

where the values of single-phase heat transfer coefficients of water and refrigerant (α_w and $\alpha_{ref,g}$ respectively) have been calculated with the empirical correlations developed by the BPHE manufacturer. When the value of the inlet wall temperature ($T_{wall,in}$, calculated with Eq. 3) is lower than the refrigerant saturation temperature calculated at the inlet pressure, the overall condensation heat transfer coefficient is evaluated as:

$$K = \frac{q}{A_{ht}\,\Delta T_{ml}} \tag{5}$$

where A_{ht} is the total heat transfer area of the corrugated plate multiplied by the number N of effective heat transfer plates. ΔT_{ml} in Eq. 5 is calculated from the measured water temperatures and the calculated saturation temperatures of the refrigerant:

$$\Delta T_{ml} = \frac{(T_{sat,out} - T_{w,in}) - (T_{sat,in} - T_{w,out})}{\ln((T_{sat,out} - T_{w,in})/(T_{sat,in} - T_{w,out}))} \tag{6}$$

Neglecting the fouling resistance on the plate surface, it is possible to obtain the average refrigerant heat transfer coefficient from the overall heat transfer coefficient (K) as:

$$HTC_{ref} = \left(\frac{1}{K} - \frac{1}{\alpha_w} - \frac{s}{\lambda_p}\right)^{-1} \tag{7}$$

where the water-side heat transfer coefficient (α_w) is calculated through the single-phase correlations supplied by the manufacturer.
If the value of the inlet wall temperature ($T_{wall,in}$) is higher than the inlet refrigerant saturation temperature and condensation process is consequently delayed, a sensible vapour de-superheating process occurs at the inlet of the condenser. In this case, it is possible to estimate the area of the plate where only sensible heat is exchanged considering the energy balance between the heat exchanged on the water and refrigerant sides within this region. The so obtained value has not been taken into consideration in the calculation of the condensation heat transfer coefficient.

3.2. Complete condensation tests

To study how the liquid subcooling influences the heat transfer during complete condensation process, a heat transfer coefficient ($HTC_{ref,tot}$) calculated as the average value over the entire heat transfer area has been considered. $HTC_{ref,tot}$ has been evaluated considering the logarithmic mean temperature difference that is defined in Eq. 8, and takes into account the de-superheating, condensation and subcooling regions.

$$\Delta T_{ml} = \frac{q_{tot}}{q_{SH}/\Delta T_{ml,SH} + q_{cond}/\Delta T_{ml,cond} + q_{SC}/\Delta T_{ml,SC}} \tag{8}$$

In Eq. 8, q_{SH}, q_{cond} and q_{SC} are the heat flow rates that refer to the de-superheating, condensation and subcooling processes respectively, and $\Delta T_{ml,SC}$, $\Delta T_{ml,cond}$ and $\Delta T_{ml,SH}$ are defined as:

$$\Delta T_{ml,SC} = \frac{(T_{sat,out} - T_{w,SC}) - (T_{ref,out} - T_{w,in})}{\ln((T_{sat,out} - T_{w,SC})/(T_{ref,out} - T_{w,in}))} \tag{9}$$

$$\Delta T_{ml,cond} = \frac{(T_{sat,out} - T_{w,SC}) - (T_{sat,in} - T_{w,SH})}{\ln((T_{sat,out} - T_{w,SC})/(T_{sat,in} - T_{w,SH}))} \tag{10}$$

$$\Delta T_{ml,SH} = \frac{(T_{ref,in} - T_{w,out}) - (T_{sat,in} - T_{w,SH})}{\ln((T_{ref,in} - T_{w,out})/(T_{sat,in} - T_{w,SH}))} \tag{11}$$

in which $T_{w,SH}$ refers to the temperature of the water at end of the de-superheating zone and $T_{w,SC}$ the temperature of the water corresponding to the onset of the sub-cooling zone. From the overall heat transfer coefficient (K) it is therefore possible to obtain the refrigerant-side heat transfer coefficient using Eq. 7, where K is defined as:

$$K = \frac{q_{tot}}{\Delta T_{ml} A_{ht}} \tag{12}$$

To assess the accuracy of the empirical correlations, a heat transfer coefficient accounting only for the de-superheating and condensation processes was calculated by subtracting the plate area in the subcooling zone from the total heat transfer area, following the procedure adopted by Zhang et al. [10].

3.3. *Uncertainty analysis*

A detailed uncertainty analysis was performed following the procedure described by Kline and McClintock [14]. Considering a coverage factor k equal to 2, the average uncertainty of the refrigerant heat transfer coefficient resulted equal to 15.5%, with a maximum value of 26.9% and a minimum value equal to 9.5%.

4. Experimental results

Regarding the partial condensation tests, a comparison between the refrigerant-side heat transfer coefficients obtained with R134a and R1234ze(E) is shown in Fig. 2, for mass fluxes equal to 20 kg m^{-2} s^{-1}, 30 kg m^{-2} s^{-1} and 40 kg m^{-2} s^{-1}. The results show that the heat transfer coefficient increases with the mass flux and with the outlet vapour quality. An increase in the refrigerant-side heat transfer coefficient equal to 43% was observed passing from G = 20 kg m^{-2} s^{-1} to G = 40 kg m^{-2} s^{-1} when considering R1234ze(E) and x_{out} equal to 0.2, while when the outlet vapour quality is set equal to 0.6 the difference becomes 60%. Regarding the comparison between the two fluids in terms of heat transfer coefficient, R134a shows higher average heat transfer coefficients at the same operative conditions. When the refrigerant mass flux is equal to 40 kg m^{-2} s^{-1} and the outlet vapour quality equal to 0.3, this difference is around 4%, whereas it is equal to 8% when the outlet vapour quality increases to 0.6. This could depend on the lower value of liquid thermal conductivity of R1234ze(E) compared to that of R134a and the consequent increase in the thermal resistance of the liquid film condensate along the plate, which reduces the heat transfer coefficient.

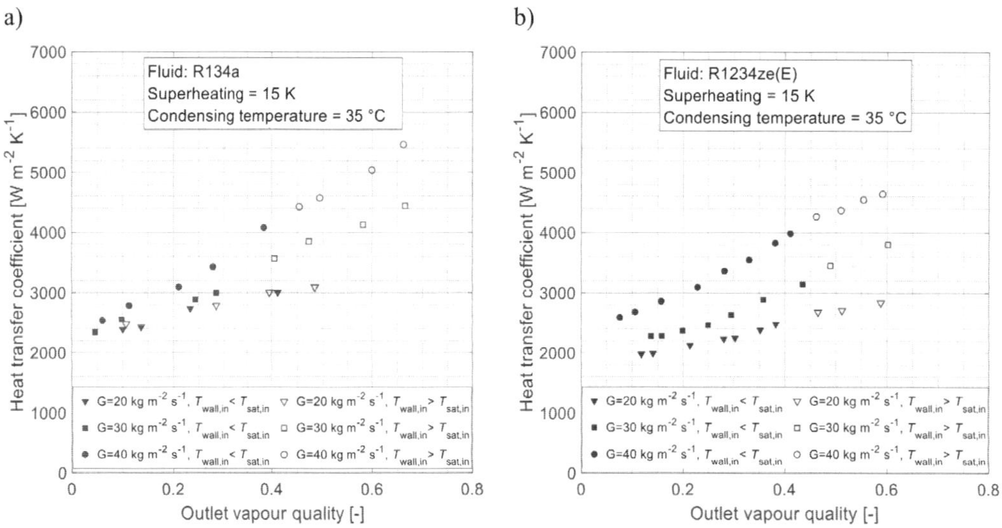

Fig. 2. R134a (a) and R1234ze(E) (b) heat transfer coefficients vs outlet vapour quality.

Fig. 3 shows the heat transfer coefficients obtained during complete condensation tests with R1234ze(E) considering two different water temperature differences (5 K and 10 K). Fig. 3a shows that the refrigerant-side heat transfer coefficient increases with the vapour superheating at the inlet of the test section. Passing from a superheating degree equal to 10 K to 30 K, the increase in the refrigerant heat transfer coefficient is around 15% both in the case of a water temperature difference equal to 5 K and 10 K. Particularly interesting is the case of water temperature difference equal to 10 K, at the lowest tested refrigerant mass flux and superheating equal to 30 K. In these conditions, the heat transfer coefficient is 38% higher with respect to the case with a superheating degree equal to 10 K. Fig. 3b shows the effect of the subcooling degree. Increasing the subcooling degree, the average heat transfer coefficient on the refrigerant side is penalized since a larger plate area is needed for the refrigerant subcooling process. At low mass fluxes and for a water temperature difference equal to 5 K, this phenomenon is highlighted since the heat transfer coefficient decreases by 2.7 times passing from a subcooling degree equal to 3 K to 8 K. In these conditions, the calculated area needed for the subcooling process is around 30% of the overall heat transfer area.

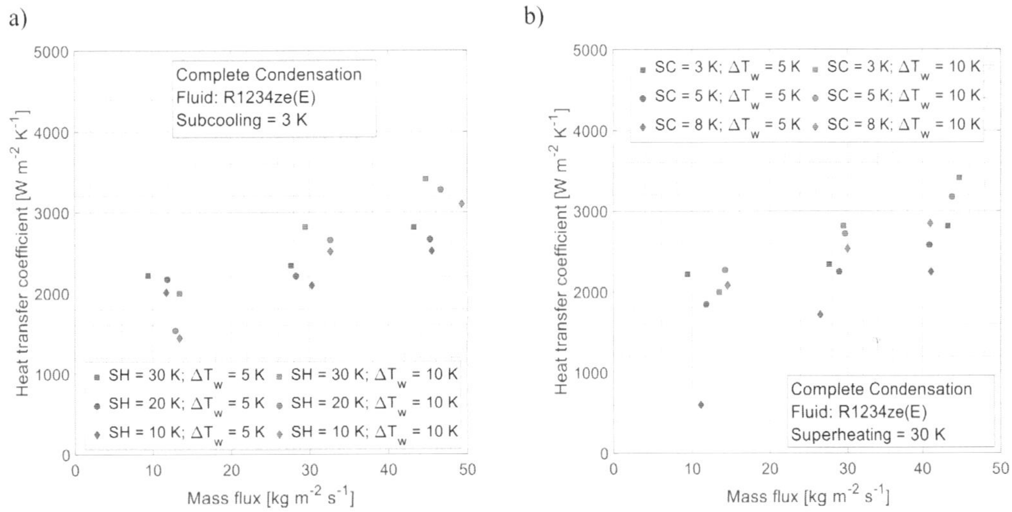

Fig. 3. Average refrigerant heat transfer coefficient vs. mass flux with R1234ze(E) and water temperature difference equal to 5 K and 10 K for different values of inlet vapour superheating (a) and outlet liquid subcooling (b).

5. Comparison against models

The present condensation heat transfer coefficients database has been used to evaluate the predictions of correlations available in the literature: Longo et al. [15], Mancin et al. [16], Mancin et al. [17] and Zhang et al. [12]. Fig. 4 depicts the results predicted by each correlation against the experimental heat transfer coefficients. It can be observed that the model proposed by Zhang et al. presents the lowest average deviations and 99% of the experimental data are predicted within ±30% error bands. The results obtained with this model are characterized by a mean absolute and a mean bias deviations equal to 13.5% and 1.61%, respectively. Zhang et al. validated their correlation using data covering HydroFluoroCarbons (R134a, R236fa and R245fa), HydroFluoroOlefins (R1234ze(E) and R1233zd(E)), and HydroCarbons, for a range of condensation temperatures that goes from 30 °C to 90 °C and mass fluxes range between 12 to 92 kg m^{-2} s^{-1}.

The model proposed by Mancin et al. [16] presents the lowest standard deviation (equal to 14.2%) among the assessed correlations. Mancin et al. [16] model has been developed using experimental data obtained with R410A and R407C and the refrigerant-side heat transfer coefficient is calculated combining the value obtained applying the Nusselt equation and a component derived from the Cavallini et al. [18] correlation, suitable for shear-stress dominated condensation inside tubes.

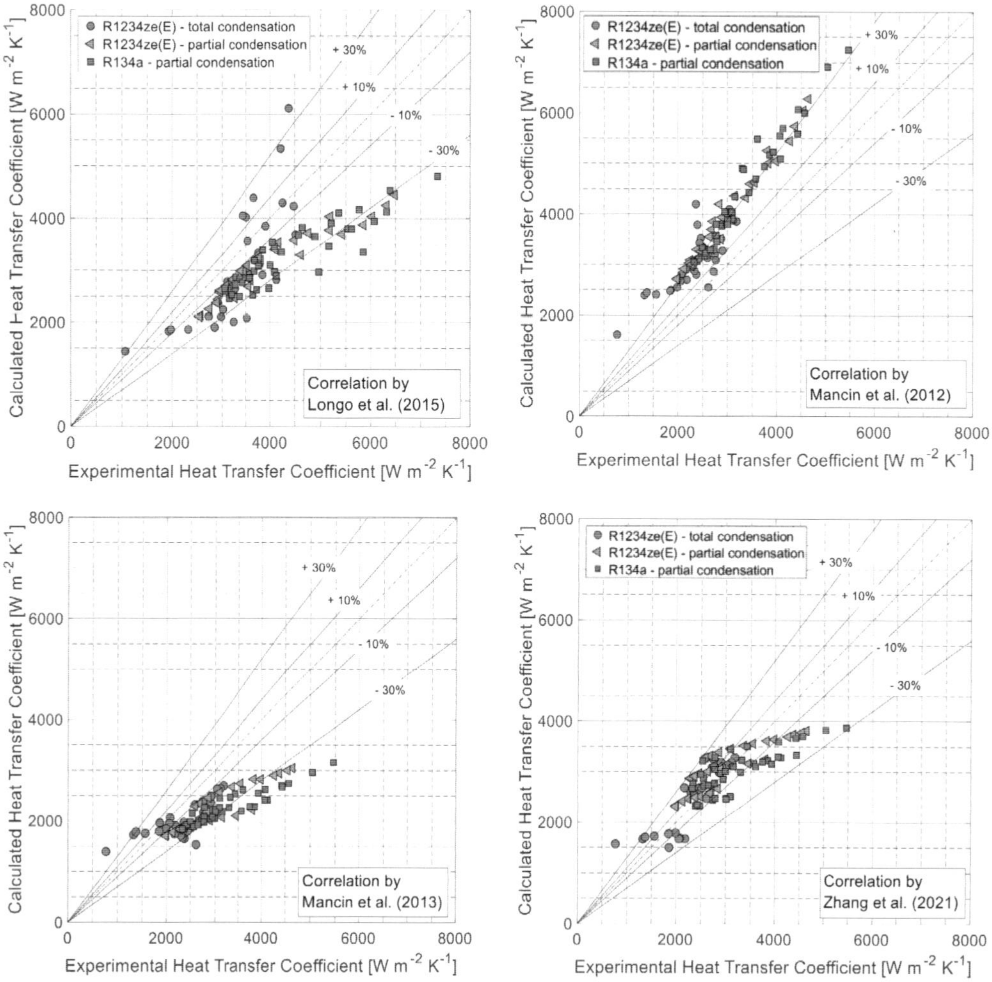

Fig. 4. Comparison between the experimental heat transfer coefficients and the predictions by the models of Longo et al. (2015), Mancin et al. (2012), Mancin et al. (2013), Zhang et al. (2021).

6. Conclusions

In this study, experimental partial condensation tests with R134a and R1234ze(E) and complete condensation tests with R1234ze(E) inside a brazed plate heat exchanger are presented. Partial condensation measurements have been performed with a superheating equal to 15 K at the inlet of the BPHE, while the outlet vapour quality ranged from 0.05 to 0.67. When the heat transfer coefficients of R134a and R1234ze(E) are compared during partial condensation tests, it results that the heat transfer coefficients of R134a are from 4% to 8% higher than those of R1234ze(E) at the same operative conditions.

Complete condensation tests have been conducted when varying the superheating degree from 10 K to 30 K and the subcooling from 3 K to 8 K, to assess the effect of these parameters on the refrigerant heat transfer coefficient. It results that increasing the superheating degree from 10 K to 30 K, the heat transfer coefficient increases by 15%. When the subcooling is increased, the average heat transfer coefficient is penalized because a larger plate area is needed for the subcooling process. At low mass fluxes, the average heat transfer coefficient decreases by 2.7 times when the subcooling changes from 3 K to 8 K.

The experimental data presented in this study have been compared with the predictions of some empirical models available in the literature. Among the assessed correlations, the one proposed by Zhang et al. [12] was found to be the most accurate in the prediction of the heat transfer coefficients, since 99% of the experimental data were predicted within ± 30%, with a mean absolute deviation equal to 13.5% and a mean bias deviation equal to 1.6%.

Acknowledgments
ALFA LAVAL S.p.A. is gratefully acknowledged for the support of this study.

Nomenclature

Symbols

A	area [m^2]	s	plate thickness [m]
$BPHE$	brazed plate heat exchanger	T	temperature [°C]
c_p	specific heat [J kg^{-1} K^{-1}]	x	vapour quality [-]
h	specific enthalpy [J kg^{-1}]		*Greek Symbols*
HTC	heat transfer coefficient [W m^{-2} K^{-1}]	α	single-phase HTC [W m^{-2} K^{-1}]
K	overall HTC [W m^{-2} K^{-1}]	ΔT_{ml}	logarithmic medium temperature difference [K]
$\dot{m}$	mass flow rate [kg s^{-1}]	λ	thermal conductivity [W m^{-1} K^{-1}]
q	heat flow rate [W]		

Subscripts

cond	condensation	*g*	gas
ht	heat transfer	*in*	inlet
L	liquid phase	*out*	outlet
p	plate	*ref*	refrigerant
sat	saturation	*SC*	subcooling
SH	superheating	*s-ph*	single phase
tot	total	*V*	vapour phase
w	water	*wall*	wall

References

[1] European Parliament and of the Council of 16 April 2014 on fluorinated greenhouse gases and repealing Regulation (EC) 2014 *European F-Gas Regulation (EU 517/2014)*
[2] Nations U *Amendment to the Montreal Protocol On Substances That Deplete The Ozone Layer*
[3] Hodnebrog, Etminan M, Fuglestvedt J S, Marston G, Myhre G, Nielsen C J, Shine K P and Wallington T J 2013 Global warming potentials and radiative efficiencies of halocarbons and related compounds: A comprehensive review *Rev. Geophys.* **51** 300–78
[4] Hossain M A, Onaka Y and Miyara A 2012 Experimental study on condensation heat transfer and pressure drop in horizontal smooth tube for R1234ze(E), R32 and R410A *Int. J. Refrig.* **35** 927–38
[5] Agarwal R and Hrnjak P 2015 Condensation in two phase and desuperheating zone for R1234ze(E), R134a and R32 in horizontal smooth tubes *Int. J. Refrig.* **50** 172–83
[6] Diani A, Liu Y, Wen J and Rossetto L 2022 Experimental investigation on the flow condensation of R450A, R515B, and R1234ze(E) in a 7.0 mm OD micro-fin tube *Int. J. Heat Mass Transf.* **196** 123260
[7] Del Col D, Bortolato M, Azzolin M and Bortolin S 2015 Condensation heat transfer and two-phase frictional pressure drop in a single minichannel with R1234ze(E) and other refrigerants *Int. J. Refrig.* **50** 87–103
[8] Azzolin M, Berto A, Bortolin S and Del Col D 2022 Condensation heat transfer of R1234ze(E) and its A1 mixtures in small diameter channels *Int. J. Refrig.* **137** 153–65
[9] Longo G A, Zilio C, Righetti G and Brown J S 2014 Condensation of the low GWP refrigerant HFO1234ze(E) inside a Brazed Plate Heat Exchanger *Int. J. Refrig.* **38** 250–9
[10] Zhang J, Kærn M R, Ommen T, Elmegaard B and Haglind F 2019 Condensation heat transfer and pressure drop characteristics of R134a, R1234ze(E), R245fa and R1233zd(E) in a plate heat exchanger *Int. J. Heat Mass Transf.* **128** 136–49
[11] Zhang J, Elmegaard B and Haglind F 2021 Condensation heat transfer and pressure drop correlations in plate heat exchangers for heat pump and organic Rankine cycle systems *Appl. Therm. Eng.* **183** 116231
[12] Jung J H, Ko Y M and Kang Y T 2021 Condensation heat transfer characteristics and energy conversion performance analysis for low GWP refrigerants in plate heat exchangers *Int. J. Heat Mass Transf.* **166** 120727
[13] Lemmon, E. W.; Huber, M. L.; McLinden M O 2018 NIST Reference Fluid Thermodynamic and Transport Properties, REFPROP
[14] Kline, S.J., McClintock F A 1953 Describing uncertainties in single-sample experiments *Mech. Eng.* **75** 3–8
[15] Longo G A, Righetti G and Zilio C 2015 A new computational procedure for refrigerant condensation inside herringbone-type Brazed Plate Heat Exchangers *Int. J. Heat Mass Transf.* **82** 530–6
[16] Mancin S, Del Col D and Rossetto L 2012 Partial condensation of R407C and R410A refrigerants inside a plate heat exchanger *Exp. Therm. Fluid Sci.* **36** 149–57
[17] Mancin S, Del Col D and Rossetto L 2013 R32 partial condensation inside a brazed plate heat exchanger *Int. J. Refrig.* **36** 601–11
[18] Cavallini A, Del Col D, Doretti L, Matkovic M, Rossetto L, Zilio C and Censi G 2006 Condensation in horizontal smooth tubes: A new heat transfer model for heat exchanger design *Heat Transf. Eng.* **27** 31–8

AUTHOR INDEX

Aimi, Alessandra........128
Allesina, Giulio........136
Ambrosini, Dario........187
Andreozzi, A........112
Aquaro, D........179
Arcasi, A........202
Arpino, F........194
Azzolin, M........226
Barletta, A........120
Berti, L........179
Bertone, M........194
Besagni, Giorgio........46
Bianco, N........112
Bianco, Nicola........1
Bocanegra, Johan Augusto........54
Bodys, Jakub........46
Bonfiglio, P........62
Bortolin, S........210
Bozzoli, Fabio........128
Brandão, P V........120
Brignoli, Vittorio........46
Buonomo, Bernardo........30
Canale, C........194
Caramello, Marco........153
Carraretto, I M........218
Castellani, Beatrice........163, 171
Cattani, Luca........128
Cattelan, G........226
Celli, M........120
Chiu, W K S........112
Ciani, F. S........62
Col, D Del........210, 226
Colombo, L P M........218
Cortellessa, G........194
Corti, Maria........70, 78, 86
Croci, Lorenzo........46
D'Alessandro, V........94
Diani, A........226
Ding, Yulong........171
Falone, M........94
Fanciulli, Carlo........86
Fingas, Rafal........46
Fragnito, Andrea........1
Fustinoni, Damiano........70, 78
Gambelli, Alberto Maria........163
Gannavarpu, Rajshekhar........187
Giambartolomei, G........179
Giammichele, L........94
Giannetti, Fabio........153
Giovannini, Michael........145
Golia, Maria Rita........30
Gramazio, Pasqualino........70, 78
Grossi, G........194
Guardasoni, Chiara........128
Haida, Michal........46
Hirata, S Da C........120
Iasiello, M........112
Iasiello, Marcello........1
Inzoli, Fabio........46
Kaur, Parminder........171
Lazzeri, R........179
Lezzi, A M........20
Lorenzini, Marco........103, 145
Lucchini, A........218
Manca, Oronzio........30
Martucci, A........210
Massarotti, Nicola........38
Mastrullo, R........202
Mauro, A W........202
Mauro, G M........112
Mauro, Gerardo Maria........1
Meglio, Armando Di........38
Melka, Bartlomiej........46
Minelli, Giorgio........163
Miotti, Davide........9
Mirafiori, M........210
Misale, Mario........54
Molinari, Martina........153
Molinaroli, L........218
Moretti, L........194
Narcisi, Vincenzo........153
Nardini, Sergio........30
Naso, V........112
Neri, M........20
Nespoli, Adelaide........86
Niro, Alfonso........70, 78, 86
Nobile, Enrico........9
Nowak, Andrzej J........46
Ottani, Filippo........136
Ouarzazi, M N........120
Palacz, Michal........46
Paoletti, Domenica........187
Passarelli, A F........202
Pedrazzi, Simone........136
Pesetti, A........179
Phan, T N........218

Piechnik, Edyta ... 46
Pittoni, P G ... 218
Piva, S. ... 62
Plomitallo, Renato Elpidio ... 30
Ramaiah, Jagadesh ... 187
Raucci, M ... 179
Ricci, R. ... 94
Rossetto, L ... 226
Rossi, Federico ... 163, 171
Rubeis, Tullio De ... 187
Sciacchitano, A. ... 194
Sciacovelli, Adriano ... 171
Smolka, Jacek ... 46
Stebel, Michal ... 46
Suzzi, Nicola ... 103
Tancon, M ... 210
Tarantino, Mariano ... 153
Tartarini, Paolo ... 136
Terra, Caterina La ... 86
Viscardi, I ... 202
Vitali, Luigi ... 70, 78
Zamolo, Riccardo ... 9

Institute of Physics
Dirac House, Temple Back
Bristol BS1 6BE UK

ISSN: 1742-6588
ISBN 978-1-7138-7465-2